Selected Topics
in Integral Geometry

Translations of
MATHEMATICAL
MONOGRAPHS

Volume 220

Selected Topics in Integral Geometry

I. M. Gelfand
S. G. Gindikin
M. I. Graev

American Mathematical Society
Providence, Rhode Island

И. М. Гельфанд, С. Г. Гиндикин, М. И. Граев

ИЗБРАННЫЕ ЗАДАЧИ ИНТЕГРАЛЬНОЙ ГЕОМЕТРИИ

ДОБРОСВЕТ, МОСКВА, 2000

Translated from the Russian by A. Shtern

2000 *Mathematics Subject Classification.* Primary 53C65; Secondary 42A38, 42B10, 43A32, 44A12, 46F12, 60D05, 60E05, 60E10, 65R10, 92C55.

For additional information and updates on this book, visit
www.ams.org/bookpages/mmono-220

Library of Congress Cataloging-in-Publication Data

Gel′fand, I. M. (Izrail′ Moiseevich)
[Izbrannye zadachi integral′noi geometrii. English]
Selected topics in integral geometry / I.M. Gelfand, S.G. Gindikin, M.I. Graev ; [translated from the Russian by A. Shtern].
p. cm. — (Translations of mathematical monographs, ISSN 0065-9282 ; v. 220)
Includes bibliographical references and index.
ISBN 0-8218-2932-7 (acid-free paper)
1. Integral geometry. I. Gindikin, S. G. (Semen Grigor′evich) II. Graev, M. I. (Mark Iosifovich) III. Title. IV. Series.

QA672.G4513 2003
516.3′62—dc21 2003052222

∞ The paper used in this book is acid-free and falls within the guidelines
established to ensure permanence and durability.
Visit the AMS home page at http://www.ams.org/

10 9 8 7 6 5 4 3 2 1 08 07 06 05 04 03

Contents

Preface to the English Edition

This is the English translation of a book published in Russia in 2000. The book is a realization of our old plan to write a small book explaining the main ideas of integral geometry in the context of several simple examples, and it follows our point of view that even now certain specific examples play a bigger role in integral geometry than general results.

For these simple examples we selected the classical Radon transform, its generalization suggested by F. John, hyperbolic versions of the Radon transform, and the horospherical transform for the group $SL(2, \mathbb{C})$. In discussing the Radon transform, which, of course, is treated in other books, we emphasize several circumstances, which are usually not considered. One example is the projective invariance of the Radon transform. This allowed us to regard the affine Radon transform, the Minkowski–Funk transform, and the geodesic hyperbolic Radon transform as different realizations of the projective Radon transform. We also considered it important to illustrate, by simple examples, the central role played in integral geometry by the operator κ, which is responsible for the universality of explicit inversion formulas.

The English edition of the book contains some modifications and corrections. In particular, we added Chapter 5 devoted to integral geometry on quadrics, or, in other words, to a conformally invariant version of the Radon transform. This approach allows us to combine the previously mentioned version of the Radon transform with the hyperbolic horospherical transform.

We are grateful to Alik Shtern for his excellent translation and to Sergei Gelfand for a number of useful comments.

April 2003 The authors

Preface

Integral geometry studies mainly integral transforms assigning to a function on a manifold X the integrals of this function over submanifolds that form a family M. It is assumed that the family M itself is endowed with the structure of a manifold. This establishes a correspondence between functions on the manifold X and functions on some manifold M of submanifolds of X. For instance, to functions on Euclidean space E^n one can assign the integrals of these functions over all possible lines; this rule defines an integral transform sending functions on E^n to functions on the manifold of lines.

Along with the integration of functions on X over submanifolds, integral geometry considers similar integral transforms of other analytic objects on X (densities, differential forms, sections of bundles, etc.). The main problems are in the description of the images and kernels of these transforms and in the construction of explicit inversion formulas recovering the original objects from their images. The first book devoted to this area of mathematics was the monograph by I. M. Gelfand, M. I. Graev, and N. Ya. Vilenkin [**8**].

Integral geometry interacts with the classical direction in geometry originating from Plücker, Klein, and Lie; the cornerstone of this direction is dualities between pairs of manifolds such that points of one manifold are realized as submanifolds of the other. Manifolds whose points have such geometrical nature, carry specific structures used in integral geometry. Transforms in integral geometry are studied in the language of these geometric structures. The typical examples of such manifolds are the four-dimensional variety of all lines in the three-dimensional projective space (the Plücker–Klein quadric) and, more generally, the Grassmann manifold of k-dimensional planes in the n-dimensional projective space, the Lie manifold of spheres, and so on. All these manifolds play a significant role in integral geometry.

The best known example of an integral geometry transform is the Radon transform which appeared in 1917 in Radon's paper [**31**] staying aloof from his other mathematical heritage. Radon considered the operator of integration over hyperplanes in Euclidean space and gave the exposition a surprisingly perfect form, which, uncommon at that time, combined analytic and geometric considerations and anticipated possible analogs of this transform in other homogeneous spaces. The attention was focused on the inversion formula reconstructing a function from its integrals over the hyperplanes. It is remarkable that the resulting formula is absolutely explicit and has principally different form depending on whether the dimension of the space is even or odd. In the odd-dimensional case the formula is local; namely, some differential operator is averaged along a family of parallel hyperplanes (and to reconstruct the function at a point, it suffices to know the integrals of this function only along the hyperplanes close to this point). In the even-dimensional case the inversion formula is nonlocal; namely, some integral operator is averaged along a family of parallel hyperplanes (in contrast to the differential operator arising in the odd-dimensional case), and we need integrals over distant planes. This corresponds to the fact that the Huygens principle holds for wave propagation in odd-dimensional spaces and fails in even-dimensional ones.

Already before Radon, Minkowski and Funk considered an analog of this transform for the spheres in which an even function on the sphere is reconstructed from the integrals of this function over the great circles. This reconstruction can be carried out by using spherical polynomials. Radon knew about these results, but he seemingly did not know that these transforms are projectively equivalent (see Chapter 1).

In 1938, F. John [**28**] considered a natural generalization of the Radon transform in which a function on three-dimensional space is integrated over all possible lines rather than over planes. A new exceptionally important feature of this construction is that the family of lines depends on four parameters. Thus, the John transform acts from functions of three variables to functions of four variables, and hence it is natural to expect that functions in the image satisfy an additional condition. The main observation of John was that these functions satisfy the ultrahyperbolic differential equation which completely describes the image. This observation, which was of great importance for the future of integral geometry, not only put overdetermined problems into circulation but also determined the relations between integral geometry and differential equations.

Starting from the 1940s, one of the central problems in mathematics was to develop an analog of the Fourier integral for noncommutative Lie groups. Among the first groups under investigation (mainly because of physical applications) was the Lorentz group $SL(2, \mathbb{C})$, i.e., the group of second-order unimodular complex matrices. For this group, I. M. Gelfand and M. A. Naimark succeeded in constructing a theory in which the role of exponential functions was played by irreducible infinite-dimensional unitary representations of the Lorentz group, and, at the same time, the main features of the classical Fourier integral were preserved. Obtaining analogs of the inversion formula and the Plancherel formula for the Fourier integral was the culmination of this theory.

Later on it became clear that the main point in the proof of these formulas was in the reconstruction of a function in $\mathbb{C}^3$ from the integrals of this function over the lines intersecting a hyperbola. Thus, a complex version of the John transform was considered; however, to recover the function, the integrals over the lines of the three-parameter family of lines intersecting the hyperbola were used rather than the integrals over all complex lines (forming a four-parameter family). This setting of the problem is natural because the reconstructed function depends on three variables. The analog of the Fourier transform on the Lorentz group is related to the above transform of integral geometry by an ordinary one-dimensional Fourier transform.

This observation opened new possibilities for integral geometry. First, the following problem arose: To what extent the phenomenon noticed in the case of Lorentz group is general? It turned out that, for a broad class of homogeneous spaces including complex semisimple Lie groups and Riemannian symmetric spaces, there is a transform of integral geometry, namely, the horospherical transform, which is related to an analog of the Fourier transform for these homogeneous spaces by means of the ordinary Fourier transform. This connection is similar to the relation between the Radon transform and the classical n-dimensional Fourier transform carried out by the one-dimensional Fourier transform. On the other hand, the context of integral geometry is much broader than that of representation theory. For instance, instead of the family of lines in $\mathbb{C}^3$ intersecting a hyperbola one may consider the family of lines intersecting any other algebraic curve and obtain a similar explicit inversion formula, although here there is no relation to groups or homogeneous spaces. This suggests the idea that perhaps the natural setting for multidimensional harmonic analysis contains not only groups or homogeneous spaces but also some more general geometric structures, in particular, "good" manifolds of submanifolds for which there are explicit inversion formulas.

It then seems natural to develop methods of geometric analysis meant for integral geometry and enabling one, in particular, to construct harmonic analysis on a broader class of homogeneous spaces including symmetric spaces. This became the mainstream for the further development of integral geometry. Much has already been done in this direction, but many things remain unclear. There is a rather complete picture for the problem of "good" (admissible) families of complex curves for which there is an explicit local inversion formula (see [**4, 1, 22**]). However, the results on families of complex submanifolds of dimension greater than one are far less complete (although the known results are sufficient to obtain the Plancherel formula for complex semisimple Lie groups by using the technique of integral geometry; see [**14**]). There are many problems that remain unsolved in the real case; this concerns especially the construction of nonlocal inversion formulas and the clarification of relation between the discrete series representations and integral geometry. Some results in this direction were obtained recently [**17**].

On the other hand, interesting and deep connections of integral geometry with multidimensional complex analysis, symplectic geometry, and nonlinear differential equations explicitly integrable by methods of the inverse problem were discovered; see [**19, 16, 12, 13**]. We also note that there are applied aspects of integral geometry, see [**2, 18, 30**]. The inversion of the Radon transform is the background of computer tomography. There are other physical problems (in astrophysics, geophysics, electron microscopy, etc.) in which the data can be interpreted as the Radon transform.

The aim of this book is quite modest. We intend to show some ideas and constructions of integral geometry by the example of the most elementary problems. We hope that the main part of the material is accessible to university students, who can see by these examples how the interaction of elementary analysis and geometry leads to beautiful and important results. The five chapters of this book are devoted to five particular transforms. We begin with the Radon transform (Chapter 1). An important point here (which is often neglected) is that the Radon transform is of projective nature, and in the projective version, it includes not only the affine Radon transform but also the Minkowski–Funk transform on the sphere. In Chapter 2 we study the John transform (the X-ray transform). It has already been noted that the inversion problem for this transform is overdetermined, and therefore it is of importance to understand how to pose this problem and describe a class of inversion formulas for appropriate three-parametric families of lines. This problem is solved here in the complex case. In Chapter 3 we present integral geometry on two-dimensional and three-dimensional hyperbolic spaces. The nonzero curvature of these spaces leads to a greater diversity of problems. In particular, there are two versions of the Radon transform, namely, geodesic and horospherical. However, the inversion formulas are similar to those in the Euclidean case, and their proof is only slightly more complicated. In general, it is typical for integral geometry that explicit inversion formulas have some standard structure and only slightly change if the geometry becomes more complicated (contrary to the formulas of group representation theory). This phenomenon has grave reasons, which have mainly been understood, but their discussion is beyond the framework of our book. We also consider the analog of the Fourier transform in the hyperbolic space; the results for this transform are obtained by using the horospherical transform. In Chapter 4 we develop integral geometry on the group $SL(2, \mathbb{C})$; as we have already mentioned, this example stimulated the modern development of integral geometry. Finally, in Chapter 5 we present a conformally invariant version of the Radon transform (the projective Radon transform). We give three variants of this transform, which are convenient in different applications; they are associated with the family of hyperplane sections of the sphere in $\mathbb{R}^{n+1}$, the family of hyperplane sections of the n-dimensional hyperboloid of two sheets, and the family of all spheres in $\mathbb{R}^n$, respectively. As in Chapter 2, we have an overdetermined problem of integral geometry and we give a universal inversion formula (the operator κ). Among interesting consequences is the conformal equivalence of the Radon transform and the horospherical transform in the hyperbolic space.

In view of the elementary nature of the book, we do not present a detailed bibliography on integral geometry and restrict ourselves to references to some publications mentioned in our exposition.

The authors are deeply indebted to M. M. Graev (Jr.) for his help in preparing the manuscript and useful comments.

CHAPTER 1

Radon Transform

1. Radon transform on the plane

1.1. Radon transform on the Euclidean plane. By the *Radon transform* on the Euclidean plane we mean the integral transform sending a function f on the plane to the integrals of this function over all possible lines (with respect to the Euclidean length on these lines).

Let us represent the Radon transform by an explicit formula. We define the lines on the plane by equations of the form

$$x_1 \cos\varphi + x_2 \sin\varphi - p = 0$$

or by the parametric equations (in which p is fixed and t is a parameter)

$$\begin{aligned} x_1 &= -t \sin\varphi + p \cos\varphi, \\ x_2 &= t \cos\varphi + p \sin\varphi; \end{aligned}$$

Euclidean measure on such a line is equal to dt. We thus introduced the parameters (φ, p) on the manifold of lines, and it is obvious that a line corresponds to two pairs (φ, p) and (φ', p') of parameters if and only if $\varphi' = \varphi + \pi k$ and $p' = (-1)^k p$ for some $k \in \mathbb{Z}$. This parametrization shows that the manifold of lines on the plane is obtained from the infinite circular cylinder with cylindrical coordinates (φ, p) by identifying the points (φ, p) and $(\varphi + \pi, -p)$.

The Radon transform of a function $f(x_1, x_2)$ is given by the following formula:

$$\mathcal{R} f(\varphi, p) = \int_{-\infty}^{+\infty} f(-t \sin\varphi + p \cos\varphi,\ t \cos\varphi + p \sin\varphi)\, dt. \tag{1.1}$$

The integral depends on the parameters φ and p; however, it follows from the formula $\mathcal{R} f(\varphi + \pi, -p) = \mathcal{R} f(\varphi, p)$ that the function $\mathcal{R} f$ can be pushed down to the manifold of lines.

The Radon transform can be considered on various function spaces. For simplicity, we first treat the transform on the space $D(\mathbb{R}^2)$ of all compactly supported infinitely differentiable functions. In this case, the Radon transform $\mathcal{R} f$ is a compactly supported infinitely differentiable function on the manifold of lines, i.e., an infinitely differentiable function of φ and p compactly supported with respect to p. In what follows we also consider this transform on the Schwartz space $S(\mathbb{R}^2)$ of all infinitely differentiable functions on $\mathbb{R}^2$ decreasing as $r \to \infty$, together with all their derivatives, more rapidly than an arbitrary negative power of r.

REMARK. Equation (1.1) is frequently represented in the form

$$\begin{aligned} \mathcal{R} f(\varphi, p) &= \int_{\mathbb{R}^2} f(x_1, x_2)\, \delta(x_1 \cos\varphi + x_2 \sin\varphi - p)\, dx_1\, dx_2 \\ &= \langle \delta(x_1 \cos\varphi + x_2 \sin\varphi - p),\ f(x_1, x_2) \rangle, \end{aligned}$$

where $\delta(\cdot)$ is the Dirac delta function of one variable. Here one must view the expression $\delta(x_1 \cos\varphi + x_2 \sin\varphi - p)$ as a distribution (a "generalized function") of two variables x_1 and x_2. To this end, we pass to the new coordinates u, v on the plane, where $u = x_1 \cos\varphi + x_2 \sin\varphi - p$, and apply the distribution $\delta(u)$ for any fixed v. One can immediately see that the result does not depend on the choice of the coordinate v.

1.2. Inversion formula. Consider the following problem: Reconstruct the function $f \in \mathcal{D}(\mathbb{R}^2)$ from its Radon transform $\mathcal{R}f$. The inversion formula was first obtained by Radon (see the Preface), and in fact we reproduce his argument. It uses the observation that the Radon transform commutes with the motions of the Euclidean plane. This implies, first, that it suffices to find a way to reconstruct the function f at a single point, say, $O = (0,0)$, and second, we can restrict ourselves to radially symmetric functions.

We proceed with the calculations. Let the function $f(x_1, x_2) = F(r)$ depend only on the distance $r = (x_1{}^2 + x_2{}^2)^{1/2}$ from the point $O = (0,0)$. One must reconstruct the value $f(0,0) = F(0)$ from the function $\mathcal{R}f(\varphi, p)$. Since the Radon transform commutes with rotations, it follows that $\mathcal{R}f$ depends only on p, $\mathcal{R}f(\varphi, p) = \widehat{F}(p)$. The problem finally reduces to the inversion problem for some integral transform $F \mapsto \widehat{F}$ for functions of one variable.

It immediately follows from (1.1) that

$$\widehat{F}(p) = \int_{-\infty}^{+\infty} F(\sqrt{t^2 + p^2})\, dt,$$

i.e.,

$$\widehat{F}(p) = 2\int_{|p|}^{\infty} \frac{F(r)}{\sqrt{r^2 - p^2}}\, r\, dr = \int_{p^2}^{\infty} \frac{F(\sqrt{t})}{\sqrt{t - p^2}}\, dt. \tag{1.2}$$

Thus, the passage (transformation) from F to $\widehat{F}$ reduces to the Abel transform (integration of order 1/2). It is known (and can readily be verified directly) that formula (1.2) implies

$$F(r) = -\frac{1}{\pi}\int_r^{\infty} \frac{\widehat{F}'(p)}{\sqrt{p^2 - r^2}}\, dp. \tag{1.3}$$

In particular,

$$F(0) = -\frac{1}{\pi}\int_0^{\infty} \frac{\widehat{F}'(p)}{p}\, dp. \tag{1.4}$$

Note that $\widehat{F}(p)$ is an even function. Therefore, since $\widehat{F}(p)$ is smooth, it follows that the integrand is a regular function, and hence the integral exists.

We pass from (1.4) to the inversion formula for an arbitrary function f. Averaging the function f over a circle centered at the point $O = (0,0)$ we obtain a function depending only on the distance r from the point O,

$$F(r) = \frac{1}{2\pi}\int_0^{2\pi} f(r \cos\varphi,\, r \sin\varphi)\, d\varphi.$$

Note that $F(0) = f(0,0)$. Let $\widehat{F}$ be the Radon transform of the function F. Since the Radon transform commutes with rotations, it follows that

$$\widehat{F}(p) = \frac{1}{2\pi} \int_0^{2\pi} \mathcal{R} f(\varphi, p)\, d\varphi,$$

i.e., $\widehat{F}$ is the mean value of the function $\mathcal{R} f$ over the set of all lines equidistant from the point O. From (1.4) we obtain

$$f(0,0) = -\frac{1}{\pi} \int_0^\infty \frac{\widehat{F}'(p)}{p}\, dp.$$

Since the Radon transform commutes with the translations, we thus obtain the following assertion.

THEOREM 1.1. *If* $\mathcal{R} f$ *is the Radon transform of a function* $f \in \mathcal{D}(\mathbb{R}^2)$, *then the following inversion formula holds*:

$$f(x_1, x_2) = -\frac{1}{\pi} \int_0^\infty \frac{\widehat{F}'_p(x_1, x_2; p)}{p}\, dp, \tag{1.5}$$

where

$$\widehat{F}(x_1, x_2, p) = \frac{1}{2\pi} \int_0^{2\pi} \mathcal{R} f(\varphi, p + x_1 \cos\varphi + x_2 \sin\varphi)\, d\varphi, \tag{1.6}$$

i.e., $\widehat{F}$ *is the mean value of the function* $\mathcal{R} f$ *over the lines equidistant from the point* $x = (x_1, x_2)$.

Using integration by parts, we can represent (1.5) as follows:

$$f(x_1, x_2) = -\frac{1}{\pi} \int_0^\infty \frac{\widehat{F}_p(x_1, x_2; p) - \widehat{F}_p(x_1, x_2; 0)}{p^2}\, dp. \tag{1.7}$$

Recall that $\widehat{F}$ is an even infinitely differentiable function of p.

To invert the Radon transform, we have used (1.3) for $r = 0$ only. For $r > 0$, this formula contains an additional information concerning the relation between the averages of the function f and the Radon transform $\mathcal{R} f$ of f. Namely,

THEOREM 1.2. *The following relation holds* (*the* Asgeirsson relation):

$$F(x_1, x_2; r) = -\frac{1}{\pi} \int_r^\infty \frac{\widehat{F}'_p(x_1, x_2; p)}{\sqrt{p^2 - r^2}}\, dp, \tag{1.8}$$

where F *is the mean value of the function* f *over the circles centered at the point* (x_1, x_2), *i.e.*,

$$F(x_1, x_2; r) = \frac{1}{2\pi} \int_0^{2\pi} f(x_1 + r\cos\varphi, x_2 + r\sin\varphi)\, d\varphi, \tag{1.9}$$

and $\widehat{F}$ *is the mean value of the function* $\mathcal{R} f$ *over the lines equidistant from the point* (x_1, x_2) (*see* (1.6)). *We also have*

$$\widehat{F}(x_1, x_2; p) = 2 \int_{|p|}^\infty \frac{F(x_1, x_2; r)}{\sqrt{r^2 - p^2}}\, r\, dr. \tag{1.10}$$

Formulas (1.10) and (1.8) can be obtained from (1.2) and (1.3), respectively, by using the corresponding translation.

COROLLARY. *For any* $x = (x_1, x_2)$ *and* $a > 0$, *the condition that* $F(x, r) = 0$ *for* $r > a$ *is equivalent to the condition that* $\widehat{F}(x, p) = 0$ *for* $|p| > a$.

1.3. Remarks. 1°. In the inversion formula one can substantially weaken the assumption that the function f is infinitely differentiable. It is clearly sufficient that the function $\mathcal{R}f$ be twice differentiable with respect to p and, to this end, it is sufficient that the function f be twice differentiable. Moreover, it suffices to assume that the function f belongs to the space C^1 of differentiable functions with continuous first partial derivatives because the Radon transform of radially symmetric functions is smoothing at the origin, namely, if $F \in C^1$, then $\widehat{F}'(p) = O(p)$ as $p \to 0$. This follows from the relation

$$\widehat{F}'(p) = 2p \int_p^\infty \frac{F'(r)}{\sqrt{r^2 - p^2}}\, dr = p \int_0^\infty \frac{F'(\sqrt{s + p^2})}{\sqrt{s}\,\sqrt{s + p^2}}\, ds.$$

The interest to the Radon transform has grown in connection with tomography problems; see [**30, 2, 18**]). To study these problems, one must consider the Radon transform not only for smooth functions but also for piecewise smooth functions whose discontinuity lines are piecewise smooth curves (for instance, the Radon transform of the characteristic function of an open set). The inversion formula works well at the points of smoothness of such functions because the Radon transform is a smooth function in a neighborhood of the lines passing through the points of smoothness. However, the inversion formula is poorly adjusted for recovering the function in a neighborhood of a point of nonsmoothness. Special regularizing procedures are used to overcome these difficulties. The development of computational methods based on the inversion formula requires getting over some additional difficulties.

2°. For the validity of the inversion formula, we do not need to assume that the function f is compactly supported. It is enough to assume that f is rapidly decreasing, and even the weaker condition $f = O((1 + x_1^2 + x_2^2)^{-1})$ is sufficient. In particular, the inversion formula holds for the functions f in the Schwartz space $S(\mathbb{R}^2)$. Recall that $S(\mathbb{R}^2)$ is the space of infinitely differentiable functions on $\mathbb{R}^2$ decreasing as $r \to \infty$, together with all derivatives, more rapidly than any power of r.

3°. For any function in C^1, one can change the order of averaging $\mathcal{R}f = \widehat{F}(x_1, x_2; p)$ and the integration over p in the inversion formula and represent the inversion formula in the following form:

$$(1.11)\quad f(x_1, x_2) = -\frac{1}{4\pi^2} \int_0^{2\pi} \int_{-\infty}^{+\infty} (\mathcal{R}f)'_p(\varphi,\, p + x_1 \cos\varphi + x_2 \sin\varphi)\, p^{-1}\, dp\, d\varphi.$$

Here the integral with respect to p must be taken in the principal value sense.

1.4. Radon transform on the affine plane. In fact, the Euclidean structure in not needed to define the Radon transform on a plane, and the affine structure is quite sufficient. Let us define the lines by the equations

$$\xi_1 x_1 + \xi_2 x_2 - p = 0$$

and endow every line $\xi_1 x_1+\xi_2 x_2-p=0$ with the measure $d\mu_\xi$ such that $dx_1\,dx_2 = d(\xi_1 x_1 + \xi_2 x_2 - p)\,d\mu_\xi$. In the coordinates x_1, x_2 this measure can be written as follows:

$$d\mu_\xi = \frac{dx_1}{|\xi_2|} = \frac{dx_2}{|\xi_1|}.$$

We define the *Radon transform* of a function f on the affine plane by the relation

$$\mathcal{R} f(\xi_1,\xi_2;p) = \int_{\xi_1 x_1+\xi_2 x_2=p} f(x_1,x_2)\,d\mu_\xi = \int_{-\infty}^{+\infty} f\Big(x_1, \frac{p-\xi_1 x_1}{\xi_2}\Big)\frac{dx_1}{|\xi_2|}. \tag{1.12}$$

The function $\mathcal{R} f$ thus obtained satisfies the condition

$$\mathcal{R} f(\lambda\,\xi_1,\,\lambda\,\xi_2;\,\lambda\,p) = |\lambda|^{-1}\,\mathcal{R} f(\xi_1,\xi_2;p). \tag{1.13}$$

Thus, this function depends not only on the line $\xi_1 x_1 + \xi_2 x_2 - p = 0$ but also on the choice of the parameters ξ_1, ξ_2, p.

In the case of $(\xi_1,\xi_2) = (\cos\varphi,\,\sin\varphi)$, the function $\mathcal{R} f$ coincides with the function $\mathcal{R} f(\varphi,p)$ introduced in the Euclidean case.

One frequently writes the expression for $\mathcal{R} f$ in the form

$$\mathcal{R} f(\xi_1,\,\xi_2;\,p) = \int_{\mathbb{R}^2} f(x_1,x_2)\,\delta(\xi_1 x_1 + \xi_2 x_2 - p)\,dx_1\,dx_2,$$

where the definition of $\delta(\xi_1 x_1 + \xi_2 x_2 - p)$ is similar to that given in 1.1.

THEOREM 1.3. *The inversion formula for the Radon transform* (1.12) *is*

$$\begin{aligned} f(x_1,x_2) &\\ = -\frac{1}{4\pi^2}\int_\Gamma \Big(\int_{-\infty}^{+\infty} (\mathcal{R} f)'_p(\xi_1,\xi_2,p+\xi_1x_1+\xi_2x_2)\,p^{-1}\,dp\Big)\,(\xi_1\,d\xi_2 - \xi_2\,d\xi_1), \end{aligned} \tag{1.14}$$

where Γ is an arbitrary contour in $\mathbb{R}^2\setminus O$ intersecting once almost every ray issuing from the point O.

The validity of this formula follows from two simple facts.

1) The differential form in the integrand can be pushed down to the manifold S of rays issuing from the point O, i.e., to the base space of the bundle $\mathbb{R}^2\setminus O \to S$. This follows from the fact that this form is homogeneous with respect to ξ of degree zero, and the form vanishes on the vectors tangent to the fibers of the bundle $\big((d\xi_1,\,d\xi_2) = (\xi_1,\,\xi_2)\,d\lambda\big)$.

2) If for Γ we take the unit circle centered at the point O, then (1.14) coincides with formula (1.11) proven above.

1.5. Relation to the Fourier transform and another proof of the inversion formula. Let $\mathcal{F}(\xi_1,\xi_2)$ be the Fourier transform of the function $f(x_1,x_2)$, i.e.,

$$\mathcal{F}(\xi_1,\xi_2) = (2\pi)^{-1}\int_{\mathbb{R}^2} f(x_1,x_2)\,e^{i\,(\xi_1 x_1+\xi_2 x_2)}\,dx_1\,dx_2. \tag{1.15}$$

To establish the relation between the Fourier transform $\mathcal{F}(\xi_1,\xi_2)$ and the Radon transform $\mathcal{R} f(\xi_1,\xi_2;p)$ of the function f, we represent the integral (1.15) for $(\xi_1,\xi_2)\neq O$ as an iterated integral, where the integration is taken first over the lines $\xi_1 x_1 + \xi_2 x_2 = p$ and then over p,

$$\mathcal{F}(\xi_1,\xi_2) = (2\pi)^{-1}\int_{-\infty}^{+\infty}\Big(\int_{\xi_1 x_1+\xi_2 x_2=p} f(x_1,x_2)\,d\mu_\xi\Big)\,e^{i\,p}\,dp.$$

Thus, we have

THEOREM 1.4. *The Radon transform $\mathcal{R} f$ and the Fourier transform $\mathcal{F}$ of a function f are related as follows:*

$$\mathcal{F}(\xi_1, \xi_2) = (2\pi)^{-1} \int_{-\infty}^{+\infty} \mathcal{R} f(\xi_1, \xi_2; p)\, e^{i\,p}\, dp \qquad \big((\xi_1, \xi_2) \neq O\big). \tag{1.16}$$

Since the function $\mathcal{R} f$ is homogeneous, one can represent (1.16) in the form

$$\mathcal{F}(\lambda\,\xi_1,\, \lambda\,\xi_2) = (2\pi)^{-1} \int_{-\infty}^{+\infty} \mathcal{R} f(\xi_1, \xi_2; p)\, e^{i\,\lambda p}\, dp. \tag{1.17}$$

Thus, the two-dimensional Fourier transform is the composition of the Radon transform on the plane and the one-dimensional Fourier transform.

Let us now present another proof of the inversion formula for the Radon transform. This approach is based on the inversion formula

$$\begin{aligned} f(x_1,\, x_2) &= (2\pi)^{-1} \int_{\mathbb{R}^2} \mathcal{F} f(\xi_1,\, \xi_2)\, e^{-i\,(\xi_1\, x_1 + \xi_2\, x_2)}\, d\xi_1\, d\xi_2 \\ &= (4\pi)^{-1} \int_{-\infty}^{+\infty} \int_0^{2\pi} \mathcal{F} f(\lambda\, \cos\varphi,\, \lambda\, \sin\varphi)\, e^{-i\lambda(x_1 \cos\varphi + x_2 \sin\varphi)}\, |\lambda|\, d\varphi\, d\lambda \end{aligned}$$

for the Fourier transform. Substitute the expression in (1.17) (in which the function $\mathcal{F} f(\lambda\, \cos\varphi,\, \lambda\, \sin\varphi)$ is represented in terms of $\mathcal{R} f(\cos\varphi,\, \sin\varphi;\, p) = \mathcal{R} f(\varphi; p)$) into this inversion formula. We obtain

$$\begin{aligned} &f(x_1,\, x_2) \\ &= \frac{1}{8\pi^2} \int_{-\infty}^{+\infty} \int_0^{2\pi} \int_{-\infty}^{+\infty} \mathcal{R} f(\varphi,\, p)\, e^{i\,\lambda\,(p - x_1 \cos\varphi - x_2 \sin\varphi)}\, |\lambda|\, dp\, d\varphi\, d\lambda \\ &= \frac{1}{4\pi} \int_{-\infty}^{+\infty} \int_{-\infty}^{+\infty} \widehat{F}(x_1,\, x_2,\, p)\, e^{i\,\lambda\, p}\, |\lambda|\, dp\, d\lambda \\ &= -\frac{1}{4\pi\, i} \int_{-\infty}^{+\infty} \int_{-\infty}^{+\infty} \widehat{F}'_p(x_1,\, x_2,\, p)\, e^{i\,\lambda\, p}\, \operatorname{sgn}\lambda\, dp\, d\lambda \\ &= -\frac{1}{2\pi} \int_{-\infty}^{+\infty} \widehat{F}'_p(x_1,\, x_2,\, p)\, p^{-1}\, dp. \end{aligned}$$

Here the function $\widehat{F}$ is given by (1.6). At the last step we used the following fact: the Fourier transform of the distribution $\operatorname{sgn}\lambda \in S'(\mathbb{R})$ is the distribution $2ip^{-1}$ (see [**10**]), i.e., if $\psi \in S(\mathbb{R})$ and $\widetilde{\psi}(\lambda) = \int_{-\infty}^{+\infty} \psi(p) e^{i\lambda p}\, dp$, then

$$\int_{-\infty}^{+\infty} \widetilde{\psi}(\lambda)\, \operatorname{sgn}\lambda\, d\lambda = 2i \int_{-\infty}^{+\infty} \psi(p)\, p^{-1}\, dp.$$

In general, the integral on the right-hand side is understood in the principal value sense. However, if ψ is a smooth odd function, then this integral converges in the ordinary sense. We finally obtain the inversion formula (1.5).

REMARK. Note that the Radon transform sends the Laplace operator $\Delta = \frac{\partial^2}{\partial x^2} + \frac{\partial^2}{\partial y^2}$ on the functions $f(x_1,\, x_2)$ to the operator $\frac{\partial^2}{\partial p^2}$ on the functions $\mathcal{R} f(\varphi; p)$. More generally, if there is a distribution $K(r^2)$, where $r^2 = x_1^2 + x_2^2$, then the function $K(\Delta)\, f \overset{\text{def}}{=} K(-r^2) * f$ becomes the convolution $K(-p^2) * \mathcal{R} f$ (with respect to the variable p). These reasonings can be used to invert the Radon transform of a nonsmooth function.

If f is a compactly supported distribution (in particular, a nonsmooth function), then we can represent f in the form $f = (1 - \Delta)^k \varphi$, where φ is a smooth function and k is a positive integer. The Radon transform of the function f can be defined as the convolution $K(1+p^2)^k * \mathcal{R}\varphi$. This definition agrees with that given above for compactly supported continuous functions. If f is a nonsmooth function, then we take the convolution of $\mathcal{R} f$ with $(1+p^2)^{-k}$ and obtain a smooth function $\mathcal{R}\varphi$. Applying the inversion formula for the Radon transform to the function $\mathcal{R}\varphi$, we find the function φ. Finally, applying the differential operator $(1-\Delta)^k$ to φ, we obtain the desired function f.

2. Radon transform in three-dimensional space

2.1. Radon transform in Euclidean space. By the Radon transform in three-dimensional Euclidean space we mean the integral transform sending a function on the space to the integrals of this function over all possible planes (with respect to the Euclidean area element on each of these planes). All considerations here are carried out along the same scheme as for the Radon transform on the plane.

Let us define the Radon transform by an explicit formula. We present every plane by an equation

$$\langle \omega,\, x\rangle - p = 0, \quad \text{where} \quad |\omega| = 1$$

($\langle \omega,\, x\rangle = \omega_1 x_1 + \omega_2 x_2 + \omega_3 x_3$), or by parametric equations

$$x = t_1\,\alpha^1 + t_2\,\alpha^2 + p\,\omega,$$

where α^1, α^2 is a pair of mutually orthogonal unit vectors, i.e., $\langle \alpha^1,\, \alpha^2\rangle = \langle \omega,\, \alpha^1\rangle = \langle \omega,\, \alpha^2\rangle = 0$. Euclidean measure on such a plane is equal to $dt_1\, dt_2$. Thus, every plane is defined by a pair (ω, p), where ω is a point of the unit sphere and p is a number, and it is clear that the planes corresponding to different pairs (ω, p) and (ω', p') coincide if and only if $(\omega', p') = -(\omega, p)$.

The Radon transform of a function $f(x)$ is given by the relation

$$\mathcal{R} f(\omega, p) = \int_{\mathbb{R}^2} f(t_1\,\alpha^1 + t_2\,\alpha^2 + p\,\omega)\, dt_1\, dt_2, \tag{2.1}$$

where α^1, α^2 are arbitrary unit vectors such that $\langle \alpha^1,\, \alpha^2\rangle = \langle \omega,\, \alpha^1\rangle = \langle \omega,\, \alpha^2\rangle = 0$ (and the integral does not depend on the choice of these vectors). Since $\mathcal{R} f(-\omega, -p) = \mathcal{R} f(\omega, p)$, it follows that the function $\mathcal{R} f$ can be pushed down to the manifold of planes.

As in the two-dimensional case, one frequently represents (2.1) in the form

$$\mathcal{R} f(\omega, p) = \int_{\mathbb{R}^3} f(x)\, \delta(\langle \omega,\, x\rangle - p)\, dx \qquad (dx = dx_1\, dx_2\, dx_3),$$

where $\delta(\cdot)$ is a function on $\mathbb{R}$.

For simplicity, we consider the Radon transform on the functions $f \in \mathcal{D}(\mathbb{R}^3)$. In this case, the functions $\mathcal{R} f$ are compactly supported and infinitely differentiable on the manifold of planes, i.e., these functions are infinitely differentiable when regarded as functions of ω and p and are compactly supported with respect to p.

Let us find the inversion formula for the Radon transform on three-dimensional Euclidean space. We first consider a function $f(x) = F(r)$ depending on $r = ({x_1}^2 + {x_2}^2 + {x_3}^2)^{1/2}$ only. The Radon transform of f does not depend on ω, i.e., $\mathcal{R}\, f(\omega, p) = \widehat{F}(p)$. According to (2.1) we have

$$\widehat{F}(p) = \int_{\mathbb{R}^2} F(\sqrt{{t_1}^2 + {t_2}^2 + p^2})\, dt_1\, dt_2 = 2\pi \int_{|p|}^{\infty} F(r)\, r\, dr. \tag{2.2}$$

Thus, in contrast to the two-dimensional case where we obtained the integral of order 1/2, here we simply obtain the integration operator. Therefore, the inverse is the differentiation operator, and hence

$$F(r) = -\frac{1}{2\pi} \frac{\widehat{F}'(r)}{r}. \tag{2.3}$$

In particular,

$$f(0) = F(0) = -\frac{1}{2\pi} \widehat{F}''(0). \tag{2.4}$$

To obtain the inversion formula for an arbitrary function f, we use the fact that, as in the two-dimensional case, the Radon transform commutes with the motions of Euclidean space. Introduce the mean value of the function f over the spheres centered at a fixed point x,

$$F(x; r) = \frac{1}{4\pi} \int_{|\omega|=1} f(x + r\omega)\, d\omega, \tag{2.5}$$

where $d\omega$ is the surface area element of the unit sphere. Further, let $\widehat{F}(x; p)$ be the mean value of the function $\mathcal{R}\, f$ over the planes equidistant from the point x, i.e.,

$$\widehat{F}(x; p) = \frac{1}{4\pi} \int_{|\omega|=1} \mathcal{R}\, f(\omega,\, p + \langle \omega,\, x\rangle)\, d\omega. \tag{2.6}$$

Then it follows from (2.4) and from the fact that the Radon transform commutes with the motions of Euclidean space that

$$f(x) = F(x, 0) = -\frac{1}{2\pi} \widehat{F}''_p(\omega; 0).$$

Thus, we have proved the following assertion.

THEOREM 2.1. *The inversion formula for the Radon transform on three-dimensional Euclidean space has the following form*:

$$f(x) = -\frac{1}{8\pi^2} \int_{|\omega|=1} (\mathcal{R}\, f)''_p(\omega; \langle \omega,\, x\rangle)\, d\omega. \tag{2.7}$$

We see that the three-dimensional case substantially differs from the two-dimensional one. In contrast to the two-dimensional case, here the inversion formula is local. In other words, to reconstruct a function f at a point x, one must know only the integrals of f over the planes close to x (in fact, over the planes in an infinitesimal neighborhood of the set of all planes passing through x).

In passing, we have obtained the following relations between the averages F

and $\widehat{F}$ of the functions f and $\mathcal{R} f$, respectively ((2.8) is the so-called *Asgeirsson relation* for three-dimensional space):

$$\widehat{F}(x;p) = 2\pi \int_{|p|}^{\infty} F(x,\, r)\, r\, dr, \tag{2.8}$$

$$F(r) = -\frac{1}{2\pi}\,\frac{\widehat{F}'_r(x;r)}{r}. \tag{2.9}$$

(These equations are obtained from (2.2) and (2.3) by translations.)

The Radon transform of a function f is related to its Fourier transform $\mathcal{F} f$ by the following formula:

$$\mathcal{F} f(\lambda\,\omega) = (2\pi)^{-3/2} \int_{-\infty}^{+\infty} \mathcal{R} f(\omega,\, p)\, e^{i\,\lambda\,p}\, dp \qquad (|\omega| = 1). \tag{2.10}$$

Starting from this equation and using the inversion formula for the Fourier transform, one can obtain another proof of the inversion formula for the Radon transform, namely,

$$\begin{aligned} f(x) &= (2\pi)^{-3/2} \int_{\mathbb{R}^3} \mathcal{F} f(\xi)\, e^{-i\langle \xi,\, x\rangle}\, d\xi \\ &= \frac{1}{2}\,(2\pi)^{-3/2} \int_{|\omega|=1} \int_{-\infty}^{+\infty} \mathcal{F} f(\lambda\,\omega)\, e^{-i\,\lambda\langle \omega,\, x\rangle}\, \lambda^2\, d\omega\, d\lambda \\ &= \frac{1}{2}\,(2\pi)^{-3/2} \int_{-\infty}^{+\infty} \int_{|\omega|=1} \int_{-\infty}^{+\infty} \mathcal{R} f(\omega,\, p)\, e^{i\,\lambda\,(p - \langle \omega,\, x\rangle)}\, \lambda^2\, dp\, d\omega\, d\lambda \\ &= \frac{1}{4\pi^2} \int_{-\infty}^{+\infty} \int_{-\infty}^{+\infty} \widehat{F}(x,\, p)\, e^{i\,\lambda\,p}\, \lambda^2\, dp\, d\lambda = -\frac{1}{2\pi}\, \widehat{F}''_p(x,\, 0). \end{aligned}$$

Comparing this proof with similar arguments for the plane (in 1.5), we note that the difference between the inversion formulas for the plane and for the space is caused by the distinction between the expressions for volume elements in the polar coordinates, namely, $|\lambda|\; d\lambda\; d\omega$ for the plane and $\lambda^2\; d\lambda\; d\omega$ for the space.

REMARK. In the two-dimensional case, the original function f is reconstructed from $\mathcal{R} f$ by a nonlocal formula. However, the function $\sqrt{\Delta}\, f$, where Δ stands for the Laplace operator, is reconstructed from $\mathcal{R} f$ by a local inversion formula:

$$\sqrt{\Delta}\, f(x_1,\, x_2) = \frac{1}{4\pi} \int_0^{2\pi} (\mathcal{R} f)''_p(\varphi,\, x_1 \cos\varphi + x_2 \sin\varphi)\, d\varphi.$$

Since $\sqrt{\Delta}$ is an elliptic pseudodifferential operator, the singularities of the functions $\sqrt{\Delta}\, f$ and f coincide. This fact is sometimes used in tomography to reconstruct the singularities of the function f from the known Radon transform $\mathcal{R} f$.

2.2. Radon transform in the affine space. To define the Radon transform in three-dimensional space, it suffices, as in the two-dimensional case, to have an affine rather than Euclidean structure. Let us define planes by equations of the form

$$\xi_1\, x_1 + \xi_2\, x_2 + \xi_3\, x_3 - p = 0$$

and equip every plane $\xi_1\, x_1 + \xi_2\, x_2 + \xi_3\, x_3 - p = 0$ by a measure $d\mu_\xi$ such that $dx_1\, dx_2 = d(\xi_1\, x_1 + \xi_2\, x_2 + \xi_3\, x_3 - p)\, d\mu_\xi$. For the coordinates on the plane one can take any pair x_i, x_j, $i, j = 1, 2, 3$, and the measure $d\mu_\xi$ in these coordinates can be written as

$$d\mu_\xi = \frac{dx_i\, dx_j}{|\xi_k|} \qquad (k \neq i, j),$$

i.e., it depends on the equation of the plane.

We define the Radon transform of a function f on the affine space by the relation

$$\begin{aligned} \mathcal{R} f(\xi; p) = \mathcal{R} f(\xi_1, \xi_2, \xi_3; p) &= \int_{\xi_1 x_1 + \xi_2 x_2 + \xi_3 x_3 = p} f(x)\, d\mu_\xi \\ &= \int f\Big(x_1,\, x_2,\, \frac{p - \xi_1 x_1 - \xi_2 x_2}{\xi_3}\Big) \frac{dx_1\, dx_2}{|\xi_3|}. \end{aligned} \tag{2.11}$$

As in the two-dimensional case, the function $\mathcal{R} f$ satisfies the condition

$$\mathcal{R} f(\lambda \xi;\, \lambda p) = |\lambda|^{-1}\, \mathcal{R} f(\xi; p).$$

Thus, the function $\mathcal{R} f$ depends not only on the plane $\xi_1 x_1 + \xi_2 x_2 + \xi_3 x_3 - p = 0$ but also on the choice of the parameters ξ_1, ξ_2, ξ_3, p.

For $|\xi| = 1$, the function $\mathcal{R} f$ coincides with the function $\mathcal{R} f(\omega, p)$ introduced earlier in the Euclidean case.

The expression for $\mathcal{R} f$ is frequently represented in the form

$$\mathcal{R} f(\xi_1,\, \xi_2, \xi_3;\, p) = \int_{\mathbb{R}^3} f(x_1, x_2, x_3)\, \delta(\xi_1 x_1 + \xi_2 x_2 + \xi_3 x_3 - p)\, dx_1\, dx_2\, dx_3,$$

where the definition of $\delta(\xi_1 x_1 + \xi_2 x_2 + \xi_3 x_3 - p)$ is similar to that in the two-dimensional case.

THEOREM 2.2. *The inversion formula for the affine Radon transform* (2.11) *is*

$$f(x) = -\frac{1}{8\pi^2} \int_\Gamma (\mathcal{R} f)''_p(\xi; p)|_{p = \langle \xi,\, x \rangle}\, \omega(\xi), \tag{2.12}$$

where

$$\omega(\xi) = \xi_1\, d\xi_2 \wedge d\xi_3 + \xi_2\, d\xi_3 \wedge d\xi_1 + \xi_3\, d\xi_1\, d\xi_2,$$

and Γ *is an arbitrary surface in* $\mathbb{R}^3 \setminus O$ *whose intersection with almost every ray starting from the origin* O *consists of one point.*

As in the two-dimensional case, the inversion formula (2.12) follows from two simple facts: first, the integral (2.12) does not depend on the choice of the surface Γ, and second, if Γ is the unit sphere, then (2.12) coincides with the inversion formula (2.7) for the Euclidean space.

2.3. Radon transform for a space of arbitrary dimension. The definition of Radon transform for $\mathbb{R}^2$ and $\mathbb{R}^3$ can be extended to spaces of arbitrary dimension. Namely, the Radon transform of a function $f \in S(\mathbb{R}^n)$ is defined by the formula

$$\mathcal{R} f(\xi,\, p) = \int_{\mathbb{R}^n} f(x)\, \delta(\langle \xi,\, x \rangle - p)\, dx, \qquad \xi \in (\mathbb{R}^n)' \setminus O, \quad p \in \mathbb{R}, \tag{2.13}$$

where $\langle \xi,\, x \rangle = \sum \xi_i x_i$, $dx = dx_1 \cdots dx_n$, and $\delta(p)$ stands for the delta function of one variable. The definition of $\delta(\langle \xi,\, x \rangle - p)$ is similar to that given above for $n = 2$ and $n = 3$.

We present without proof the inversion formulas for the Radon transform. If $\varphi = \mathcal{R} f$, then

$$f(x) = \frac{(-1)^{\frac{n-1}{2}}}{2\,(2\pi)^{n-1}} \int_\gamma \frac{\partial^{n-1}\varphi(\xi, s)}{\partial s^{n-1}}\bigg|_{s=(\xi,x)} \sigma(\xi) \tag{2.14}$$

for odd n, where $\sigma(\xi) = \sum_{i=1}^n (-1)^{i-1}\xi_i \wedge_{j\neq i} d\xi_j$ and $\gamma \subset \mathbb{R}^n$ is an arbitrary surface intersecting once every ray in $\mathbb{R}^n$ issuing from the origin O;

$$f(x) = \frac{(-1)^{n/2}}{(2\pi)^n} \int_\gamma \int_{-\infty}^{+\infty} \frac{\partial^{n-1}\varphi(\xi, s)}{\partial s^{n-1}}\bigg|_{s=(\xi,x)+p} p^{-1}\, dp \wedge \sigma(\xi) \tag{2.15}$$

for even n, where the integral over p is understood in the principal value sense. Thus, similarly to the cases $n = 2$ and $n = 3$, the inversion formula is local for odd n and nonlocal for even n.

The inversion formulas (2.14) and (2.15) can be combined by using the distribution $(p - i0)^{-1}$. By definition, this distribution is the limit as $s \to +0$ of the distribution $u_s(p) = (p - is)^{-1}$. It is known [**10**] that the function $(p - i0)^{-1}$ can be represented as a linear combination of the distributions p^{-1} and $\delta(p)$,

$$(p - i0)^{-1} = p^{-1} + i\,\pi\,\delta(p). \tag{2.16}$$

For any n we have the following inversion formula:

$$f(x) = \left(\frac{i}{2\pi}\right)^n \int_\gamma \int_{-\infty}^{+\infty} \frac{\partial^{n-1}\varphi(\xi, p)}{\partial p^{n-1}} (p - (\xi, x) - i0)^{-1}\, dp \wedge \sigma(\xi). \tag{2.17}$$

Indeed, it follows from (2.16) that relation (2.17) coincides with (2.14) for odd n and with (2.15) for even n.

Using the integration by parts $(n-1)$ times with respect to p, we can rewrite this formula as follows:

$$f(x) = (n-1)! \left(\frac{i}{2\pi}\right)^n \int_\gamma \int_{-\infty}^{+\infty} \varphi(\xi, p)\,(p - (\xi, x) - i0)^{-n}\, dp \wedge \sigma(\xi). \tag{2.18}$$

We can represent the inversion formula differently by passing from the Radon transform $\mathcal{R}$ to the "complex" Radon transform $\widetilde{\mathcal{R}}$. Namely, define the operator $\widetilde{\mathcal{R}}$ by the following formula:

$$\widetilde{\mathcal{R}}f(\xi, p) = \int_{\mathbb{R}^n} f(x)\,(\langle \xi,\, x\rangle - p - i0)^{-1}\, dx. \tag{2.19}$$

One can readily see that the functions $\widetilde{\mathcal{R}}f$ and $\mathcal{R} f$ are related as follows:

$$\widetilde{\mathcal{R}}f(\xi, p) = \int_{-\infty}^{+\infty} \mathcal{R} f(\xi, s)\,(s - p - i0)^{-1}\, ds.$$

This, together with (2.17), implies the inversion formula for the "complex" Radon transform $\widetilde{\mathcal{R}}$: if $\psi = \widetilde{\mathcal{R}}f$, then

$$f(x) = \left(\frac{i}{2\pi}\right)^n \int_{|\omega|=1} \frac{\partial^{n-1}\,\psi(\omega, s)}{\partial s^{n-1}}\bigg|_{s=\langle\omega,\, x\rangle} d\sigma. \tag{2.20}$$

Note that this formula is local for any n.

3. Wave equation and the Huygens principle

Here we shall show how one can obtain the classical formulas for the solutions of the wave equation by using the Radon transform. The proof is based on the method of plane waves.

3.1. Two-dimensional case. For the two-dimensional wave equation

$$\frac{\partial^2 u}{\partial t^2} = \frac{\partial^2 u}{\partial {x_1}^2} + \frac{\partial^2 u}{\partial {x_2}^2},$$

consider the following Cauchy problem:

$$u(0,\, x) = 0, \quad u'_t(0,\, x) = f(x), \tag{3.1}$$

and assume for simplicity that $f \in \mathcal{D}(\mathbb{R}^2)$.

The method of plane waves is based on the following fact: if $\varphi(s)$ is an arbitrary smooth function of one variable, then the function $\Phi_\omega(t,\, x) = \varphi(t + \langle \omega,\, x\rangle)$, where $|\omega| = 1$, is a solution of the wave equation for any ω (this fact can be verified directly). The solutions of this kind are called the *plane waves.* The method of plane waves consists in representing an arbitrary solution of the wave equation as a superposition of plane waves.

We proceed with constructing a solution of the Cauchy problem (3.1). Consider the following function:

$$u(t,\, x) = \frac{1}{4\pi^2} \int_{-\infty}^{+\infty} \int_0^{2\pi} \mathcal{R} f(\varphi,\, p - t + x_1 \cos\varphi + x_2 \sin\varphi)\, p^{-1}\, d\varphi\, dp, \tag{3.2}$$

where $\mathcal{R} f(\varphi,\, p)$ is the Radon transform of the function f. Since the function u is obtained as a superposition of plane waves, it follows that this function is a solution of the wave equation. Let us show that u is the solution of the Cauchy problem (3.1). Indeed, for $t = 0$, the integrand in (3.2) changes sign under the change of variables $(\varphi,\, p) \mapsto (\varphi + \pi,\, -p)$, and hence $u(0, x) = 0$. Furthermore, we have

$$\begin{aligned} u'_t(0,\, x) &= -\frac{1}{4\pi^2} \int_{-\infty}^{+\infty} \int_0^{2\pi} (\mathcal{R} f)'_p(\varphi,\, p + x_1 \cos\varphi + x_2 \sin\varphi)\, p^{-1}\, d\varphi\, dp \\ &= f(x), \end{aligned}$$

where the last equality follows from formula (1.11) for the Radon transform on the plane.

We intend to return in (3.2) from the function $\mathcal{R} f$ to the original function f. To this end, we use relation (1.10) for the mean values of the functions f and $\mathcal{R} f$,

$$\begin{aligned} &\int_0^{2\pi} \mathcal{R} f(\varphi,\, p + x_1 \cos\varphi + x_2 \sin\varphi)\, d\varphi \\ &\qquad = 2 \int_{|p|}^{\infty} \int_0^{2\pi} \frac{f(x_1 + r\cos\varphi,\, x_2 + r\sin\varphi)}{\sqrt{r^2 - p^2}}\, r\, d\varphi\, dr \\ &\qquad = 2 \int_{|y|^2 > p^2} \frac{f(x+y)}{\sqrt{|y|^2 - p^2}}\, dy \qquad (dy = dy_1\, dy_2), \end{aligned}$$

where $|y|^2 = {y_1}^2 + {y_2}^2$. We obtain

$$u(t,\, x) = \frac{1}{2\pi^2} \int_{\mathbb{R}^2} K(t,\, y)\, f(x+y)\, dy,$$

where

$$K(t, y) = \int_{t-|y|}^{t+|y|} \frac{dp}{p\sqrt{|y|^2 - (p-t)^2}}$$

(for $t < |y|$, the last integral must be understood in the principal value sense). Elementary manipulations show that $K(t, y) = 0$ for $t < |y|$ and $K(t, y) = \frac{\pi}{\sqrt{t^2 - |y|^2}}$ for $t > |y|$. Finally, we see that

$$u(t, x) = \frac{1}{2\pi^2} \int_{|y|<t} \frac{f(x+y)}{\sqrt{t^2 - |y|^2}} \, dy \tag{3.3}$$

(the Poisson formula).

Thus, the value of a solution of the wave equation at a point x and at a moment t depends only on the initial data in the disk of radius t centered at x. This is the so-called effect of "finite domain of dependence" for the existence of the forward front of the wave; namely, the signal moves with unit speed, and at time t it can come to x only from the points whose distance from x is not greater than t.

3.2. Three-dimensional case. Now we find the solution of the Cauchy problem (cf. (3.1)) for the three-dimensional wave equation,

$$\frac{\partial^2 u}{\partial t^2} = \frac{\partial^2 u}{\partial {x_1}^2} + \frac{\partial^2 u}{\partial {x_2}^2} + \frac{\partial^2 u}{\partial {x_3}^2}, \qquad u(0, x) = 0, \quad u_t'(0, x) = f(x). \tag{3.4}$$

Consider the following function:

$$u(t, x) = -\frac{1}{8\pi^2} \int_{|\omega|=1} (\mathcal{R} f)_p'(\omega, t + \langle \omega, x \rangle) \, d\omega, \tag{3.5}$$

where $\mathcal{R} f(\omega, p)$ is the Radon transform of the function $f(x) = f(x_1, x_2, x_3)$. This function is a superposition of plane waves, and hence it satisfies the wave equation. We show that it is the solution of the Cauchy problem (3.4). Indeed, for $t = 0$, the integrand in (3.5) is an odd function on the sphere $|\omega| = 1$. Hence, $u(0, x) = 0$. Furthermore,

$$u_t'(0, x) = -\frac{1}{8\pi^2} \int_{|\omega|=1} (\mathcal{R} f)_p''(\omega, \langle \omega, x \rangle) \, d\omega \; = \; f(x),$$

where the last equality follows from the inversion formula for the Radon transform on three-dimensional space.

We return in (3.5) from the function $\mathcal{R} f$ to the original function f. To this end, we use relation (2.9) between the mean values of the functions f and $\mathcal{R} f$:

$$\int_{|\omega|=1} f(x + t\,\omega) \, d\omega = -\frac{1}{2\pi\, t} \int_{|\omega|=1} (\mathcal{R} f)_p'(\omega, t + \langle \omega, x \rangle) \, d\omega.$$

We obtain

$$u(t, x) = \frac{t}{4\pi} \int_{|\omega|=1} f(x + t\,\omega) \, d\omega \tag{3.6}$$

(the Kirchhoff formula).

We see that the value of a solution of the three-dimensional wave equation at a point x and at a moment t depends only on the initial data at the points whose distance from x is equal to t (the Huygens principle). Thus, here we face not only the effect of finite domain of dependence but also a stronger effect, namely, the wave has a sharp back front: at the moment t the signal issued by any point placed at a distance less than t from the point x had already passed through the point x at some previous moment and cannot be observed at this point. This distinction between the two-dimensional and the three-dimensional case corresponds to the fact that the inversion formula for the Radon transform is local in the three-dimensional case and nonlocal in the two-dimensional case.

4. Cavalieri's conditions and Paley–Wiener theorems for the Radon transform

4.1. Cavalieri's conditions for rapidly decreasing functions. Consider the Radon transform $\mathcal{R}f(\xi_1, \xi_2; p)$ (in the affine version) on the functions $f \in S(\mathbb{R}^2)$; see (1.12). We are interested in the family of functions $\mathcal{R}f$ obtained in this way.

Recall that the functions $\mathcal{R}f$ satisfy the *homogeneity condition* (of degree -1)

$$\mathcal{R}f(\lambda\xi, \lambda p) = |\lambda|^{-1}\,\mathcal{R}f(\xi, p). \tag{4.1}$$

Further, these functions are infinitely differentiable on $(\mathbb{R}^2 \setminus O) \times \mathbb{R}$ and rapidly decreasing with respect to p as $p \to \infty$ together with all derivatives, and this decrease is uniform with respect to ξ, $|\xi| = 1$. The following question arises: Are there other conditions on the functions $\mathcal{R}f$ in the image of $S(\mathbb{R}^2)$?

First we note that

$$\int_{-\infty}^{+\infty} \mathcal{R}f(\xi, p)\,dp = \int_{\mathbb{R}^2} f(x)\,dx, \tag{4.2}$$

i.e., the left-hand side does not depend on ξ. This is closely related to the classical Cavalieri principle, which enables one to calculate the area of a plane figure in terms of the lengths of the intersections of this figure with a pencil of parallel lines of any given direction. Namely, if f is a characteristic function of a plane figure, then $\mathcal{R}f(\xi, p)$ is equal to the length of the intersection of this figure with the line $\xi_1 x_1 + \xi_2 x_2 - p = 0$ for any $|\xi| = 1$, and thus relation (4.2) expresses the area of a plane figure in terms of the lengths of the intersections of this figure with a pencil of parallel lines. In particular, the left-hand side of this relation does not depend on the direction of the pencil. For this reason, we refer to both the condition that the integral $\int_{-\infty}^{+\infty} \mathcal{R}f(\xi, p)\,dp$ does not depend on ξ and some generalizations of this condition as Cavalieri's conditions.

Consider the moments of the function $\mathcal{R}f(\xi, p)$ with respect to p,

$$I_k(\xi) = \int_{-\infty}^{+\infty} \mathcal{R}f(\xi, p)\,p^k\,dp, \quad k = 0, 1, 2, \dots . \tag{4.3}$$

Substituting the expression for $\mathcal{R}f$ in terms of the original function $f(x)$ into (4.3), we obtain

$$\int_{-\infty}^{+\infty} \mathcal{R}f(\xi, p)\,p^k\,dp = \int_{\mathbb{R}^2} f(x)\langle \xi, x\rangle^k\,dx. \tag{4.4}$$

Relation (4.4) implies the following property.

Cavalieri's conditions. *The integral $I_k(\xi)$ is a homogeneous polynomial of degree k in ξ_1, ξ_2 $(k = 0, 1, 2, \dots)$.*[1]

[1]These conditions (for affine spaces of arbitrary dimension) were first formulated in [**6**].

We give an interpretation of the integrals $I_k(\xi)$ in the language of the Fourier transform $\mathcal{F}f$ of the function f. It was shown in § 1 that the functions $\mathcal{F}f(\xi)$ and $\mathcal{R}f(\xi,p)$ are related as follows:

$$\mathcal{F}f(\lambda\xi) = (2\pi)^{-1}\int_{-\infty}^{+\infty} \mathcal{R}f(\xi,\, p)\, e^{i\lambda p}\, dp \qquad (\xi \neq 0).$$

This implies the formula

$$I_k(\xi) = 2\pi\, i^{-k}\, \frac{\partial^k \mathcal{F}f(\lambda\xi)}{\partial\lambda^k}\bigg|_{\lambda=0}. \tag{4.5}$$

Assume now that the function $\mathcal{F}f$ is rapidly decreasing and infinitely differentiable at the point 0 in any direction, i.e., the function $\mathcal{F}f(\lambda\xi)$ is infinitely differentiable with respect to λ at the point $\lambda = 0$ for any $\xi \neq 0$. We show that Cavalieri's conditions are equivalent to the condition that the function $\mathcal{F}f(\xi)$ is infinitely differentiable at the point $\xi = 0$.

Indeed, let $\mathcal{F}f(\xi)$ be infinitely differentiable at the point $\xi = 0$. Then the right-hand side of (4.5) is equal to the homogeneous component of degree k in the Taylor expansion of $\mathcal{F}f$ in a neighborhood of the point $\xi = 0$; thus, relations (4.5) imply Cavalieri's conditions. Conversely, assume that the function $\mathcal{F}f$ is infinitely differentiable at the point 0 in every direction and that Cavalieri's conditions are satisfied. Then it follows from the Taylor formula for $\mathcal{F}f(\lambda\xi)$ regarded as a function of λ that

$$\mathcal{F}f(\lambda\xi) = \sum_{k=0}^{n} \frac{\partial^k \mathcal{F}f(\lambda\xi)}{\partial\lambda^k}\bigg|_{\lambda=0} + o(|\lambda|^n) = (2\pi)^{-1}\sum_{k=0}^{n} \frac{i^k\lambda^k}{k!}\, I_k(\xi) + o(|\lambda|^n)$$

for any positive integer n and any ξ with $|\xi| = 1$. Since $I_k(\xi)$ is a homogeneous polynomial of degree k in ξ_1, ξ_2, this equation can be represented in the form

$$\mathcal{F}f(\xi) = (2\pi)^{-1}\sum_{k=0}^{n} \frac{i^k}{k!}\, I_k(\xi) + o(|\xi|^n). \tag{4.6}$$

We finally obtain a Taylor formula of order k for the function $\mathcal{F}f(\xi)$ at the origin O. Hence, $\mathcal{F}f(\xi)$ is infinitely differentiable with respect to ξ_1, ξ_2 at the point O.

4.2. Paley–Wiener theorem for the space $S(\mathbb{R}^2)$. The assertions describing the images of various function spaces under the Radon transform are often called *Paley–Wiener theorems* for this transform.

THEOREM 4.1. *A function $\varphi(\xi_1,\, \xi_2;\, p)$ on $(\mathbb{R}^2 \setminus O) \times \mathbb{R}$ is the Radon transform of some function $f \in S(\mathbb{R}^2)$ if and only if*

1) *φ satisfies the homogeneity condition* (4.1)*;*

2) *φ is infinitely differentiable;*

3) *every derivative of the function φ decreases for $|\xi| = 1$ as $p \to \infty$ more rapidly than any negative power of $|p|$;*

4) *φ satisfies Cavalieri's conditions.*

PROOF. We must prove the sufficiency of these conditions only. Let a function $\varphi(\xi,\, p)$ satisfy the conditions of the theorem. Define the function $F(\xi)$ on $\mathbb{R}^2 \setminus O$ by the formula

$$F(\xi) = \frac{1}{2\pi}\int_{-\infty}^{+\infty} \varphi(\xi,\, p)\, e^{ip}\, dp.$$

It follows from the homogeneity condition 1) that this relation is equivalent to the formula

$$F(\lambda\,\xi) = \frac{1}{2\pi}\int_{-\infty}^{+\infty} \varphi(\xi,\, p)\, e^{i\,\lambda\, p}\, dp.$$

Conditions 2) and 3) imply that the function F is infinitely differentiable at any point $\xi \neq 0$ and infinitely differentiable in any direction at $\xi = 0$; hence, it is infinitely differentiable at the point $\xi = 0$ by Cavalieri's conditions. Further, by conditions 2) and 3), the function F rapidly decreases together with all its derivatives. Thus, $F \in S(\mathbb{R}^2)$, and hence there is a function $f \in S(\mathbb{R}^2)$ such that $F = \mathcal{F}\, f$. Then $\varphi(\xi,\, p)$ coincides with the Radon transform $\mathcal{R}\, f(\xi,\, p)$ of the function f because the Fourier transforms of these functions with respect to p are equal to the same function $\mathcal{F}\, f(\lambda\,\xi)$.

4.3. Paley–Wiener theorem for the space $\mathcal{D}(\mathbb{R}^2)$ of compactly supported infinitely differentiable functions.

THEOREM 4.2. *A function $\varphi(\xi_1,\, \xi_2;\, p)$ on $(\mathbb{R}^2 \setminus O) \times \mathbb{R}$ is the Radon transform of a function $f \in \mathcal{D}(\mathbb{R}^2)$ if and only if*

1) *φ satisfies the homogeneity condition* (4.1)*;*

2) *φ is infinitely differentiable on $(\mathbb{R}^2 \setminus O) \times \mathbb{R}$ and compactly supported with respect to p;*

3) *φ satisfies Cavalieri's conditions.*

PROOF. If $f \in \mathcal{D}(\mathbb{R}^2)$, then the Radon transform $\mathcal{R}\, f$ of f clearly satisfies conditions 1) and 2). Cavalieri's conditions follow from the relation $\mathcal{D}(\mathbb{R}^2) \subset S(\mathbb{R}^2)$.

Conversely, let a function φ satisfy the conditions of the theorem. Then φ satisfies the conditions of Theorem 4.1 as well, and hence φ is the Radon transform of some function $f \in S(\mathbb{R}^2)$. It remains to prove that the function f is compactly supported. Let us present the proof which is due to Helgason [**26**].

By assumption, φ is a compactly supported function, i.e.,

$$\mathcal{R}\, f(\theta,\, p) = \mathcal{R}\, f(\cos\theta,\, \sin\theta;\, p) = 0 \quad \text{if} \quad |p| > a \tag{4.7}$$

for some $a > 0$. We prove that

$$F(x,\, r) \equiv \int_0^{2\pi} \mathcal{R}\, f(x_1 + r\,\cos\theta,\, x_2 + r\,\sin\theta;\, p)\, d\theta = 0 \quad \text{if} \quad r > |x| + a. \tag{4.8}$$

Indeed, let $\widehat{F}(x,\, p)$ be the mean value of the function φ on the set of lines placed at the distance $|p|$ from the point x. It follows from (4.7) that $\widehat{F}(x,\, p) = 0$ for $|p| > |x| + a$. Then $F(x,\, r) = 0$ for $r > |x| + a$ by the Asgeirsson relation (§ 1.2).

Next, let us prove that if the function f satisfies condition (4.8), then the functions $x_i\, f$, $i = 1, 2$, also satisfy this condition.

Indeed, it follows from (4.8) that

$$\int_{|y|>r} f(x + y)\, dy = 0 \quad \text{for} \quad r > |x| + a,$$

and hence

$$\int_{|y|>r} f'_{x_i}(x + y)\, dy = 0 \quad \text{for} \quad r > |x| + a, \quad i = 1, 2.$$

By the Stokes theorem, this implies that

$$\int_{|y|>r} f'_{x_i}(x+y)\,dy = r\int_0^{2\pi} f(x+r\,\omega)\,\omega_i\,d\theta,$$

where $\omega = (\cos\theta,\,\sin\theta)$. Thus,

$$\int_0^{2\pi} f(x+r\,\omega)\,(x+r\,\omega)_i\,d\theta$$

$$\equiv x_i\int_0^{2\pi} f(x+r\,\omega)\,d\theta + r\int_0^{2\pi} f(x+r\,\omega)\,\omega_i\,d\theta = 0 \quad \text{for} \quad r > |x|+a,$$

as required.

It follows from the above that the function $P(x)\,f(x)$ has zero mean value over the circle (centered at the point 0) of every radius $r > a$ for any polynomial $P(x)$. Hence, $f = 0$ on this circle, and thus f vanishes in the entire domain $|x| > a$.

4.4. Inversion of the Radon transform of a function $f \in \mathcal{D}(\mathbb{R}^2)$ using the moments. Let f be a compactly supported function. Then the Fourier transform $\mathcal{F}\,f$ of f is an entire function and, by (4.6), the Taylor series of the function $\mathcal{F}\,f$ is given by the formula

$$\mathcal{F}\,f(\xi) = (2\pi)^{-1}\sum_{k=0}^{\infty}\frac{i^k}{k!}\,I_k(\xi), \tag{4.9}$$

where $I_k(\xi)$ is the kth moment of the function $\mathcal{R}\,f(\xi,\,p)$ with respect to p, which is a polynomial of degree k. Taking the inverse Fourier transform of $\mathcal{F}\,f$, we obtain the way to reconstruct the function f from the moments I_k.

Thus, for a function $f \in \mathcal{D}(\mathbb{R}^2)$, we obtain an inversion formula using the moments I_k only. This formula substantially differs from that in § 1.

Since the moments I_k are homogeneous polynomials, in order to reconstruct the function f, it suffices to know the values of I_k on an arbitrarily small arc of the circle $|\xi| = 1$ or even on an infinite sequence of points (in general position).

REMARK. It is worth mentioning the difference between various inversion formulas. In an inversion formula we have an operator defined on some space (which is generally larger than the image of the Radon transform) and determined by some kernel (generally, a distribution kernel). The operators in different inversion formulas coincide on the image of the Radon transform. It follows from the Paley–Wiener theorem for the space $S(\mathbb{R}^2)$ that every functional vanishing on the image of $S(\mathbb{R}^2)$ is obtained from Cavalieri's conditions, and the inversion formula for $S(\mathbb{R}^2)$ is essentially unique. However, the image of $\mathcal{D}(\mathbb{R}^2)$ consists of compactly supported functions, and the family of annihilating functionals is richer in this case. Therefore, for $\mathcal{D}(\mathbb{R}^2)$, it is possible to construct inversion formulas different from that in § 1. A version of such a formula, which uses the moments, was obtained in this subsection; for another version, see 5.1.

4.5. Reconstruction of unknown directions from known values of $\mathcal{R}\,f$. Here we present another example in which Cavalieri's conditions are used when inverting the Radon transform. If the Radon transform $\mathcal{R}\,f(\varphi,\,p)$ of a compactly supported function f is known for some set of angles φ_i, $i = 1, 2, \ldots, n$, then the function f can be reconstructed approximately.

The following situation arises in electron microscopy of ribosomes: the values of $\mathcal{R} f(\varphi, p)$ are known for some angles $\varphi_1, \varphi_2, \dots, \varphi_n$ and for all p, i.e., n functions of p are given, but the values of the angles φ_i themselves are unknown. Therefore, to reconstruct f, one must first reconstruct the unknown angles, or, to be more precise, to reconstruct these angles up to a common rotation.

Goncharov [**21**] showed that using the fact that f is compactly supported and applying Cavalieri's conditions, one can solve this problem in the generic case provided that n is sufficiently large.

Let f be a compactly supported function, let $\mathcal{R} f$ be the Radon transform of f, and let $I_k(\varphi)$ be the moments of the function $\mathcal{R} f$ $(k = 1, 2, \dots)$. By Cavalieri's conditions, the moment I_k is a homogeneous polynomial of degree k in $\cos\varphi$, $\sin\varphi$; in particular,

$$\begin{aligned} I_1(\varphi) &= \lambda_1 \cos\varphi + \lambda_2 \sin\varphi, \\ I_2(\varphi) &= \mu_1 \cos^2\varphi + \mu_2 \cos\varphi \sin\varphi + \mu_3 \sin^2\varphi. \end{aligned} \tag{4.10}$$

Note that the expressions for I_1 and I_2 can be represented in the form

$$\begin{aligned} I_1(\varphi) &= a \cos(\varphi - \alpha), \\ I_2(\varphi) &= b \cos 2(\varphi - \beta) + c. \end{aligned} \tag{4.11}$$

All parameters in (4.10) and (4.11) depend on f.

Suppose that $I_1(\varphi) \not\equiv 0$ and $I_2(\varphi) \not\equiv 0$ and consider the following parametric curve Γ_f in the plane (y_1, y_2):

$$y_1 = I_1(\varphi), \qquad y_2 = I_2(\varphi).$$

LEMMA. *Γ_f is an algebraic curve of degree four.*

Indeed, the degree of the curve Γ_f is determined by the number of points in the intersection of this curve with a line in general position $A y_1 + B y_2 + C = 0$, i.e., the number of solutions of the system of equations

$$\begin{cases} A(\lambda_1 t_1 + \lambda_2 t_2) + B(\mu_1 t_1^2 + \mu_2 t_1 t_2 + \mu_3 t_2^2) + C = 0, \\ t_1^2 + t_2^2 = 1. \end{cases}$$

We emphasize that different curves correspond to different functions f.

A curve of degree four is determined by 15 points in general position on this curve (the number of coefficients of a polynomial $P(y_1, y_2)$ of degree four is 15). Therefore, if the functions $\mathcal{R} f(\varphi, p)$ are known for 15 values of φ, then 15 points on the curve Γ_f are known, and hence the curve is known. For a known curve Γ_f, the coefficients a, b, c in (4.11) are defined by the formulas

$$a = (y_1)_{\max}, \quad b = \frac{1}{2}((y_2)_{\max} - (y_2)_{\min}), \quad c = \frac{1}{2}((y_2)_{\max} + (y_2)_{\min}).$$

Since a shift of the parameter φ does not change the curve Γ_f, we can assume that $\varphi_1 = 0$. Thus, we know $\cos\alpha$, $\cos 2\beta$ and $\cos(\varphi_i - \alpha)$, $\cos 2(\varphi_i - \beta)$ for $i = 2, 3, \dots, 15$. Hence, we can readily see that

if α, β, and φ_i are in general position, then the angles $\varphi_2, \varphi_3, \dots, \varphi_{15}$ are determined uniquely.

REMARK. The number of angles $n = 15$ is excessive because the above construction can give only curves Γ_f of special form rather than all curves Γ of degree four. In fact, it is enough to know only five angles $\varphi_1, \varphi_2, \dots, \varphi_5$ $(\varphi_1 = 0)$ for the following reason. Since the angle φ_1 is equal to 0, it follows that the coefficients λ_1 and μ_1 in (4.11) are known. As a result, we obtain an overdetermined system of $2\cdot 4 = 8$ equations in seven unknowns λ_2, μ_2, μ_3 and φ_2, φ_3, φ_4, φ_5. One can show that this system has a unique solution in the generic case.

REMARK. The results of this section can be immediately extended to the multidimensional Radon transform.

5. Poisson formula for the Radon transform, and the discrete Radon transform[2]

5.1. Poisson formula for the Radon transform on a plane. Along with Fourier integral, we have its discrete analog, namely, Fourier series. We shall see in this section that there is a discrete analog of the Radon transform, which is related to Fourier series in just the same way as the Radon transform is related to the Fourier transform.

Recall that the Fourier series can be obtained from the Poisson formula for the Fourier integral. Namely, let f be a smooth compactly supported function on $\mathbb{R}^2$ and let $\mathcal{F}$ be the Fourier transform of f. Then, by the Poisson formula,

$$\sum_{k_1,\,k_2=-\infty}^{+\infty} f(x_1+k_1,\, x_2+k_2) = 2\pi \sum_{k_1,\,k_2=-\infty}^{+\infty} \mathcal{F}f(2\pi k_1,\, 2\pi k_2)\, e^{-2\pi\, i\,(k_1\,x_1+k_2\,x_2)} \tag{5.1}$$

(both series are absolutely convergent). If a function f is supported on the square $0 < x_1,\, x_2 < 1$, then only the term with $(k_1,\, k_2) = (0,\, 0)$ on the left-hand side can be nonzero for points $(x_1,\, x_2)$ in this square, and we obtain a representation of the function f by a Fourier series,

$$f(x_1,\, x_2) = 2\pi \sum_{k_1,\,k_2=-\infty}^{+\infty} \mathcal{F}f(2\pi k_1,\, 2\pi k_2)\, e^{-2\pi\, i\,(k_1\,x_1+k_2\,x_2)}. \tag{5.2}$$

Thus, a function supported on the above square can be represented by both the Fourier integral and the Fourier series.

Let us present an analog of the Poisson formula for the Radon transform. Introduce the subset A of points $(k_1,\, k_2) \in \mathbb{Z}^2$ such that either $k_1 > 0$ and the numbers k_1, k_2 are coprime or $(k_1,\, k_2) = (0,\, 1)$.

THEOREM 5.3. *Let f be a smooth compactly supported function on the plane $\mathbb{R}^2$ and let $\mathcal{R}\, f(\xi_1, \xi_2; p)$ be the Radon transform of f. Then*

$$\begin{aligned}\sum_{k_1,\,k_2=-\infty}^{+\infty} & f(x_1+k_1,\, x_2+k_2) \\ &= I(\mathcal{R}\, f) + \sum_{(k_1,\,k_2)\in A} \Big(\sum_{m=-\infty}^{+\infty} \mathcal{R}\, f(k_1,\, k_2; k_1\,x_1 + k_2\,x_2 + m) - I(\mathcal{R}\, f)\Big),\end{aligned} \tag{5.3}$$

where

$$I(\mathcal{R}\, f) = \int_{-\infty}^{+\infty} \mathcal{R}\, f(\xi_1, \xi_2; p)\, dp. \tag{5.4}$$

---

[2] See [**11**] and the computational applications in [**20**].

PROOF. Recall that the Radon transform is obtained as a composition of the two-dimensional and one-dimensional Fourier transforms, namely,

$$\mathcal{R}f(\xi_1,\xi_2;p) = \int_{-\infty}^{+\infty} \mathcal{F}f(\lambda\,\xi)\,e^{-i\,\lambda\,p}\,d\,\lambda, \tag{5.5}$$

where $\mathcal{F}f$ is the Fourier transform of f. Therefore, formula (5.3) can be obtained by combining the Poisson formulas for the two-dimensional and one-dimensional Fourier transforms. Note first that each nonzero point of $\mathbb{Z}^2$ can be represented in the form $m(k_1,\,k_2)$, where $m \in \mathbb{Z}\setminus 0$ and $(k_1,\,k_2) \in A$. Thus, the Poisson formula (5.1) can be rewritten as

$$\begin{aligned}\sum_{k_1,\,k_2=-\infty}^{+\infty} f(x_1+k_1,\,x_2+k_2) &= 2\pi\Big\{\mathcal{F}f(0) \\ &+ \sum_{(k_1,\,k_2)\in A}\Big(\sum_{m=-\infty}^{+\infty}\mathcal{F}f(2\pi\,m\,k_1,\,2\pi\,m\,k_2)\,e^{-2\pi\,i\,m\,(k_1\,x_1+k_2\,x_2)} - \mathcal{F}f(0)\Big)\Big\}.\end{aligned} \tag{5.6}$$

It follows from (5.5) and the Poisson formula for the one-dimensional Fourier transform that

$$\sum_{m=-\infty}^{+\infty} \mathcal{R}f(\xi;\,p+m) = \sum_{m=-\infty}^{+\infty} \mathcal{F}f(2\pi\,m\,\xi)\,e^{-2\pi\,i\,m\,p}.$$

Hence,

$$\begin{aligned}&2\pi\sum_{m=-\infty}^{+\infty}\mathcal{F}f(2\pi\,m\,k_1,\,2\pi\,m\,k_2)\,e^{-2\pi\,i\,m\,(k_1\,x_1+k_2\,x_2)} \\ &\qquad = \sum_{m=-\infty}^{+\infty}\mathcal{R}f(k_1,\,k_2;\,k_1\,x_1+k_2\,x_2+m).\end{aligned} \tag{5.7}$$

Since

$$\mathcal{F}f(0) = (2\pi)^{-1}\int_{-\infty}^{+\infty}\mathcal{R}f(\xi,p)\,dp$$

for any ξ, relation (5.3) follows from (5.6) and (5.7).

COROLLARY. *If a function f is supported on the unit square $0 < x_1,\,x_2 < 1$, then*

$$f(x_1,\,x_2) = I(\mathcal{R}f) + \sum_{((k_1,\,k_2)\in A}\Big(\sum_{m=-\infty}^{+\infty}\mathcal{R}f(k_1,\,k_2;\,k_1\,x_1+k_2\,x_2+m) - I(\mathcal{R}f)\Big) \tag{5.8}$$

for any point $(x_1,\,x_2)$ of this square, and the series on the right-hand side is absolutely convergent.

Thus, we have obtained another inversion formula for a smooth function f supported on the unit square. Moreover, if $I(\mathcal{R}f) = 0$, then the function f can be reconstructed at a point x provided that only the integrals of f over the lines with rational angular coefficients passing through x and the parallel translations of these lines corresponding to the integral values of Δp are known. Note that the distance between the parallel lines $k_1\,x_1 + k_2\,x_2 = p$ and $k_1\,x_1 + k_2\,x_2 = p+1$ is equal to $({k_1}^2 + {k_2}^2)^{-1/2}$ (in the ordinary Euclidean metric). The presence of integrals over shifted lines reflects the fact that the inversion formula is nonlocal.

REMARK. Formula (5.8) can be used to numerically invert the Radon transform for not necessarily smooth functions.

5.2. Discrete Radon transform; relation to the Fourier series. Let f be a function supported on the unit square $0 < x_1, x_2 < 1$ and let $\mathcal{R} f$ be the Radon transform of f. Starting from (5.8), it is natural (by analogy with the Fourier series) to refer to the function F on $A \times \mathbb{R}$ given by the formula

$$F(k_1, k_2; p) = \sum_{m=-\infty}^{+\infty} \mathcal{R} f(k_1, k_2; p+m) - I(\mathcal{R} f) \tag{5.9}$$

as the *discrete Radon transform* of f. This function is periodic in p with period 1.

By (5.8), the function f can be expressed in terms of its discrete Radon transform as follows:

$$f(x_1, x_2) = I(\mathcal{R} f) + \sum_{(k_1, k_2)\in A} F(k_1, k_2; k_1 x_1 + k_2 x_2), \quad 0 < x_1, x_2 < 1. \tag{5.10}$$

The function F is related to the Fourier series of the function f in exactly the same way in which the Radon transform $\mathcal{R} f$ is related to the Fourier transform $\mathcal{F} f$. Namely, let

$$f(x_1, x_2) = \sum_{n_1, n_2=-\infty}^{+\infty} a(n_1, n_2)\, e^{2\pi i (n_1 x_1 + n_2 x_2)}$$

be the expansion of f in the Fourier series on $0 < x_1, x_2 < 1$ and let

$$F(k_1, k_2; p) = \sum_{m=-\infty}^{+\infty} b_m(k_1, k_2)\, e^{2\pi i m p}$$

be the expansion of F in the Fourier series with respect to p on the interval $0 < p < 1$. Note that $b_0(k_1, k_2) = 0$. Substituting these series into (5.10), we obtain

$$\sum_{n_1, n_2=-\infty}^{+\infty} a(n_1, n_2)\, e^{2\pi i (n_1 x_1 + n_2 x_2)} = I(\mathcal{R} f) + \sum_{(k_1, k_2)\in A} \sum_{m\neq 0} b_m(k_1, k_2)\, e^{2\pi i m (k_1 x_1 + k_2 x_2)}.$$

Hence,

$$a(m k_1, m k_2) = b_m(k_1, k_2) \quad \text{for} \quad m \neq 0, \quad (k_1, k_2) \in A, \tag{5.11}$$

$$a(0,0) = I(\mathcal{R} f). \tag{5.12}$$

Relation (5.11) is a discrete analog of (5.5).

5.3. Problem of integral geometry on the torus. One can interpret two-dimensional Fourier series in two ways, namely, as a representation on the square $0 < x_1, x_2 < 1$ of a function supported on this square and as a representation of a periodic function f on the entire $\mathbb{R}^2$, i.e., a function f such that $f(x+n) = f(x)$ for any $n \in \mathbb{Z}^2$. In the latter case, the Fourier series can be interpreted as an expansion of a function defined in the torus $T = \mathbb{R}^2/\mathbb{Z}^2$. We give the same interpretation for the discrete Radon transform.

The square $0 < x_1, x_2 < 1$ can be viewed as an unfolding of the torus $T = \mathbb{R}^2/\mathbb{Z}^2$. In this case, for any $k = (k_1, k_2) \in A$ and $p \in \mathbb{R}$, the union of the segments of the lines

$$k_1 x_1 + k_2 x_2 = p + m, \qquad m \in \mathbb{Z},$$

contained in the square corresponds to a closed geodesic $\gamma(k, p)$ on the torus T. Conversely, every closed geodesic on the torus T can be represented as the union of segments of this kind. Thus, for the restriction of a periodic function f to the square $0 < x_1,\, x_2 < 1$, the discrete Radon transform $F(k_1,\, k_2,\, p)$ can be interpreted as follows:

$$F(k_1,\, k_2,\, p) = ({k_1}^2 + {k_2}^2)^{-1/2}\, \widehat{f}(k_1,\, k_2;\, p) - I(\widehat{f}),$$

where $\widehat{f}(k_1,\, k_2;\, p)$ stands for the integral of the function f on the torus T over the closed geodesic $\gamma(k, p)$ (taken with respect to the Euclidean length on T), and $I(\widehat{f})$ for the integral of the function $\widehat{f}$ over the family of parallel closed geodesics. Relation (5.10) implies the following assertion.

THEOREM 5.4. *Every smooth function f on the torus $T = \mathbb{R}^2/\mathbb{Z}^2$ is reconstructed from the integrals $\widehat{f}(k_1,\, k_2;\, p)$ over the closed geodesics $\gamma(k, p)$ as follows:*

$$f(x) = I(\widehat{f}) + \sum_{(k_1,\, k_2) \in A} \Big(({k_1}^2 + {k_2}^2)^{-1/2}\, \widehat{f}(k_1,\, k_2;\, k_1 x_1 + k_2 x_2) - I(\widehat{f}) \Big). \tag{5.13}$$

REMARKS. 1°. The length of the geodesic $\gamma(k, p)$ is equal to $({k_1}^2 + {k_2}^2)^{-1/2}$. Therefore, if one renormalizes the measures on the geodesics in such a way that the lengths of all geodesics become equal to one and denotes by $\widehat{f}'(k_1,\, k_2;\, p)$ the integral over the geodesic $\gamma(k, p)$ with respect to the new measure, then (5.13) simplifies to

$$f(x) = I(\widehat{f}) + \sum_{(k_1,\, k_2) \in A} \Big(\widehat{f}'(k_1,\, k_2;\, k_1 x_1 + k_2 x_2) - I(\widehat{f}') \Big). \tag{5.14}$$

2°. Formula (5.13) takes even a simpler form for functions f with mean value zero on the torus, namely,

$$f(x) = \sum_{(k_1,\, k_2) \in A} \widehat{f}'(k_1,\, k_2;\, k_1 x_1 + k_2 x_2).$$

In this case, the inversion formula is local. To reconstruct the function f at a point x on the torus, one must know only the integrals of this function over the closed geodesics passing through the point x.

6. Minkowski–Funk transform

This section is devoted to an analog of the Radon transform in which the plane $\mathbb{R}^2$ is replaced by the unit sphere $S^2 \subset \mathbb{R}^3$ and the lines in $\mathbb{R}^2$ by the great circles on S^2.

Let f be an even smooth function on the unit sphere $S \subset \mathbb{R}^3$. By the *Minkowski–Funk transform* we mean the integral transform sending the function f to the integrals of this function over all great circles on the sphere S. The following problem arises: Reconstruct the function f from its Minkowski–Funk transform. This problem had first been solved by Minkowski and Funk even before the Radon paper appeared. A similar problem can be formulated for functions on the three-dimensional unit sphere $S^3 \subset \mathbb{R}^4$.

REMARK. We consider only even functions on the sphere because the integral of an odd function over a great circle equals zero.

We represent the Minkowski–Funk transform by an explicit formula. Every great circle on the sphere S is the intersection of this sphere with a plane passing through the point 0,

$$\langle \xi,\, x\rangle = 0, \qquad |\xi| = 1. \tag{6.1}$$

Thus, the manifold of great circles is parametrized by the points ξ of the unit sphere S' in the dual space $(\mathbb{R}^3)'$, and the same great circle corresponds to antipodal points ξ and $-\xi$. To write out the integral over the circle (6.1), we present this circle parametrically,

$$x = \cos\varphi \cdot \alpha + \sin\varphi \cdot \beta, \qquad 0 \le \varphi < 2\pi, \tag{6.2}$$

where α and β are mutually orthogonal unit vectors such that $\langle \xi,\, \alpha\rangle = \langle \xi,\, \beta\rangle = 0$. The Minkowski–Funk transform is given by the following relation:

$$\mathcal{M}f(\xi) = \int_0^{2\pi} f(\cos\varphi \cdot \alpha + \sin\varphi \cdot \beta)\, d\varphi, \qquad |\xi| = 1. \tag{6.3}$$

It is clear that the integral does not depend on the choice of vectors α and β and that $\mathcal{M}f(-\xi) = \mathcal{M}f(\xi)$.

Thus, the Minkowski–Funk transform sends an even function on the sphere $S \subset \mathbb{R}^3$ into an even function on the sphere $S' \subset (\mathbb{R}^3)'$.

In order to identify functions on S and S', we identify the spheres S and S'. To be precise, we assign to any point $x \in S$ the point $\xi \in S'$ corresponding to the equatorial great circle with the pole x.

It follows from the definition that the Minkowski–Funk transform commutes with rotations of the sphere. In other words, for any rotation $g\,\omega \mapsto \omega\, g$ of the sphere S, the Minkowski–Funk transform $\mathcal{M}f_g(\xi)$ of the function $f_g(\xi) = f(\xi g)$ is equal to $\mathcal{M}f(\xi g)$. We shall use this property to reconstruct a function on the sphere from the Minkowski–Funk transform of this function.

Consider the Hilbert space $\mathcal{H}$ of even functions on S whose absolute values are square-integrable (with respect to the invariant measure on S). We construct the spectral decomposition of the operator $\mathcal{M}$ on $\mathcal{H}$.

Introduce a filtration of $\mathcal{H}$ by subspaces invariant under rotations. Namely, let $\mathcal{H}_{2n} \subset \mathcal{H}$ be the space of spherical polynomials of degree $2n$, i.e., the subspace of functions on S obtained by restricting the homogeneous polynomials of degree $2n$ to $S \subset \mathbb{R}^3$. We obtain a filtration

$$\mathcal{H}_0 \subset \mathcal{H}_2 \subset \cdots \subset \mathcal{H}_{2n} \subset \cdots \tag{6.4}$$

of the space $\mathcal{H}$ because the restriction to S of a homogeneous polynomial $P(x)$ of degree $2n$ coincides with the restriction to S of the polynomial $({x_1}^2 + {x_2}^2 + {x_3}^2)\, P(x)$ of degree $2n+2$. Moreover, the closure of the space $\bigcup \mathcal{H}_{2n}$ coincides with $\mathcal{H}$. Obviously, all spaces $\mathcal{H}_{2n}$ are finite-dimensional and rotation invariant, and the operator $\mathcal{M}$ is defined on each subspace $\mathcal{H}_{2n}$. Note that the restriction to S of a nonzero homogeneous polynomial of degree $2n$ is a nonzero function on S. This readily implies that

$$\dim \mathcal{H}_{2n} = (n+1)(2n+1). \tag{6.5}$$

We construct the grading of invariant subspaces of $\mathcal{H}$ associated to the filtration (6.4):

$$\mathcal{H} = \bigoplus_{n=0}^{\infty} H_{2n}, \tag{6.6}$$

where $H_0 = \mathcal{H}_0$ is the subspace of constants, and H_{2n}, $n > 0$, is the orthogonal complement of $\mathcal{H}_{2n-2}$ in $\mathcal{H}_{2n}$, i.e.,

$$\mathcal{H}_{2n} = \mathcal{H}_{2n-2} \oplus H_{2n}.$$

It follows from (6.5) that $\dim H_{2n} = 4n + 1$.

The space H_{2n} $(n = 0, 1, 2, \dots)$ admits the following intrinsic description. It consists of the functions on the sphere S that are restrictions to S of the homogeneous harmonic polynomials of degree $2n$ (i.e., the polynomials P satisfying the Laplace equation $\frac{\partial^2 P}{\partial x_1{}^2} + \frac{\partial^2 P}{\partial x_2{}^2} + \frac{\partial^2 P}{\partial x_3{}^2} = 0$).

The main fact is that *each H_{2n} is an eigenspace of the Minkowski–Funk operator $\mathcal{M}$.* This assertion immediately follows from the fact that the operator $\mathcal{M}$ commutes with rotations of the sphere S, whereas the subspaces H_{2n} are irreducible (i.e., contain no proper invariant subspaces) and are pairwise nonequivalent.

To describe $\mathcal{M}$, it remains to compute the eigenvalue λ_n of the operator $\mathcal{M}$ on each subspace H_{2n}. To this end, we consider the function $f \in H_{2n}$ obtained by restricting the polynomial $(x_2 + i\, x_3)^{2n}$ to S. Since $(x_2 + i\, x_3)^{2n}$ is obviously a harmonic polynomial, it follows that $f \in H_{2n}$. Hence,

$$\mathcal{M} f = \lambda_n\, f.$$

We write out the expression for f in the spherical coordinates $\theta,\ \varphi$:

$$\omega_1 = \cos\theta,\ \omega_2 = \sin\theta\, \cos\varphi,\ \omega_3 = \sin\theta\, \sin\varphi\ (0 < \theta < \pi,\ 0 < \varphi < 2\pi).$$

We have

$$f(\theta,\, \varphi) = e^{2\, i\, n\, \varphi} \sin^{2n}\theta.$$

In particular, $f(\frac{\pi}{2},\, 0) = 1$, and therefore $\mathcal{M} f(\frac{\pi}{2},\, 0) = \lambda_n$. On the other hand, $\mathcal{M} f(\frac{\pi}{2},\, 0)$ is the integral of the function f over the great circle on which $\varphi = \pm\frac{\pi}{2}$, i.e.,

$$\mathcal{M} f(\frac{\pi}{2},\, 0) = 2(-1)^n \int_0^{\pi} \sin^{2n}\theta\, d\theta = 2\sqrt{\pi}(-1)^n \frac{\Gamma(n+\frac{1}{2})}{\Gamma(n+1)}.$$

Hence,

$$\lambda_n = 2\sqrt{\pi}(-1)^n \frac{\Gamma(n+\frac{1}{2})}{\Gamma(n+1)}. \tag{6.7}$$

It follows from the description of $\mathcal{M}$ that to reconstruct a function f on S from its image $\varphi = \mathcal{M} f$, it suffices to decompose φ over the eigenspaces of the operator $\mathcal{M}$, namely, $\varphi = \sum_{n=0}^{\infty} \varphi_n$, where $\varphi_n \in H_{2n}$. Then $f = \sum_{n=0}^{\infty} \lambda_n^{-1}\, \varphi_n$.

In § 8 we shall find another inversion formula for the Minkowski–Funk transform. This formula will be based on a simple relation between this transform and the Radon transform for the projective plane.

7. Radon transform of differential forms

The Radon transform can be defined not only for functions but also for other objects. In this section we consider the Radon transform of differential forms on the plane and in three-dimensional space. The main problem still is to describe the image of the transform and to reconstruct the original form from its Radon transform. More general discussion can be found in [**3**].

7.1. Radon transform of 1-forms on the plane. Consider an arbitrary differential 1-form on $\mathbb{R}^2$,

$$\omega = f_1(x_1, x_2)\, dx_1 + f_2(x_1, x_2)\, dx_2,$$

whose coefficients belong to the Schwartz space $S(\mathbb{R}^2)$. One can take the integral of this form over any oriented line in the plane $\mathbb{R}^2$. This defines a function on the manifold H of all oriented lines, and we refer to this function as the Radon transform of the original 1-form ω and denote it by $\mathcal{R}\,\omega$. Thus, by definition,

$$\mathcal{R}\,\omega(l) = \int_l \omega, \qquad l \in H. \tag{7.1}$$

We write out an explicit expression for the function $\mathcal{R}\,\omega$ in coordinates on H. As in § 1, we present the lines on the plane by parametric equations,

$$\begin{aligned} x_1 &= -t\sin\varphi + p\cos\varphi, \\ x_2 &= t\cos\varphi + p\sin\varphi. \end{aligned} \tag{7.2}$$

For the positive direction on the line (7.2) we take the direction in which t increases. Thus, the parameters (φ, p) define not only the line but also the orientation of this line, and hence they can be taken for the coordinates on the manifold H of all oriented lines. Note that the same line with opposite orientation has the coordinates $(\varphi + \pi,\ -p)$.

In the coordinates (φ, p), the expression for $\mathcal{R}\,\omega$ becomes

$$\begin{aligned} \mathcal{R}\,\omega(\varphi, p) = \int_{-\infty}^{+\infty} \big(&-\sin\varphi\ f_1(-t\sin\varphi + p\cos\varphi,\ t\cos\varphi + p\sin\varphi) \\ &+ \cos\varphi\ f_2(-t\sin\varphi + p\cos\varphi,\ t\cos\varphi + p\sin\varphi)\big)\, dt. \end{aligned} \tag{7.3}$$

We wish to reconstruct the differential form ω from the Radon transform $\mathcal{R}\,\omega$. It is impossible to obtain a unique solution of this problem because of the following simple assertion.

LEMMA. *If the differential* 1*-form* ω *is closed, i.e.,* $d\omega = 0$, *then* $\mathcal{R}\,\omega = 0$.

Indeed, every closed form $\omega = f_1\, dx_1 + f_2\, dx_2$ with $f_1,\ f_2 \in S(\mathbb{R}^2)$ is exact, i.e., representable as $\omega = du$, where $u \in S(\mathbb{R}^2)$. Hence, the integral of the closed form ω over any line vanishes.

We solve another geometric problem. Assuming that the integrals of a 1-form $\omega = f_1\, dx_1 + f_2\, dx_2$ over all lines in $\mathbb{R}^2$ are known, i.e., for the given Radon transform of ω, find the exterior differential of ω, i.e., the 2-form

$$d\omega = \Big(\frac{\partial f_2}{\partial x_1} - \frac{\partial f_1}{\partial x_2}\Big)\, dx_1 \wedge dx_2 = F(x_1, x_2)\, dx_1 \wedge dx_2,$$

or, briefly, reconstruct the function F from $\mathcal{R}\,\omega$.

PROPOSITION 7.1. *To reconstruct F, one must apply the inversion formula for the Radon transform on the plane to the function* $\frac{\partial}{\partial p}\mathcal{R}\,\omega(\varphi, p)$.

The proposition results from the following lemma.

LEMMA. *The Radon transform $\mathcal{R}\,\omega$ of the form $\omega = f_1\,dx_1 + f_2\,dx_2$ is related to the Radon transform $\mathcal{R}\,F$ of the function $F = \frac{\partial f_2}{\partial x_1} - \frac{\partial f_1}{\partial x_2}$ by the formula*

$$\mathcal{R}\,F\,(\varphi,\,p) = \frac{\partial}{\partial p}\,\mathcal{R}\,\omega\,(\varphi,\,p). \tag{7.4}$$

PROOF OF THE LEMMA. Obviously, for a fixed coordinate $\varphi = \varphi_0$, the integral of the function $\mathcal{R}\,F$ with respect to p from p_1 to p_2 coincides with the integral of the 2-form $d\omega$ over the strip $\pi \subset \mathbb{R}^2$ between the lines $l_1 = (\varphi_0,\,p_1)$ and $l_2 = (\varphi_0,\,p_2)$. Hence, by the Stokes formula,

$$\int_{p_1}^{p_2} \mathcal{R}\,F\,(\varphi_0,\,p)\,dp = \int_\pi d\omega = \int_{l_2}\omega - \int_{l_1}\omega = \mathcal{R}\,\omega\,(\varphi_0,\,p_2) - \mathcal{R}\,\omega\,(\varphi_0,\,p_1).$$

This implies (7.4).

COROLLARY. *The Radon transform $\mathcal{R}\,\omega$ of a differential 1-form ω with coefficients in the space $S(\mathbb{R}^2)$ is identically zero if and only if this form is closed.*

7.2. Radon transform of 2-forms on the plane. In analysis, it is customary to integrate differential k-forms over k-dimensional submanifolds. This corresponds to the contents of 7.1, where differential 1-forms were integrated over lines on $\mathbb{R}^2$. However, if a family of submanifolds of dimension k is chosen, then one can integrate an m-form with $m \geq k$ over this family. If $m > k$, then the result is a differential $(m-k)$-form on the variety of submanifolds rather than a numerical function.

In this subsection we consider the integration of a 2-form ω over the lines on $\mathbb{R}^2$; the result of this integration is a 1-form $\mathcal{R}\,\omega$ on the manifold of lines. The informal idea to evaluate $\mathcal{R}\,\omega$ is as follows. To find the value $\mathcal{R}\,\omega(l, \tau)$, where l is a line and τ is a tangent vector to the manifold of lines, one must infinitesimally shift the line l in the direction of τ and integrate ω over the corresponding infinitesimal 2-dimensional domain. The same idea works in the general situation as well.

We proceed with the corresponding computations. First we define the Radon transform of a differential 2-form

$$\omega = f(x_1, x_2)\,dx_1 \wedge dx_2, \qquad f \in S(\mathbb{R}^2).$$

For that, we introduce the three-dimensional flag manifold A whose elements are pairs formed by an oriented line on the plane and a point on this line.

Since points on the plane are given by coordinates (x_1, x_2), and oriented lines by coordinates (φ, p), it follows that one can equip A with two coordinate systems. The first system is given by the coordinates (x_1, x_2, φ), where (x_1, x_2) determine a point on the plane and φ gives the direction of a line passing through the point, and the other system is given by the triple (φ, p, t), where (φ, p) determine an oriented line on the plane and t gives a point on this line. The coordinates (x_1, x_2, φ) are expressed in terms of the coordinates (φ, p, t) by relations (7.2). The inverse relations look as follows:

$$t = x_2\cos\varphi - x_1\sin\varphi, \qquad p = x_1\cos\varphi + x_2\sin\varphi.$$

Let us view the 2-form $\omega = f\,dx_1 \wedge dx_2$ as a differential form on the manifold A with coordinates (x_1, x_2, φ). To define the Radon transform of ω, we first represent the form ω in the coordinates (φ, p, t) by using (7.2) for x_1, x_2 and the corresponding expressions for dx_1, dx_2,

$$dx_1 = -\sin\varphi\,dt - t\,\cos\varphi\,d\varphi - p\,\sin\varphi\,d\varphi + \cos\varphi\,dp,$$
$$dx_2 = \cos\varphi\,dt - t\,\sin\varphi\,d\varphi + p\,\cos\varphi\,d\varphi + \sin\varphi\,dp.$$

We obtain

$$\begin{aligned}\omega = &-f(-t\,\sin\varphi + p\,\cos\varphi,\ t\,\cos\varphi + p\,\sin\varphi)\,dp \wedge dt\\ &- t\,f(-t\,\sin\varphi + p\,\cos\varphi,\ t\,\cos\varphi + p\,\sin\varphi)\,d\varphi \wedge dt + \cdots,\end{aligned}$$

where the dots stand for the terms that do not contain dt.

We denote by ω_0 the differential form obtained from ω by deleting all terms that do not contain dt. For fixed values of φ, p, $d\varphi$, and dp, the resulting form ω_0 is a differential 1-form on the oriented line corresponding to (φ, p). Integrating the form over the line, we obtain a differential 1-form on the manifold of oriented lines; this form is called the Radon transform of the original 2-form ω and is denoted by $\mathcal{R}\,\omega$. Thus, we obtain the following definition.

DEFINITION. The Radon transform $\mathcal{R}\,\omega$ of a differential 2-form

$$\omega = f(x_1, x_2)\,dx_1 \wedge dx_2$$

on the plane is the differential 1-form

$$\mathcal{R}\,\omega = a(\varphi, p)\,dp + b(\varphi, p)\,d\varphi \tag{7.5}$$

on the manifold H of oriented lines on the plane, where

$$\begin{aligned} a(\varphi, p) &= \int_{-\infty}^{+\infty} f(-t\sin\varphi + p\cos\varphi,\ t\cos\varphi + p\sin\varphi)\,dt,\\ b(\varphi, p) &= -\int_{-\infty}^{+\infty} t\,f(-t\sin\varphi + p\cos\varphi,\ t\cos\varphi + p\sin\varphi)\,dt.\end{aligned} \tag{7.6}$$

We establish some properties of the differential 1-form $\mathcal{R}\,\omega$. It follows from the explicit expressions for the functions a and b that both of them are infinitely differentiable and rapidly decreasing together with all their derivatives on the manifold H of oriented lines, i.e., these functions belong to the Schwartz space $S(H)$. Further, the function $a(\varphi, p)$ is the Radon transform of the function f, $a = \mathcal{R}\,f$, and hence this function additionally satisfies Cavalieri's conditions (see § 4). Finally, one can immediately see that

$$\frac{\partial a}{\partial \varphi} = \frac{\partial b}{\partial p},$$

and hence the differential form $\mathcal{R}\,\omega$ is closed.

PROPOSITION 7.2. *A differential* 1*-form* $\Omega = a\,dp + b\,d\varphi$ *with rapidly decreasing coefficients* $a, b \in S(H)$ *that is defined on the manifold* H *of oriented lines in* $\mathbb{R}^2$ *belongs to the image of the Radon transform of the* 2*-forms if and only if this form is closed and the function* a *satisfies Cavalieri's conditions.*

PROOF. In one direction the assertion has already been proved. Conversely, let $\Omega = a\,dp + b\,d\varphi$ be any form with the above-listed properties on the manifold $H = S^1 \times \mathbb{R}^1$. Then $a = \mathcal{R}\,f$ for some function $f \in S(\mathbb{R}^2)$. By setting $\omega = f\,dx_1 \wedge dx_2$, we obtain $\Omega - \mathcal{R}\,\omega = c(\varphi, p)\,d\varphi$. Since this 1-form is closed, it follows that the function c does not depend on p, and it follows from the rapid decrease that $c = 0$. Thus, $\Omega = \mathcal{R}\,\omega$.

REMARK. The fact that the form $\mathcal{R}\,\omega$ is closed can also be derived from the general fact that the operator $\mathcal{R}$ commutes with the operator d of exterior differentiation. We only need to note that the form ω itself is closed as a form of maximal degree.

To reconstruct the differential form $\omega = f\,dx_1 \wedge dx_2$ from its Radon transform $\mathcal{R}\,\omega = a\,dp + b\,d\varphi$, or, equivalently, to reconstruct the function f, one must apply the inversion formula for the Radon transform on the plane (see § 1) to the function $a(\varphi, p)$.

We see that the Radon transform of 2-forms has trivial kernel, in contrast to the case of 1-forms. We also note that the problem of reconstructing a differential 2-form ω from its Radon transform $\mathcal{R}\,\omega$ is overdetermined because one must know only one coefficient of the form $\mathcal{R}\,\omega$ to find ω completely.

7.3. Radon transform of 2-forms in three-dimensional space. Consider an arbitrary differential 2-form in $\mathbb{R}^3$,

$$\omega = f_1(x)dx_2 \wedge dx_3 + f_2(x)dx_3 \wedge dx_1 + f_3(x)dx_1 \wedge dx_2, \qquad x = (x_1, x_2, x_3), \tag{7.7}$$

with coefficients in the Schwartz space $S(\mathbb{R}^3)$. Integrating the form ω over all possible orientable planes in $\mathbb{R}^3$, we obtain a function on the manifold of planes. We refer to this function as the *Radon transform* of ω and denote it by $\mathcal{R}\,\omega$.

We present the expression for $\mathcal{R}\,\omega$ in the coordinates on the manifold of planes in $\mathbb{R}^3$. We define the planes by equations of the form

$$x_3 = \alpha_1 x_1 + \alpha_2 x_2 + \beta. \tag{7.8}$$

On the manifold of planes we take the coefficients $\alpha_1, \alpha_2, \beta$ of these equations for the local coordinates. Further, we take (x_1, x_2) for the coordinates on each plane given by (7.8), and thus define an orientation on this plane. In this case, the restriction of the form ω to the plane (7.8) can be obtained by substituting the expressions $x_3 = \alpha_1 x_1 + \alpha_2 x_2 + \beta$ and $dx_3 = \alpha_1\,dx_1 + \alpha_2\,dx_2$ in (7.7). We finally obtain

$$\begin{aligned}\mathcal{R}\,\omega\,(\alpha_1, \alpha_2, \beta) \\ &= \int_{-\infty}^{+\infty}\int_{-\infty}^{+\infty} \big[(f_3 - \alpha_1 f_1 - \alpha_2 f_2)(x_1,\, x_2,\, \alpha_1 x_1 + \alpha_2 x_2 + \beta)\big]\,dx_1\,dx_2.\end{aligned}$$

As in the similar problem for 1-forms on the plane (see 7.1), the form ω can be reconstructed from its Radon transform $\mathcal{R}\,\omega$ only up to a summand which is a closed differential 2-form. Therefore, we formulate another problem: Reconstruct the exterior differential of the form ω from the function $\mathcal{R}\,\omega$,

$$d\omega = \Big(\frac{\partial f_1}{\partial x_1} + \frac{\partial f_2}{\partial x_2} + \frac{\partial f_3}{\partial x_3}\Big)\,dx_1 \wedge dx_2 \wedge dx_3,$$

or, equivalently, reconstruct the function $F = \frac{\partial f_1}{\partial x_1} + \frac{\partial f_2}{\partial x_2} + \frac{\partial f_3}{\partial x_3}$ from $\mathcal{R}\,\omega$.

PROPOSITION 7.3. *In order to reconstruct the function F from the function $\mathcal{R}\,\omega$, one must apply the inversion formula for the Radon transform of functions in the three-dimensional space to the function $\frac{\partial}{\partial\beta}\,\mathcal{R}\,\omega(\alpha_1, \alpha_2, \beta)$.*

The kernel of the Radon transform of 2-forms in $\mathbb{R}^3$ with rapidly decreasing coefficients consists of the closed forms ω.

The proposition results from the following lemma.

LEMMA. *The Radon transform of a 2-form*

$$\omega = f_1\, dx_2 \wedge dx_3 + f_2\, dx_3 \wedge dx_1 + f_3\, dx_1 \wedge dx_2$$

is related to the Radon transform of the function $F = \frac{\partial f_1}{\partial x_1} + \frac{\partial f_2}{\partial x_2} + \frac{\partial f_3}{\partial x_3}$ by the formula

$$\mathcal{R}\, F(\alpha_1, \alpha_2, \beta) = \frac{\partial}{\partial \beta} \mathcal{R}\,\omega(\alpha_1, \alpha_2, \beta). \tag{7.9}$$

The proof of (7.9) essentially coincides with that of (7.4), with the strip between two parallel lines in $\mathbb{R}^2$ replaced by a domain in $\mathbb{R}^3$ between two parallel planes.

REMARK. The results of 7.1 and 7.3 can be readily extended to differential $(n-1)$-forms in $\mathbb{R}^n$.

7.4. Radon transform of 3-forms in three-dimensional space. The Radon transform of a differential 3-form in $\mathbb{R}^3$,

$$\omega = f(x_1, x_2, x_3)\, dx_1 \wedge dx_2 \wedge dx_3, \qquad f \in S(\mathbb{R}^3),$$

is defined similarly to that of a differential 2-form in $\mathbb{R}^2$; see 7.2. Namely, introduce the 5-dimensional flag manifold A whose elements are the pairs given by a plane in $\mathbb{R}^3$ and a point on this plane. Since points in $\mathbb{R}^3$ are given by the coordinates (x_1, x_2, x_3), and the planes by the local coordinates $(\alpha_1, \alpha_2, \beta)$, it follows that one can equip A with two coordinate systems. The first coordinate system is $(x_1, x_2, x_3; \alpha_1, \alpha_2)$, where the triple (x_1, x_2, x_3) determines a point in $\mathbb{R}^3$ and the pair (α_1, α_2) gives a plane passing through this point. The other coordinate system is $(\alpha_1, \alpha_2, \beta; x_1, x_2)$, where the triple $(\alpha_1, \alpha_2, \beta)$ determines a plane in $\mathbb{R}^3$ and the pair (x_1, x_2) gives a point on this plane. These two coordinate systems are related by (7.8).

Consider a differential form $\omega = f\, dx_1 \wedge dx_2 \wedge dx_3$ on A in coordinates $(x_1, x_2, x_3; \alpha_1, \alpha_2)$. To define the Radon transform of this form, we first express ω in the coordinates $(\alpha_1, \alpha_2, \beta; x_1, x_2)$. To do this, one must replace x_3 with $\alpha_1 x_1 + \alpha_2 x_2 + \beta$ and dx_3 with $\alpha_1\, dx_1 + \alpha_2\, dx_2 + x_1\, d\alpha_1 + x_2\, d\alpha_2 + d\beta$ in the expression for ω. We obtain

$$\omega = f(x_1,\, x_2,\, \alpha_1 x_1 + \alpha_2 x_2 + \beta)(d\beta + x_1\, d\alpha_1 + x_2\, d\alpha_2) \wedge dx_1 \wedge dx_2.$$

For chosen values of $\alpha_1, \alpha_2, \beta, d\beta$ and for integration variables x_1, x_2, this 3-form can be regarded as a differential 2-form on the plane with the coordinates $\alpha_1, \alpha_2, \beta$. Integrating the 2-form over this plane, we obtain a 1-form at the point $(\alpha_1, \alpha_2, \beta)$ of the manifold of planes in $\mathbb{R}^3$. We call this 1-form the *Radon transform* of the original differential 3-form ω and denote it by $\mathcal{R}\,\omega$. Let us summarize.

DEFINITION. By the *Radon transform* of a differential 3-form $\omega = f\, dx_1 \wedge dx_2 \wedge dx_3$ on $\mathbb{R}^3$ we mean the following differential 1-form $\mathcal{R}\,\omega$ on the manifold of planes in $\mathbb{R}^3$:

$$\mathcal{R}\,\omega = a_1(\alpha_1, \alpha_2, \beta)\, d\alpha_1 + a_2(\alpha_1, \alpha_2, \beta)\, d\alpha_2 + b(\alpha_1, \alpha_2, \beta)\, d\beta, \tag{7.10}$$

where
(7.11)

$$a_i\,(\alpha_1,\alpha_2,\beta) = \int_{-\infty}^{+\infty}\int_{-\infty}^{+\infty} x_i\, f(x_1,x_2,\alpha_1x_1+\alpha_2x_2+\beta)\,dx_1dx_2, \qquad i=1,2,$$

$$b\,(\alpha_1,\alpha_2,\beta) = \int_{-\infty}^{+\infty}\int_{-\infty}^{+\infty} f(x_1,x_2,\alpha_1x_1+\alpha_2x_2+\beta)\,dx_1\,dx_2.$$

The next assertion immediately follows from the definition of the 1-form $\mathcal{R}\,\omega$.

1) The coefficients of the form $\mathcal{R}\,\omega$ satisfy the relations

$$\frac{\partial a_1}{\partial \alpha_2} = \frac{\partial a_2}{\partial \alpha_1}, \qquad \frac{\partial b}{\partial \alpha_i} = \frac{\partial a_i}{\partial \beta}, \qquad i=1,2.$$

Hence, the form $\mathcal{R}\,\omega$ is closed.

2) The coefficients a_1, a_2, b of the form $\mathcal{R}\,\omega$ are infinitely differentiable functions, which rapidly decrease together with all their derivatives as $|\beta|\to\infty$.

3) The function b is the Radon transform of the function $f\in S(\mathbb{R}^3)$, i.e., $b=\mathcal{R}\,f$.

One can readily see that conversely, if a differential 1-form $\Omega = a_1\,d\alpha_1 + a_2\,d\alpha_2 + b\,d\beta$ on the manifold of planes in $\mathbb{R}^3$ satisfies conditions 1), 2), and 3), then this form is the Radon transform of a certain differential 3-form $\omega = f\,dx_1\wedge dx_2\wedge dx_3$, $f\in S(\mathbb{R}^3)$ (cf. a similar assertion in 7.2 in the case of $\mathbb{R}^2$).

It also follows from 3) that to reconstruct the form ω from its Radon transform $\mathcal{R}\,\omega = a_1\,d\alpha_1 + a_2\,d\alpha_2 + b\,d\beta$, one must apply the inversion formula for the Radon transform of functions in $\mathbb{R}^3$ to the coefficient b of the form $\mathcal{R}\,\omega$.

8. Radon transform for the projective plane and projective space

We have started from the Euclidean definition of the Radon transform; later on, we have established that this transform is affine-invariant. In this section we show that the Radon transform is in fact projective-invariant [**3**]. In this respect, the Radon transform substantially differs from the Fourier transform because the latter has no projective version.

We consider the Radon transform for the real projective plane $\mathbb{P}^2$ and the real projective space $\mathbb{P}^3$. The definitions of the Radon transforms for $\mathbb{P}^2$ and $\mathbb{P}^3$ are similar; however, as in the case of affine spaces, the inversion formulas for these transforms are substantially different: the formula is local for $\mathbb{P}^3$ and nonlocal for $\mathbb{P}^2$. Therefore, we begin with the case $\mathbb{P}^3$ because it is simpler.

8.1. Spaces $\mathbb{P}^3$ and $(\mathbb{P}^3)'$. As usual, points in $\mathbb{P}^3$ are given by homogeneous coordinates x_1,x_2,x_3,x_4, i.e., by vectors $x\in\mathbb{R}^4\setminus O$ defined up to a common factor. In other words, $\mathbb{P}^3$ is defined as the base of the natural bundle

$$\mathbb{R}^4\setminus O\to\mathbb{P}^3, \qquad \lambda x\sim x, \qquad \lambda\in\mathbb{R}\setminus 0.$$

Since the preimages of points in $\mathbb{P}^3$ are one-dimensional subspaces in $\mathbb{R}^4$, it follows that $\mathbb{P}^3$ can be interpreted as the manifold of one-dimensional subspaces in $\mathbb{R}^4$.

In the homogeneous coordinates x_1,x_2,x_3,x_4, each plane in $\mathbb{P}^3$ is given by an equation of the form

$$\langle\xi,\,x\rangle \equiv \xi_1\,x_1+\xi_2\,x_2+\xi_3\,x_3+\xi_4\,x_4=0,$$

where the coefficients ξ_i are defined up to a common factor. Thus, the manifold of planes in $\mathbb{P}^3$ is itself a three-dimensional projective space with homogeneous coordinates $\xi_1, \xi_2, \xi_3, \xi_4$. This space is said to be dual to $\mathbb{P}^3$ and is denoted by $(\mathbb{P}^3)'$.

8.2. Radon transform for $\mathbb{P}^3$. Denote by $\mathcal{F}(m)$ the space of infinitely differentiable functions on $\mathbb{R}^4 \setminus O$ satisfying the homogeneity condition

$$f(\lambda x) = |\lambda|^m f(x) \quad \text{for any} \quad \lambda \neq 0. \tag{8.1}$$

The first task in passing from the affine case to the projective one is to select an appropriate function space on which the Radon transform can be constructed. As such a function space we take the space $\mathcal{F}(-3)$.

The reason why one should take the space $\mathcal{F}(-3)$ rather than of a space of functions on the projective space $\mathbb{P}^3$ itself is related to the fact that there is no projective-invariant volume form on $\mathbb{P}^3$, i.e., there is no differential 3-form invariant with respect to the action of the group $SL(4, \mathbb{R})$ of projective transformations on $\mathbb{P}^3$. On the other hand, such a 3-form exists on the one-dimensional bundle $\mathbb{R}^4 \setminus O$ over $\mathbb{P}^3$. This is the Leray differential form

$$\begin{aligned} \omega(x) = {} & x_1\, dx_2 \wedge dx_3 \wedge dx_4 \\ & + (\text{forms obtained by nontrivial cyclic permutations of } 1, 2, 3, 4), \end{aligned} \tag{8.2}$$

which can conveniently be written as the determinant with commuting and anti-commuting variables,

$$\omega(x) = \frac{1}{6} \begin{vmatrix} x_1 & dx_1 & dx_1 & dx_1 \\ x_2 & dx_2 & dx_2 & dx_2 \\ x_3 & dx_3 & dx_3 & dx_3 \\ x_4 & dx_4 & dx_4 & dx_4 \end{vmatrix}.$$

In what follows, we write, for brevity,

$$\omega(x) = [x,\, dx,\, dx,\, dx].$$

Let $\xi \in (\mathbb{R}^4)' \setminus O$ and let

$$h_\xi : \ \langle \xi,\, x \rangle = 0$$

be the subspace in $\mathbb{R}^4$ corresponding to ξ. Denote by $\sigma_\xi(x)$ the interior product of the Leray form $\omega(x)$ and the form $d(\langle \xi,\, x\rangle)$ on $\mathbb{R}^4$,

$$\sigma_\xi(x) = d(\langle \xi,\, x\rangle) \,\rfloor\, \omega(x).$$

By definition, $\sigma_\xi(x)$ is a differential 2-form on $\mathbb{R}^4$ satisfying the relation

$$d(\langle \xi,\, x\rangle) \wedge \sigma_\xi(x) = \omega(x). \tag{8.3}$$

It is known that such a form on the entire space $\mathbb{R}^4$ is not unique; however, the restriction of such a form to the subspace h_ξ is uniquely defined. We can set, for instance,

$$\sigma_\xi(x) = \frac{[x,\, u,\, dx,\, dx]}{\langle \xi,\, u\rangle},$$

where $u \in \mathbb{R}^4$ is an arbitrary (but fixed) vector such that $\langle \xi,\, u\rangle \neq 0$.

Let $f \in \mathcal{F}(-3)$. Then the differential 2-form

$$f(x)\, \sigma_\xi(x)$$

on the space h_ξ is homogeneous of degree zero, and this form is orthogonal to the fibers of the bundle $h_\xi \to Ph_\xi$, where Ph_ξ is the projectivization of h_ξ. Hence, this differential form can be pushed down from h_ξ to Ph_ξ, i.e., to a plane in $\mathbb{P}^3$.

DEFINITION. By the Radon transform of a function $f \in \mathcal{F}(-3)$ we mean the function $\mathcal{R} f$ on $(\mathbb{R}^4)' \setminus O$ given by the formula

$$\mathcal{R} f(\xi) = \int_{\langle \xi,\, x\rangle = 0} f(x)\, \sigma_\xi(x). \tag{8.4}$$

It follows from the definition that $\varphi = \mathcal{R} f$ is an infinitely differentiable function on $(\mathbb{R}^4)' \setminus O$ satisfying the homogeneity condition $\varphi(\lambda\, \xi) = |\lambda|^{-1}\, \varphi(\xi)$. Hence, $\varphi \in \mathcal{F}(-1)$. Thus, the Radon transform defines a mapping

$$\mathcal{R} : \mathcal{F}(-3) \to \mathcal{F}(-1).$$

It is convenient to represent the expression for $\mathcal{R} f$ in the following form:

$$\mathcal{R} f(\xi) = \int_{\mathbb{P}^3} f(x)\, \delta(\langle \xi,\, x\rangle)\, \omega(x), \tag{8.5}$$

where $\delta(t)$ is the delta function on $\mathbb{R}$. The integral must be understood as follows. First we represent the form in the integrand in an arbitrary system of variables that includes $t = \langle \xi,\, x\rangle$ as one of the variables, and then apply the distribution $\delta(t)$ to the variable t (i.e., take only terms containing dt, delete dt in these terms, and then set $t = 0$). It is remarkable that the differential form thus obtained on the plane $\langle \xi,\, x\rangle = 0$ does not depend on the choice of the original variables.

8.3. Relation to the affine Radon transform for $\mathbb{R}^3$ and to the Minkowski–Funk transform for the three-dimensional sphere. Both the Radon transform for $\mathbb{R}^3$ and the Minkowski–Funk transform for the three-dimensional sphere are special cases of the projective Radon transform. Namely, one must replace homogeneous functions $f \in \mathcal{F}(-3)$ by their restrictions to appropriate sections of the bundle $\mathbb{R}^4 \setminus O \to \mathbb{P}^3$.

To obtain the affine Radon transform, we take the hyperplane $x_4 = 1$ for the section and set

$$f_0(x_1, x_2, x_3) = f(x_1, x_2, x_3, 1).$$

Since $\omega(x)\Big|_{x_4=1} = -dx_1 \wedge dx_2 \wedge dx_3$, formula (8.5) becomes

$$\mathcal{R} f(\xi_1, \xi_2, \xi_3, \xi_4) = -\int_{\mathbb{R}^3} f_0(x_1, x_2, x_3)\; \delta(\xi_1 x_1 + \xi_2 x_2 + \xi_3 x_3 + \xi_4)\; dx_1\, dx_2\, dx_3,$$

i.e., $\mathcal{R} f$ is the affine Radon transform of the function f_0 on $\mathbb{R}^3$.

To obtain the Minkowski–Funk transform, one must replace the functions f and $\varphi = \mathcal{R} f$ by their restrictions to the sphere $|x| = 1$ in $\mathbb{R}^4$ and to the sphere $|\xi| = 1$ in $(\mathbb{R}^4)'$, respectively. In this case, the transform (8.5) coincides with the Minkowski–Funk transform $\mathcal{M}$ for the functions on the sphere $|x| = 1$ in $\mathbb{R}^4$, namely, $\mathcal{M} f(\xi) = \mathcal{R} f(\xi)$ for $|\xi| = 1$.

8.4. Inversion formula for the Radon transform on $\mathbb{P}^3$. Our problem is to reconstruct a function $f \in \mathcal{F}(-3)$ from its Radon transform $\mathcal{R} f$. Let us immediately present the answer,

$$f(x) = -(2\,\pi)^{-2} \int_\Gamma \mathcal{R} f(\xi)\; \delta''(\langle \xi,\, x\rangle)\; \omega(\xi), \tag{8.6}$$

where $\delta''(t)$ is the second derivative of the delta function on $\mathbb{R}$; the integral is taken over an arbitrary section Γ of the bundle $\mathbb{R}^4 \setminus O \to \mathbb{P}^3$.

Formally, the differential 3-form in the integrand,

$$\Omega_x = \mathcal{R} f(\xi)\, \delta''(\langle \xi, x\rangle)\, \omega(\xi),$$

is homogeneous of degree zero and is orthogonal to the fibers of the bundle $\mathbb{R}^4 \setminus O \to \mathbb{P}^3$. Therefore, the integral does not depend on the choice of the section Γ.

We must make sense of the integral (8.6). To this end, we set

$$\xi = \eta + p\,\zeta,$$

where $\eta \in (\mathbb{R}^4)' \setminus O$ satisfies the condition $\langle \eta, x\rangle = 0$, p belongs to $\mathbb{R}$, and $\zeta \in (\mathbb{R}^4)' \setminus O$ is an arbitrary (but fixed) vector such that $\langle \zeta, x\rangle \neq 0$.

In the expression for Ω_x we pass from variables $\xi \in \mathbb{R}^4$ to variables $\eta \in \mathbb{R}^4$ and $p \in \mathbb{R}$ and then delete the terms that do not contain dp. Since in the new variables we have

$$\omega(\xi) = [\eta + p\,\zeta,\, d\eta + \zeta\, dp,\, d\eta + \zeta\, dp,\, d\eta + \zeta\, dp],$$
$$\delta''(\langle \xi, x\rangle) = |\langle \zeta, x\rangle|^{-3}\, \delta''(p),$$

after deleting the terms without dp we obtain

$$\Omega_x' = |\langle \zeta, x\rangle|^{-3}\, \mathcal{R} f(\eta + p\,\zeta)\, \delta''(p)\, [\eta, \zeta, d\eta, d\eta] \wedge dp.$$

Integrating the form Ω_x' with respect to p yields a differential 2-form on the subspace $h_\xi : \langle \eta, x\rangle = 0$ in $(\mathbb{R}^4)'$, namely,

$$|\langle \zeta, x\rangle|^{-3} (L^2 \varphi)\,(\eta)\, [\eta, \zeta, d\eta, d\eta],$$

where

$$L = \zeta\, \frac{\partial}{\partial \eta} = \sum \zeta_i\, \frac{\partial}{\partial \eta_i}.$$

This form is homogeneous of degree zero and is orthogonal to the fibers of the bundle $h_\xi \to P(h_\xi)$, where $P(h_\xi)$ is the projectivization of h_ξ. Thus, we can regard L as a form on $P(h_\xi)$.

Hence, we have reduced the inversion formula (8.6) to the following formula.

THEOREM 8.1. *If $\varphi = \mathcal{R} f$, where $f \in \mathcal{F}(-3)$, then*

$$f(x) = -(2\pi)^{-2} |\langle \zeta, x\rangle|^{-3} \int_{\langle \eta, x\rangle = 0} \Bigg(\sum_{1 \le i,j \le 4} \zeta_i \zeta_j\, \frac{\partial^2 \varphi(\eta)}{\partial \eta_i\, \partial \eta_j} \Bigg) [\eta, \zeta, d\eta, d\eta], \tag{8.7}$$

where $\zeta \in \mathbb{R}^4$ is an arbitrary (but fixed) vector such that $\langle \zeta, x\rangle \neq 0$.

REMARK. In contrast to (8.5), the restriction of the differential form $\Omega_x = \mathcal{R} f(\xi)\, \delta''(\langle \xi, x\rangle)\, \omega(\xi)$ to the subspace $\langle \xi, x\rangle = 0$ (this restriction was introduced in (8.7)) is not unique because it depends on the choice of the vector ζ. However, the difference of any two forms obtained in this way is an exact form.

PROOF OF THE INVERSION FORMULA (8.7). Denote by $\mathcal{F}_i$, $i = 1, 2, 3, 4$, the subspace of functions $f \in \mathcal{F}(-3)$ that vanish on the subspace $x_i = 0$ together with all their derivatives. Since the sum of the subspaces $\mathcal{F}_i$ is the entire space $\mathcal{F}(-3)$, it suffices to prove the validity of (8.7) only for functions belonging to these subspaces. To be definite, assume that $f \in \mathcal{F}_4$. Then the function on $\mathbb{R}^3$ given by the formula

$$f_0(x_1, x_2, x_3) = f(x_1, x_2, x_3, 1)$$

belongs to the Schwartz space $S(\mathbb{R}^3)$, and it follows from 8.3 that $\varphi = \mathcal{R} f$ is the affine Radon transform of the function f_0. Thus, by the inversion formula for the affine Radon transform,

$$f_0(x_1, x_2, x_3) = -(2\pi)^{-2} \int_\gamma \Big(\frac{\partial^2 \varphi}{\partial \eta_4^2} \Big|_{\eta_4 = -\eta_1 x_1 - \eta_2 x_2 - \eta_3 x_3} \Big) \, \omega_0(\eta),$$

where $\omega_0(\eta) = \eta_1 \, d\eta_2 \wedge d\eta_3 - \eta_2 \, d\eta_1 \wedge d\eta_3 + \eta_3 \, d\eta_1 \wedge d\eta_2$. Hence,

$$f(x_1, x_2, x_3, x_4) = |x_4|^{-3} f_0\Big(\frac{x_1}{x_4}, \frac{x_2}{x_4}, \frac{x_3}{x_4} \Big) = -(2\pi)^{-2} |x_4|^{-3} \int_{\langle \eta,\, x \rangle = 0} \frac{\partial^2 \varphi}{\partial \eta_4} \, \omega_0(\eta).$$

The equation thus obtained coincides with (8.7) in the case of $\zeta = (0,0,0,1)$. Thus, the inversion formula is proved for $\zeta = (0,0,0,1)$. The case of arbitrary vector ζ for which $\langle \zeta, x \rangle \neq 0$ can be reduced to the above case by a transformation of coordinates in $\mathbb{R}^4$ and $(\mathbb{R}^4)'$.

REMARK. One could drop the assumption that ζ is constant. Then a more general inversion formula than (8.7) is obtained; the integration in this new formula is over a two-dimensional cycle in the manifold of pairs η, ζ.

8.5. On the inversion formulas for the affine Radon transform on $\mathbb{R}^3$ and the Minkowski–Funk transform for S^3. We have already seen in 8.4 that the inversion formula proved in § 2 for the affine Radon transform is obtained from the projective inversion formula (8.7) for $\zeta = (0,0,0,1)$. If for ζ we take an arbitrary (but fixed) vector in $\mathbb{R}^4$ such that

$$\zeta_1 x_1 + \zeta_2 x_2 + \zeta_3 x_3 + \zeta_4 = c > 0,$$

then, using (8.7), we obtain a new inversion formula for the affine Radon transform, which differs from the inversion formula in § 2. Namely, let $\pi_x \subset (\mathbb{R}^4)'$ be the subspace orthogonal to $(x_1, x_2, x_3, 1) \in \mathbb{R}^4$, $P(\pi_x)$ the projectivization of this subspace, and γ_x an arbitrary section of the bundle $\pi_x \to P(\pi_x)$. If $f \in S(\mathbb{R}^3)$ and if $\varphi = \mathcal{R} f$ is the affine Radon transform of the function $f \in S(\mathbb{R}^3)$, then

$$f(x) = -(2\pi)^{-2} \, c^{-3} \int_{\gamma_x} \sum_{i,j=1}^{4} \zeta_i \zeta_j \frac{\partial^2 \varphi(\eta)}{\partial \eta_i \partial \eta_j} \, [\eta, \zeta, d\eta, d\eta].$$

REMARK. A more general class of inversion formulas for the affine Radon transform can be obtained by taking a smooth vector field $\eta \mapsto \zeta \in (\mathbb{R}^4)' \setminus O$ transversal to $\pi_x \setminus O$ (i.e., such that $\langle x, \zeta(\eta) \rangle > 0$ for any $\eta \in \pi_x \setminus O$) at the points $\eta \in \pi_x \setminus O$. To every such field there corresponds an inversion formula for the affine Radon transform.

Let us now turn to the Minkowski–Funk transform for the three-dimensional sphere. Interpreting the elements of $\mathcal{F}(-3)$ as even functions on the sphere $|x| = 1$ in $\mathbb{R}^4$, we see that the projective Radon transform $\mathcal{R} f$ and the Minkowski–Funk transform $\mathcal{M} f$ of a function $f \in \mathcal{F}(-3)$ are related as follows:

$$\mathcal{R} f(\xi) = |\xi|^{-1} \mathcal{M} f\Big(\frac{\xi}{|\xi|} \Big).$$

Hence, the inversion formula for the Minkowski–Funk transform $\mathcal{M}$ is obtained from the inversion formula (8.7) for the Radon transform $\mathcal{R}$ by replacing $\mathcal{R} f(\xi)$ with $|\xi|^{-1} \mathcal{M} f\big(\frac{\xi}{|\xi|} \big)$.

8.6. Description of the image of the Radon transform for $\mathbb{P}^3$.

THEOREM 8.2. *The Radon transform $\mathcal{R}$ defines a vector space isomorphism*

$$\mathcal{R} : \mathcal{F}(-3) \to \mathcal{F}(-1).$$

PROOF. Introduce the operator $\widetilde{\mathcal{R}}$ on $\mathcal{F}(-1)$ by the formula

$$(\widetilde{\mathcal{R}}\varphi)(x) = -(2\pi)^{-2} \int_{(\mathbb{P}^3)'} \varphi(\xi)\, \delta''(\langle \xi, x\rangle)\, \omega(\xi).$$

This definition implies that $\widetilde{\mathcal{R}}\varphi \in \mathcal{F}(-3)$ for $\varphi \in \mathcal{F}(-1)$.

It follows from what was proved in 8.4 that the operator $\mathcal{R}$ is injective, and the composition $\widetilde{\mathcal{R}}\mathcal{R}$ is the identity operator on $\mathcal{F}(-3)$. Therefore, to prove the theorem, it suffices to show that the operator $\widetilde{\mathcal{R}}$ has zero kernel.

Thus, let $\varphi \in \mathcal{F}(-1)$ and $\widetilde{\mathcal{R}}\varphi = 0$. Note that all derivatives $\frac{\partial^2 \varphi}{\partial \xi_i \partial \xi_j}$ of the function φ belong to $\mathcal{F}(-3)$, and hence the Radon transform of each of these derivatives is well defined.

Since $\frac{\partial^2}{\partial \xi_i \partial \xi_j}\, \delta(\langle \xi, x\rangle) = x_i x_j\, \delta''(\langle \xi, x\rangle)$, integration by parts shows that

$$\int_{\mathbb{P}^3} \frac{\partial^2 \varphi(\xi)}{\partial \xi_i \partial \xi_j}\, \delta(\langle \xi, x\rangle)\, \omega(\xi) = x_i x_j \int_{\mathbb{P}^3} \varphi(\xi)\, \delta''(\langle \xi, x\rangle)\, \omega(\xi) = 0.$$

Since the Radon transform is injective, $\frac{\partial^2 \varphi(\xi)}{\partial \xi_i \partial \xi_j} = 0$, $i, j = 1, 2, 3, 4$, i.e., φ is a polynomial. On the other hand, φ is of homogeneity degree -1, which is possible for $\varphi = 0$ only.

8.7. Radon transform for the projective plane $\mathbb{P}^2$. The description of the Radon transform for $\mathbb{P}^2$ is similar to that in the above case of the projective space $\mathbb{P}^3$; so we describe the transform for $\mathbb{P}^2$ more briefly. The only substantial difference between the cases of $\mathbb{P}^3$ and $\mathbb{P}^2$ is that the inversion formula is local for $\mathbb{P}^3$ and nonlocal for $\mathbb{P}^2$.

By analogy with the case of projective space $\mathbb{P}^3$, denote by $\mathcal{F}(m)$ the space of C^∞ functions on $\mathbb{R}^3 \setminus O$ satisfying the homogeneity condition (8.1).

The Radon transform for $\mathbb{P}^2$ is defined on the function space $\mathcal{F}(-2)$.

Let $\omega(x)$ be the Leray differential form on $\mathbb{R}^3$,

$$\omega(x) = x_1\, dx_2 \wedge dx_3 - x_2\, dx_1 \wedge dx_3 + x_3\, dx_1 \wedge dx_2,$$

or, briefly, $\omega(x) = [x, dx, dx]$. On every subspace

$$h_\xi : \langle \xi, x\rangle \equiv \xi_1 x_1 + \xi_2 x_2 + \xi_3 x_3 = 0$$

in $\mathbb{R}^3$ we introduce the differential 1-form $\sigma_\xi(x)$ by the rule

$$\sigma_\xi(x) = d(\langle \xi, x\rangle) \rfloor \omega(x),$$

i.e.,

$$d(\langle \xi, x\rangle) \wedge \sigma_\xi(x) = \omega(x).$$

If $f \in \mathcal{F}(-2)$, then the differential form $f(x)\sigma_\xi(x)$ is of homogeneity degree zero, and this form is orthogonal to the fibers of the bundle $h_\xi \to Ph_\xi$, where Ph_ξ is the projectivization of h_ξ. Hence, this form can be pushed down to Ph_ξ, i.e., to a line in $\mathbb{P}^2$.

DEFINITION. By the Radon transform of a function $f \in \mathcal{F}(-2)$ we mean a function $\mathcal{R} f$ on $\mathbb{R}^3 \setminus O$ given by the relation

$$\mathcal{R} f(\xi) = \int_{\langle \xi,\, x\rangle = 0} f(x)\, \sigma_\xi(x). \tag{8.8}$$

It follows from the definition that the function $\varphi = \mathcal{R} f$ belongs to $\mathcal{F}(-1)$; thus, the Radon transform yields a mapping

$$\mathcal{R} : \mathcal{F}(-2) \to \mathcal{F}(-1).$$

Similarly to the case of projective space $\mathbb{P}^3$, it is convenient to represent the expression for $\mathcal{R} f$ in the form

$$\mathcal{R} f(\xi) = \int_{\mathbb{P}^2} f(x)\, \delta(\langle \xi,\, x\rangle)\, \omega(x), \tag{8.9}$$

where $\delta(t)$ is the delta function on $\mathbb{R}$.

We now proceed with the construction of the inversion formula for $\mathcal{R}$.

As in the case of $\mathbb{P}^3$, we first present the formal answer,

$$f(x) = -\frac{1}{4\pi^2} \int_\Gamma \mathcal{R} f(\xi)\, \langle \xi,\, x\rangle^{-2}\, \omega(\xi), \tag{8.10}$$

where Γ is an arbitrary section of the bundle $\mathbb{R}^3 \setminus O \to \mathbb{P}^2$.

To make sense of this integral, we repeat the arguments used in 8.4 for $\mathbb{P}^3$. Namely, set $\xi = \eta + p\,\zeta$, where $\eta \in (\mathbb{R}^3)' \setminus O$ satisfies the condition $\langle \eta,\, x\rangle = 0$, p belongs to $\mathbb{R}$, and $\zeta \in (\mathbb{R}^3)' \setminus O$ is an arbitrary (but fixed) vector such that $\langle \zeta,\, x\rangle \neq 0$.

We pass from the variables $\xi \in \mathbb{R}^3$ to the variables $\eta \in \mathbb{R}^3$ and $p \in \mathbb{R}$ in the expression for the differential form

$$\Omega_x = \mathcal{R} f(\xi) \langle \xi, x\rangle^{-2}\, \omega(\xi)$$

and then delete the terms that do not contain the differential dp. We obtain

$$\Omega'_x = \langle \zeta,\, x\rangle^{-2}\, \mathcal{R} f(\eta + p\zeta)\, p^{-2}\, [\eta, \zeta, d\eta] \wedge dp.$$

We interpret the integral of this form as follows. One first takes the integral over p in accordance with the definition of the distribution p^{-2}, i.e.,

$$\int_{-\infty}^{+\infty} F(p)\, p^{-2}\, dp = \int_{-\infty}^{+\infty} F'(p)\, p^{-1}\, dp,$$

where the integral on the right-hand side must be understood in the principal value sense. We obtain a differential 1-form on the line and integrate the form over the line.

Hence, the inversion formula (8.10) is reduced to the following formula.

THEOREM 8.3. *If $\varphi = \mathcal{R} f$, where $f \in \mathcal{F}(-2)$, then*

$$f(x) = -\frac{1}{4\pi^2}\, \langle \zeta,\, x\rangle^{-2} \int_{\langle \eta,\, x\rangle = 0} \psi(\eta, \zeta)\, [\eta, \zeta, d\eta], \tag{8.11}$$

where

$$\psi(\eta, \zeta) = \int_{-\infty}^{+\infty} \Big(\sum_{i=1}^{3} \zeta_i \frac{\partial \varphi}{\partial \xi_i}\Big|_{\xi = \eta + p\zeta} \Big)\, p^{-1}\, dp, \tag{8.12}$$

and $\zeta \in \mathbb{R}^3 \setminus O$ *is an arbitrary (but fixed) vector satisfying the condition* $\langle \zeta, x \rangle \neq 0$. *The integral* (8.12) *must be understood in the principal value sense.*

As in the case of projective space (see 8.4), the proof of the inversion formula (8.11), (8.12) is reduced to the inversion formula in § 1 for the affine Radon transform on the plane.

THEOREM 8.4. *The Radon transform* $\mathcal{R}$ *for the projective plane* $\mathbb{P}^2$ *defines a vector space isomorphism*

$$\mathcal{R} : \mathcal{F}(-2) \to \mathcal{F}(-1).$$

PROOF. Let

$$\widetilde{\mathcal{R}} : \mathcal{F}(-1) \to \mathcal{F}(-2)$$

be the operator defined by (8.10). It follows from what was proved above that the composition $\widetilde{\mathcal{R}}\mathcal{R}$ is the identity operator on $\mathcal{F}(-2)$. Therefore, to prove the theorem, it remains to show that the operator $\widetilde{\mathcal{R}}$ is injective.

Let us use the fact that the function $\widetilde{\mathcal{R}}\varphi\Big|_{x_3=1}$ can be represented in the form

$$\widetilde{\mathcal{R}}\varphi(x_1, x_2, 1) = -\frac{1}{4\pi^2} \int_\Gamma \psi(\xi)\, \delta'(\xi_1 x_1 + \xi_2 x_2 + \xi_3)\, \omega(\xi), \tag{8.13}$$

where

$$\psi(\xi) = \int_{-\infty}^{+\infty} \varphi(\xi_1, \xi_2, p)\, (p - \xi_3)^{-1}\, dp. \tag{8.14}$$

Indeed, substituting (8.14) in (8.13) and integrating with respect to ξ_3, we obtain the original expression for $\widetilde{\mathcal{R}}\varphi\Big|_{x_3=1}$.

Equation (8.14) defines the Hilbert transform of the function $\varphi(\xi_1, \xi_2, \xi_3)$ with respect to ξ_3. This transform is known to be injective. Further, if $\varphi \in \mathcal{F}(-1)$, then ψ is an odd homogeneous C^∞ function on $(\mathbb{R}^3)' \setminus O$ whose homogeneity degree -1 is equal to that of φ. Therefore, every derivative $\frac{\partial \psi}{\partial \xi_i}$ belongs to $\mathcal{F}(-2)$.

Suppose that $\widetilde{\mathcal{R}}\varphi = 0$. Then, using the integration by parts, we obtain

$$\int_\Gamma \frac{\partial \psi}{\partial \xi_i}\, \delta(\xi_1 x_1 + \xi_2 x_2 + \xi_3 x_3)\, \omega(\xi)$$
$$= -\xi_i \int_\Gamma \psi(\xi)\, \delta'(\xi_1 x_1 + \xi_2 x_2 + \xi_3 x_3)\, \omega(\xi) = 0, \qquad i = 1, 2, 3,$$

i.e., the Radon transform of the function $\frac{\partial \psi}{\partial \xi_i}$ is zero. Since the Radon transform is injective, it follows that $\frac{\partial \psi}{\partial \xi_i} = 0$, $i = 1, 2, 3$, and hence ψ is constant. Since the function ψ is of homogeneity degree -1, we have $\psi = 0$. Since the transform (8.14) is injective, $\varphi = 0$.

As in the three-dimensional case, the inversion formulas (8.11), (8.12) for the projective Radon transform give new inversion formulas for the Radon transform related to the affine plane and for the Minkowski–Funk transform related to the two-dimensional sphere.

To obtain an inversion formula for the Minkowski–Funk transform, it suffices to interpret elements of $\mathcal{F}(-2)$ and $\mathcal{F}(-1)$ as functions on the two-dimensional sphere S^2. If $\varphi = \mathcal{M}f$ is the Minkowski–Funk transform of a function f on S^2, then the inversion formula for f is obtained from (8.11), (8.12) by replacing $\varphi(\xi)$ with $|\xi|^{-1}\varphi(\frac{\xi}{|\xi|})$.

8.8. Radon transform for the projective space of an arbitrary dimension. The Radon transforms for $\mathbb{P}^2$ and $\mathbb{P}^3$ introduced above can be extended to projective spaces of arbitrary dimension.

The Radon transform for $\mathbb{P}^n$ is defined on the space $\mathcal{F}(-n)$ of C^∞ functions $f(x)$ on $\mathbb{R}^{n+1}\setminus O$ satisfying the homogeneity condition

$$f(\lambda x) = |\lambda|^{-n} f(x).$$

Similarly to the cases of $\mathbb{P}^2$ and $\mathbb{P}^3$, this transform is defined by the formula

$$\mathcal{R} f(\xi) = \int_{\mathbb{P}^n} f(x)\,\delta((\xi,x))\,\omega(x), \qquad \xi\in\mathbb{R}^{n+1}\setminus O, \tag{8.15}$$

where $\delta(t)$ stands for the delta function on $\mathbb{R}$, $\langle\xi,x\rangle = \sum_{i=1}^{n+1}\xi_i x_i$, and $\omega(\xi) = \sum_{i=1}^{n+1}(-1)^{i-1}\xi_i \wedge_{j\neq i} d\xi_j$. The integral over $\mathbb{P}^n$ should be understood as the integral over an arbitrary section of the bundle $\mathbb{R}^{n+1}\setminus O \to \mathbb{P}^n$; this integral does not depend on the choice of the section.

By this definition, $\varphi = \mathcal{R} f$ is a C^∞ function on $\mathbb{R}^{n+1}\setminus O$ satisfying the homogeneity condition

$$\varphi(\lambda\xi) = |\lambda|^{-1}\varphi(\xi).$$

Let us present without proof the inversion formulas for the projective Radon transform. If $\varphi = \mathcal{R} f$, then

$$f(x) = \frac{(-1)^{\frac{n-1}{2}}}{2(2\pi)^{n-1}}\int_\Gamma \varphi(\xi)\,\delta^{(n-1)}((\xi,x))\,\omega(\xi) \tag{8.16}$$

for odd n, and

$$f(x) = \frac{(-1)^{n/2}\,(n-1)!}{(2\pi)^n}\int_\Gamma \varphi(\xi)\,((\xi,x))^{-n}\,\omega(\xi) \tag{8.17}$$

for even n. The meaning of these expressions was discussed in detail for $n=2$ and $n=3$. Similarly to these cases, the inversion formula is local for odd n and nonlocal for even n.

As in the case of affine Radon transform, these inversion formulas can be combined by using the distribution

$$(t-i0)^{-n} = t^{-n} + \frac{(-1)^{n-1}\pi i}{(n-1)!}\,\delta^{(n-1)}(t).$$

The following inversion formula holds for any n:

$$f(x) = (n-1)!\left(\frac{i}{2\pi}\right)^n \int_{\mathbb{P}^n}\varphi(\xi)\,((\xi,x)-i0)^{-n}\,\omega(\xi). \tag{8.18}$$

9. Radon transform on the complex affine space

The definitions and results of this chapter can be extended from the real case to the complex case. In this section we briefly present the definition of the Radon transform for the complex affine space, the relation between this transform and the Fourier transform, and the inversion formula for the Radon transform.

In essence, the only distinction between the Radon transforms in the real and complex spaces is that the inversion formula for the Radon transform in the complex case is always local and has the same structure for the space of arbitrary dimension. Therefore, we present the results for spaces of arbitrary dimension $n > 1$. The statements of the theorems are given without proofs.

9.1. Definition of the Radon transform. We represent hyperplanes in $\mathbb{C}^n$ by equations of the form

$$\langle \xi,\, x\rangle - p = 0,$$

where $\langle \xi,\, x\rangle = \xi_1\, x_1 + \xi_2\, x_2 + \cdots + \xi_n\, x_n$.

The Radon transform of a function f on $\mathbb{C}^n$ is defined by the following formula:

$$\mathcal{R} f(\xi,\, p) = \left(\frac{i}{2}\right)^{n-1} \int_{\langle \xi,\, x\rangle = p} f(x)\, \sigma(x) \wedge \overline{\sigma(x)}, \tag{9.1}$$

where σ is a holomorphic differential $(n-1)$-form on the hyperplane $\langle \xi,\, x\rangle = p$; the form σ is defined by the relation

$$dx = d\langle \xi,\, x\rangle \wedge \sigma \qquad (dx = dx_1 \wedge dx_2 \wedge \cdots \wedge dx_n).$$

In the coordinates $x_1, \ldots, x_n$, the form σ is given as follows:

$$\sigma = (-1)^{k-1}\, \xi_k^{-1}\, dx_1 \wedge \cdots \wedge \widehat{dx_k} \wedge \cdots \wedge dx_n \qquad (k = 1, 2, \ldots, n).$$

It is also convenient to use the expression for $\mathcal{R} f$ of the form

$$\mathcal{R} f(\xi, p) = \left(\frac{i}{2}\right)^{n} \int_{\mathbb{C}^n} f(x)\, \delta(\langle \xi,\, x\rangle - p)\, dx\, d\overline{x}, \tag{9.2}$$

where $\delta(\cdot)$ is the delta function on $\mathbb{C}^n$.

The function $\mathcal{R} f$ satisfies the homogeneity condition

$$\mathcal{R} f(\lambda\, \xi,\, \lambda p) = |\lambda|^{-2}\, \mathcal{R} f(\xi,\, p), \qquad \lambda \in \mathbb{C} \setminus 0. \tag{9.3}$$

9.2. Relation to the Fourier transform. The Fourier transform $\mathcal{F} f$ of a function f on $\mathbb{C}^n$ is defined as follows:

$$\mathcal{F} f(\xi) = (2\pi)^{-n} \left(\frac{i}{2}\right)^{n} \int_{\mathbb{C}^n} f(x)\, e^{i \operatorname{Re} \langle \xi,\, x\rangle}\, dx\, d\overline{x}.$$

This transform is related to the Radon transform by the formula

$$\mathcal{F} f(\xi) = (2\pi)^{-n}\, \frac{i}{2} \int \mathcal{R} f(\xi,\, p)\, e^{i \operatorname{Re} p}\, dp\, d\overline{p}. \tag{9.4}$$

Using the homogeneity (9.3) of $\mathcal{R} f$, one can represent the relation between $\mathcal{R} f$ and $\mathcal{F} f$ in the form

$$\mathcal{F} f(\lambda\, \xi) = (2\pi)^{-n}\, \frac{i}{2} \int \mathcal{R} f(\xi,\, p)\, e^{i \operatorname{Re} \lambda\, p}\, dp\, d\overline{p}. \tag{9.5}$$

Hence, by the inversion formula for the one-dimensional Fourier transform we obtain

$$\mathcal{R} f(\xi,\, p) = (2\pi)^{n-2}\, \frac{i}{2} \int \mathcal{F} f(\lambda\, \xi) e^{-i \operatorname{Re} \lambda\, p}\, d\lambda\, d\overline{\lambda}.$$

9.3. Inversion formula for the Radon transform. It is known that the operator $\mathcal{F}^{-1}$ inverse to $\mathcal{F}$ is

$$\mathcal{F}^{-1}\varphi(x) = (2\pi)^{-n} \int_{\mathbb{C}^n} \varphi(\xi)\, e^{-i\,\mathrm{Re}\,\langle \xi,\, x\rangle}\, d\xi\, d\bar{\xi}. \tag{9.6}$$

Using relations (9.6) and (9.4), we obtain the following inversion formula for the Radon transform $\mathcal{R}$:

$$f(x) = (-1)^{n-1}\,\pi^{-2n+2}\Big(\frac{i}{2}\Big)^{n-1} \int_{\Gamma} \frac{\partial^{2n-2}}{\partial p^{n-1}\,\partial \bar{p}^{n-1}}\, \mathcal{R}\, f(\xi,\, \langle \xi,\, x\rangle)\, \omega(\xi) \wedge \overline{\omega(\xi)}, \tag{9.7}$$

where $\omega(\xi) = \sum_{k=1}^{n} (-1)^{k-1}\, \xi_k\, d\xi_1 \wedge \cdots \wedge \widehat{d\xi_k} \wedge \cdots \wedge d\xi_n$, and the integral is taken over an arbitrary surface $\Gamma \subset \mathbb{C}^n \setminus O$ intersecting once almost every complex line passing through the point 0.

9.4. Case $n = 2$. For $n = 2$, the inversion formula for the Radon transform becomes

$$f(x_1,\, x_2) = -\pi^{-2}\,\frac{i}{2} \int_{\Gamma} (\mathcal{R}\, f)''_{p,\bar{p}}(\xi_1,\, \xi_2\,;\, \xi_1\, x_1 + \xi_2\, x_2) \times (\xi_1\; d\xi_2 - \xi_2\; d\xi_1) \wedge (\overline{\xi_1\; d\xi_2 - \xi_2\; d\xi_1}). \tag{9.8}$$

We represent this formula in another system of coordinates on the manifold of lines. Let us define lines in $\mathbb{C}^2$ by equations of the form

$$x_2 = \alpha\, x_1 + \beta \tag{9.9}$$

and take the coefficients $\alpha,\ \beta$ of these equations as local coordinates on the manifold of lines.

Define the integral of a function f on $\mathbb{C}^2$ over the line (9.9) by the formula

$$\widehat{\mathcal{R}}f(\alpha,\, \beta) = \frac{i}{2} \int f(t,\, \alpha\, t + \beta)\, dt\, d\bar{t}. \tag{9.10}$$

Obviously, the functions $\widehat{\mathcal{R}}f(\alpha,\, \beta)$ and $\mathcal{R}\, f(\xi_1,\, \xi_2;\, p)$ are related as follows:

$$\mathcal{R}\, f(\xi_1,\, \xi_2;\, p) = |\xi|^{-2}\, \widehat{\mathcal{R}}f\Big(-\frac{\xi_1}{\xi_2},\, \frac{p}{\xi_2}\Big).$$

Substituting this expression for $\mathcal{R}\, f$ into (9.8) and taking the line $\xi_2 = 1$ for Γ, we obtain the inversion formula

$$f(x_1,\, x_2) = -\pi^{-2}\,\frac{i}{2} \int_{\mathbb{C}} (\widehat{\mathcal{R}}f)''_{p,\bar{p}}(\alpha,\, x_2 - \alpha\, x_1)\, d\alpha\, d\bar{\alpha}. \tag{9.11}$$

9.5. Relation to Paley–Wiener theorems for the affine Radon transform in $\mathbb{R}^2$ and $\mathbb{R}^3$. In 4.2 we described the image of the space $S(\mathbb{R}^2)$ of rapidly decreasing functions on the plane under the affine Radon transform (the Paley–Wiener theorem). The proof was based on the relation between the Radon transform and the Fourier transform on the plane. A similar description of the image holds for the three-dimensional Radon transform.

We obtain this description as a consequence of the description of the image of the projective Radon transform; see Theorem 8.4 in the case of projective plane $\mathbb{P}^2$ and Theorem 8.2 in the case of projective space $\mathbb{P}^3$. For simplicity, we restrict ourselves to the case of projective space.

First we establish a relation between homogeneous functions $f \in \mathcal{F}(-3)$ on $\mathbb{R}^4 \setminus O$ and functions $f \in S(\mathbb{R}^3)$.

PROPOSITION 9.1. *A function $f_0(x_1, x_2, x_3)$ on $\mathbb{R}^3$ belongs to the space $S(\mathbb{R}^3)$ if and only if it is obtained as the restriction to the hyperplane $x_4 = 1$ of a function $f(x_1, x_2, x_3, x_4) \in \mathcal{F}(-3)$ satisfying the following condition*:

Condition A. *The function f vanishes for $x_4 = 0$ together with all its derivatives.*

PROOF. Let $f(x_1, x_2, x_3, x_4) \in \mathcal{F}(-3)$ and let f_0 be the restriction of the function f to the hyperplane $x_4 = 1$, i.e.,

$$f_0(x_1, x_2, x_3) = f(x_1, x_2, x_3, 1).$$

Then

$$f_0(r\xi_1, r\xi_2, r\xi_3) = r^{-3} f(\xi_1, \xi_2, \xi_3, 1/r) \quad \text{for any} \quad r > 0 \quad \text{and} \quad |\xi| = 1.$$

If the function f satisfies Condition A, then it follows from this relation that $f_0 \in S(\mathbb{R}^3)$.

Conversely, let $f_0 \in S(\mathbb{R}^3)$. We define the function f on $\mathbb{R}^4 \setminus 0$ by

$$f(x_1, x_2, x_3, x_4) = |x_4|^{-3} f_0(x_1/x_4, x_2/x_4, x_3/x_4) \quad \text{for} \quad x_4 \neq 0;$$
$$f(x_1, x_2, x_3, 0) = 0.$$

Then $f \in \mathcal{F}(-3)$, and the function f_0 is the restriction of f to the hyperplane $x_4 = 0$.

Note that the projective Radon transform $\mathcal{R}f$ of a function $f \in \mathcal{F}(-3)$ and the affine Radon transform of the restriction f_0 of f to the hyperplane $x_4 = 1$ coincide. Thus, by Proposition 9.1, the description of the image, under the affine Radon transform, of the space $S(\mathbb{R}^3)$ reduces to the description of the image, under the projective Radon transform, of the subspace of functions $f \in \mathcal{F}(-3)$ satisfying Condition A. The answer is given by the following proposition.

PROPOSITION 9.2. *A function $f \in \mathcal{F}(-3)$ satisfies Condition A if and only if its projective Radon transform*

$$\mathcal{R}f(\xi_1, \xi_2, \xi_3, p) = \int_{P^3} f(x)\, \delta(\langle \xi, x\rangle + px_4)\, \omega(x),$$

where $\langle \xi, x\rangle = \xi_1 x_1 + \xi_2 x_2 + \xi_3 x_3$, rapidly decreases with respect to p as $p \to \infty$ uniformly with respect to ξ and the Cavalieri conditions hold, i.e.,

$$I_k(\xi) = \int_{-\infty}^{+\infty} \mathcal{R}f(\xi, p)\, p^k\, dp, \quad k = 0, 1, 2, \dots,$$

is a homogeneous polynomial of degree k in ξ_1, ξ_2, ξ_3.

PROOF. Suppose that a function $f \in \mathcal{F}(-3)$ satisfies Condition A. For any k, $k = 0, 1, 2, \dots$, represent the product $\mathcal{R}f(\xi, p)\, p^k$ in the form

$$\mathcal{R}f(\xi, p)\, p^k = \int_{P^3} f(x) \left(-\frac{\langle \xi, x\rangle}{x_4}\right)^k \delta(\langle \xi, x\rangle + px_4)\, \omega(x). \tag{9.12}$$

Note that by Condition A, the function $f(x)\,(-\frac{\langle \xi, x\rangle}{x_4})^k$ belongs to the space $\mathcal{F}(-3)$ for any chosen ξ. Hence, the expression on the right-hand side of (9.12) is bounded with respect to p; thus, the function $\mathcal{R}f(\xi, p)$ rapidly decreases with respect to p as $p \to \infty$.

Integrating (9.12) with respect to p, we obtain

$$\int_{-\infty}^{+\infty} \mathcal{R}f(\xi,p)\,p^k\,dp = \int_{P^3} f(x)\left(-\frac{\langle \xi, x\rangle}{x_4}\right)^k |x_4|^{-1}\,\omega(x).$$

Hence, the integral $\int_{-\infty}^{+\infty} \mathcal{R}f(\xi,p)\,p^k\,dp$ is a homogeneous polynomial of degree k in ξ. Thus, the function $\mathcal{R}f(\xi,p)$ satisfies the hypothesis of the proposition.

Conversely, assume that the Radon transform of the function $f \in \mathcal{F}(-3)$ satisfies the hypothesis of the proposition. We show that in this case the function f satisfies Condition A, i.e., the function and all its derivatives vanish for $x_4 = 0$.

To this end, we use the inversion formula (8.6) for the projective Radon transform, where for Γ we take the cylindrical surface $\gamma:\ \xi_1^2+\xi_2^2+\xi_3^2 = 1$ in $\mathbb{R}^4$,

$$f(x) = -(2\pi)^{-2}\int_\gamma\int_{-\infty}^{+\infty} \mathcal{R}f(\xi,p)\,\delta''(\langle\xi,x\rangle + px_4)\,dp\wedge\omega_0(\xi)$$

for $\omega_0(\xi) = \xi_1\,d\xi_2\wedge d\xi_3 - \xi_2\,d\xi_1\wedge d\xi_3 + \xi_3\,d\xi_1\wedge d\xi_2$. Differentiating this relation with respect to x_1, x_2, x_3, x_4 and then substituting $x_4 = 0$, we obtain

$$\prod_{i=1}^4\left(\frac{\partial}{\partial x_i}\right)^{k_i} f(x)|_{x_4=0}$$
$$= -(2\pi)^{-2}\int_\gamma\int_{-\infty}^{+\infty} \mathcal{R}f(\xi,p)\,p^{k_4}\xi_1^{k_1}\xi_2^{k_2}\xi_3^{k_3}\,\delta^{(k+2)}(\langle\xi,x\rangle)\,dp\wedge\omega_0(\xi),$$

where $k = \sum k_i$. Integrating with respect to p, we see by the Cavalieri conditions that

$$\prod_{i=1}^4\left(\frac{\partial}{\partial x_i}\right)^{k_i} f(x)|_{x_4=0} = -(2\pi)^{-2}\int_\gamma P_{k_4}(\xi)\xi_1^{k_1}\xi_2^{k_2}\xi_3^{k_3}\,\delta^{(k+2)}(\langle\xi,x\rangle)\,\omega_0(\xi),$$

where $P_{k_4}(\xi)$ is a homogeneous polynomial of degree k_4. Since the degree of the polynomial $P_{k_4}(\xi)\xi_1^{k_1}\xi_2^{k_2}\xi_3^{k_3}$ is equal to k, it follows that this integral vanishes for any $(x_1,x_2,x_3) \neq 0$. Thus, the function f satisfies Condition A.

CHAPTER 2

John Transform

In this chapter we study the John transform (also called the X-ray transform) $\mathcal{J}$ of functions on $\mathbb{R}^3$ and analogs of this transform for functions on the real projective space $\mathbb{P}^3$ and the complex space $\mathbb{C}^3$.

To each compactly supported or rapidly decreasing function f on $\mathbb{R}^3$, the John transform assigns the integrals of this function over all possible lines in $\mathbb{R}^3$. This gives a function $\varphi = \mathcal{J}f$ on the manifold of lines. The transform $\mathcal{J}$ was first studied by F. John [**28**], who established a close relation between this transform and the solutions of the ultrahyperbolic differential equation.

As in the case of Radon transform $\mathcal{R}$, we consider the following two main problems: Describe the image of $\mathcal{J}$ and reconstruct a function f from its image $\mathcal{J}f$. However, the solution of these problems for $\mathcal{J}$ is substantially different. The main distinction between $\mathcal{J}$ and $\mathcal{R}$ (which also distinguishes $\mathcal{J}$ among the transforms treated in the subsequent chapters) is that the John transform sends a function of three variables to a function of four variables (because the manifold of lines in $\mathbb{R}^3$ is of dimension four). First, this implies that a function belonging to the image of the John transform satisfies an additional differential relation (the Fritz John equation). Second, the problem to reconstruct a function f on $\mathbb{R}^3$ from the John transform $\mathcal{J}f$ is overdetermined. Namely, in general, to reconstruct the function f, it suffices to define $\mathcal{J}f$ on a three-dimensional submanifold K of the four-dimensional manifold of lines H rather than on the entire manifold H. Certainly, both the existence of an inversion formula and the explicit form of this formula may substantially depend on the structure of the submanifold K.

We now briefly describe the contents of this chapter. In § 1, the definition of the John transform of a function on $\mathbb{R}^3$ is given. It is shown that one of the main special functions in analysis, namely, the Gauss hypergeometric function, naturally arises as the John transform of a certain function on $\mathbb{R}^3$. The main problem of this section is to describe the image of the Schwartz space $S(\mathbb{R}^3)$ under the John transform, and the solution of this problem is given in 1.3–1.6. We do not consider here the nonlocal problem of the inversion of the John transform. However, we will consider the complex version of the problem.

Following the scheme of Chapter 1, we then consider the John transform of differential forms on $\mathbb{R}^3$ in § 2. This transform is well defined not only for 1-forms but also for the forms of degrees 2 and 3 (cf. the case of Radon transform). In each of these cases we give the description of the image of the John transform and obtain a formula reconstructing the original differential form from its John transform.

In § 3, the main definitions and results of § 1 are extended to the case of projective space $\mathbb{P}^3$. The passage from the affine space to the projective one has two important advantages. First, the main definitions and results have a simpler and more invariant form in the projective case as compared to those in the affine case.

Second, since both the projective space $\mathbb{P}^3$ itself and the manifold of lines in $\mathbb{P}^3$ are compact, with these manifolds one can associate more natural function spaces as compared with those in the affine case. In § 3, the description of the image of the John transform is given in a projective-invariant form. A simple group-theoretic interpretation of the John transform is also presented.

The last two sections of Chapter 2 are devoted to the analog of the John transform for the complex space $\mathbb{C}^3$. The merit of the complex version is that the complex case admits local inversion formulas; namely, to reconstruct a function f at a point $x \in \mathbb{C}^3$ from the John transform $\varphi = \mathcal{J}f$, it suffices to know the function φ only on a set of lines infinitesimally close to x.

A fundamental role here is played by the operator κ introduced by I. M. Gelfand, M. I. Graev, and Z. Ya. Shapiro [**7**] to solve a more general problem; to every function φ on H this operator assigns a differential (1,1)-form $\kappa\varphi$ on the manifold of lines H. In terms of the operator κ we give a description of the image of $S(\mathbb{C}^3)$ under the John transform and construct an explicit local inversion formula for this transform in § 4. It turns out that the differential form $\kappa\varphi$, where $\varphi = \mathcal{J}f$, is closed on the submanifold $H^x \subset H$ of all lines passing through a given point x for any $x \in \mathbb{C}^3$; the function f can be reconstructed by the following inversion formula:

$$\int_{\gamma^x} \kappa\varphi = c(\gamma^x)\, f(x),$$

where $\gamma^x \subset H^x$ is an arbitrary cycle of real dimension two, and the factor $c(\gamma^x)$ depends on the homology class of the cycle γ^x only.

Note that an analog of the operator κ was introduced in § 1 for the real case; however, in that case the operator κ could be used only to describe the image of the John transform.

In § 5 we first consider the submanifolds K of lines in $\mathbb{C}^3$ that intersect an arbitrary algebraic curve $\Lambda \subset \mathbb{C}^3$. For these submanifolds we give an explicit local inversion formula reconstructing a function on $\mathbb{C}^3$ from the integrals of this function over the lines in K. This example naturally leads to the following problem: Describe all "good" three-dimensional submanifolds K (admissible complexes in the manifold H of all lines) for which there is an explicit local inversion formula reconstructing a function on $\mathbb{C}^3$ from the integrals of this function over the lines in K. The geometric structure of these admissible complexes is studied; it is shown that each admissible complex is either the manifold of lines intersecting a complex curve or the manifold of lines tangent to a complex surface.

1. John transform in the real affine space

1.1. John transform in $\mathbb{R}^3$. We present lines in $\mathbb{R}^3$ by parametric equations

$$x = \alpha t + \beta, \quad t \in \mathbb{R},$$

where $\beta = (\beta_1, \beta_2, \beta_3)$ is a point on the line in question and $\alpha = (\alpha_1, \alpha_2, \alpha_3),\ \alpha \neq 0$, is the direction vector of the line.

DEFINITION. We define the integral of a function $f \in S(\mathbb{R}^3)$ over an arbitrary line $x = \alpha t + \beta$ by the formula

$$\varphi(\alpha, \beta) = \int_{-\infty}^{+\infty} f(\alpha t + \beta)\, dt; \tag{1.1}$$

the function $\varphi(\alpha,\beta)$ is called the *John transform* of the function f and is denoted by $\mathcal{J}f$.

According to the definition, φ is a function on the manifold E of all pairs $(\alpha,\beta)\in(\mathbb{R}^3\setminus O)\times\mathbb{R}^3$. Obviously, the function φ does not depend on the choice of a point β on the line, i.e.,

$$\varphi(\alpha,\beta+\alpha t_0)=\varphi(\alpha,\beta)\quad\text{for any}\quad t_0\in\mathbb{R}. \tag{1.2}$$

Thus, for any given α, φ is a function on the quotient space $\mathbb{R}^3/\{\alpha\}$, where $\{\alpha\}$ is the subspace spanned by the vector α.

It is also clear that the function φ satisfies the homogeneity condition

$$\varphi(\lambda\alpha,\beta)=|\lambda|^{-1}\,\varphi(\alpha,\beta),\qquad \lambda\neq 0. \tag{1.3}$$

Therefore, the function φ depends only on the line and on the admissible measure on the line defined by the vector α, i.e., this is a function on the one-dimensional bundle over the manifold of lines. Moreover, if $\mathbb{R}^3$ is also equipped with the Euclidean metric, then in the definition (1.1) one can assume that $|\alpha|=1$; in this case, φ becomes a function on the manifold of lines itself.

REMARK. Conditions (1.2) and (1.3) are equivalent to the following system of equations:

$$\sum_{i=1}^{3}\alpha_i\frac{\partial\varphi}{\partial\beta_i}=0,\qquad \sum_{i=1}^{3}\alpha_i\frac{\partial\varphi}{\partial\alpha_i}=-\varphi;\qquad \varphi(-\alpha,\beta)=\varphi(\alpha,\beta). \tag{1.4}$$

Sometimes we shall present lines in $\mathbb{R}^3$ by equations

$$x_1=\alpha_1x_3+\beta_1,\qquad x_2=\alpha_2x_3+\beta_2,$$

and the John transform by the formula

$$\psi(\alpha_1,\alpha_2,\beta_1,\beta_2)=\int_{-\infty}^{+\infty}f(\alpha_1x_3+\beta_1,\,\alpha_2x_3+\beta_2,\,x_3)\,dx_3. \tag{1.5}$$

It is clear that the functions ψ and φ, where the latter is defined by (1.1), are related as follows:

$$\psi(\alpha_1,\alpha_2,\beta_1,\beta_2)=\varphi(\alpha_1,\alpha_2,1,\beta_1,\beta_2,0);$$

$$\varphi(\alpha_1,\alpha_2,\alpha_3;\beta_1,\beta_2,\beta_3)=|\alpha_3|^{-1}\,\psi\Big(\frac{\alpha_1}{\alpha_3},\frac{\alpha_2}{\alpha_3},\,\beta_1-\frac{\alpha_1}{\alpha_3}\,\beta_3,\,\beta_2-\frac{\alpha_2}{\alpha_3}\,\beta_3\Big).$$

1.2. John transform and the Gauss hypergeometric function. The integrals (1.1) and (1.5) define the John transform not only for rapidly decreasing functions $f\in S(\mathbb{R}^3)$ but also for any functions f on $\mathbb{R}^3$ for which these integrals converge or can be understood in some sense. In particular, the John transform ψ_λ of the following function on $\mathbb{R}^3$ is well defined:

$$f_\lambda(x_1,x_2,x_3)=(x_1)_+^{\lambda_1-1}\,(x_2)_+^{\lambda_2-1}\,(x_3)_+^{\lambda_3-1},$$

where $t_+^\lambda=t^\lambda$ for $t>0$ and $t_+^\lambda=0$ for $t<0$; the λ_i's stand for arbitrary complex numbers. For $\operatorname{Re}\lambda>0$ the function t_+^λ is defined as a classical function; it extends to a distribution meromorphic in λ (see [**10**]). We establish a relation between this John transform and the Gauss hypergeometric function $F(a,b,c;x)$.

By definition, the function ψ_λ is given by the integral

$$\psi_\lambda(\alpha_1, \alpha_2, \beta_1, \beta_2) = \int_0^\infty (\alpha_1 x_3 + \beta_1)_+^{\lambda_1 - 1} (\alpha_2 x_3 + \beta_2)_+^{\lambda_2 - 1} x_3^{\lambda_3 - 1} \, dx_3.$$

The integral converges if $\operatorname{Re}\lambda_i > 0$, $i = 1, 2, 3$, and $\operatorname{Re}(\lambda_1 + \lambda_2 + \lambda_3) < 2$, and in this domain the integral defines an analytic function of $\lambda_1, \lambda_2, \lambda_3$. Therefore, the integral can be defined for any values of the λ_i's as the analytic continuation of this function with respect to the λ_i's.

Note that the interval on which the integrand is nonzero depends on the signs of the numbers α_i and β_i, $i = 1, 2$. To be definite, consider the domain on which $\alpha_i < 0 < \beta_i$, $i = 1, 2$, in the manifold of lines H. In this domain, changing the variable $x_3 = -\frac{\beta_2}{\alpha_2} t$, we can reduce the integral to the form

$$\psi_\lambda(\alpha_1, \alpha_2, \beta_1, \beta_2) = \beta_1^{\lambda_1 - 1} \beta_2^{\lambda_2 + \lambda_3 - 1} |\alpha_2|^{-\lambda_3} \int_0^1 (1 - xt)_+^{\lambda_1 - 1} (1 - t)^{\lambda_2 - 1} t^{\lambda_3 - 1} \, dt,$$

where $x = \frac{\alpha_1 \beta_2}{\alpha_2 \beta_1}$.

Compare this expression with the well-known representation of the Gauss hypergeometric function $F(a, b, c; x)$ for $|x| < 1$ as the Euler integral,

$$F(a, b, c; x) = \frac{\Gamma(c)}{\Gamma(b)\,\Gamma(c - b)} \int_0^1 t^{b-1} (1 - t)^{c-b-1} (1 - xt)^{-a} \, dt;$$

this integral converges for $\operatorname{Re} c > \operatorname{Re} b > 0$, and for the other values of b and c it must be treated as the analytic continuation with respect to b and c.

We conclude that, for $\alpha_i < 0 < \beta_i$, $i = 1, 2$, and $\frac{\alpha_1 \beta_2}{\alpha_2 \beta_1} < 1$, the function ψ_λ can be expressed in terms of the hypergeometric function,

$$\begin{aligned} &\psi_\lambda(\alpha_1, \alpha_2, \beta_1, \beta_2) \\ &\quad = \frac{\Gamma(\lambda_2)\,\Gamma(\lambda_3)}{\Gamma(\lambda_2 + \lambda_3)} \beta_1^{\lambda_1 - 1} \beta_2^{\lambda_2 + \lambda_3 - 1} |\alpha_2|^{-\lambda_3} F\left(-\lambda_1 + 1, \lambda_3, \lambda_2 + \lambda_3; \frac{\alpha_1 \beta_2}{\alpha_2 \beta_1}\right). \end{aligned}$$

This implies that

$$F(a, b, c; x) = \frac{\Gamma(c)}{\Gamma(b)\,\Gamma(c - b)} \psi_\lambda(-x, -1, 1, 1) \quad \text{for} \quad 0 < x < 1,$$

where $\lambda = (1 - a,\, c - b,\, b)$.

Similar expressions for the function ψ_λ in terms of the hypergeometric hypergeometric function can be obtained for other combinations of the signs of α_i and β_i, $i = 1, 2$.

1.3. Theorem on the image of the operator $\mathcal{J}$. This subsection, together with the next three, is devoted to the description of the image of the space $S(\mathbb{R}^3)$ of Schwartz functions under the John transform $\mathcal{J}$.

First, it is clear that the functions in the image of the space $S(\mathbb{R}^3)$ satisfy conditions of smoothness and rapid decreasing. Further, since the John transform acts from a space of functions of three variables to a space of functions of four variables, one can naturally expect that the functions in the image satisfy additional conditions.

We prove that each function φ in the image of the John transform satisfies the following system of differential equations:

$$\frac{\partial^2 \varphi}{\partial \alpha_i \, \partial \beta_j} = \frac{\partial^2 \varphi}{\partial \alpha_j \, \partial \beta_i}, \quad i, j = 1, 2, 3. \tag{1.6}$$

Indeed, if $\varphi = \mathcal{J} f$, then it follows from (1.1) that

$$\frac{\partial^2 \varphi(\alpha, \beta)}{\partial \alpha_i \, \partial \beta_j} = \int_{-\infty}^{+\infty} t \, f''_{x_i, x_j}(\alpha t + \beta) \, dt.$$

Thus, the function $\frac{\partial^2 \varphi}{\partial \alpha_i \, \partial \beta_j}$ is symmetric with respect to the transposition of the indices i and j.

We claim that relations (1.5) are not only necessary but also sufficient for a function φ to belong to the image of the space $S(\mathbb{R}^3)$. Let us proceed with detailed definitions and statements.

Denote by H the manifold of lines in $\mathbb{R}^3$ and by $G \simeq P^2$ the manifold of one-dimensional subspaces in $\mathbb{R}^3$, i.e., the submanifold of lines passing through the point 0. The manifold H is equipped with the structure of a canonical vector bundle over G. Namely, the projection $\pi : H \to G$ takes every line $h \in H$ to the parallel subspace $a \in G$. Thus, the fiber of the bundle π over a is the quotient space $H_a = \mathbb{R}^3 / a$.

Since H is a vector bundle with compact base space, it follows that the notion of rapidly decreasing function on S is well defined, namely, a function on H is called *rapidly decreasing* if this function rapidly decreases on each fiber of the bundle $H \to G$.

DEFINITION. Denote by $S(H)$ the space of functions $\varphi(\alpha, \beta)$ on E satisfying conditions (1.2) and (1.3) and such that the function $\varphi(\alpha, \beta)$, $|\alpha| = 1$, regarded as a function on H, is infinitely differentiable and rapidly decreasing together with all its derivatives.

It immediately follows from the definition of the operator $\mathcal{J}$ that the image of the space $S(\mathbb{R}^3)$ under the John transform belongs to $S(H)$,

$$\mathcal{J} : S(\mathbb{R}^3) \to S(H).$$

THEOREM 1.1. *A function $\varphi \in S(H)$ belongs to the image of the space $S(\mathbb{R}^3)$ under the John transform $\mathcal{J}$, i.e., is representable in the form $\varphi = \mathcal{J} f$, where $f \in S(\mathbb{R}^3)$, if and only if this function satisfies system* (1.6).

The proof is given in 1.4–1.6.

REMARK. It is interesting to compare this description of the image of the John translform with the description of the image of the Radon transform given in Chapter 1. In the latter case, the main component in the description of the image is the nonlocal Cavalieri conditions, whereas in the former case, there are only local conditions (1.6). Of course, it is possible to write down analogs of Cavalieri's conditions for the John translfrom, but they will be consequences of the differential equations (1.6).

We now present another equivalent formulation of the image theorem. Introduce the operator κ which takes every smooth function $\varphi(\alpha, \beta)$ on E to the following differential 1-form on E:

$$\kappa\varphi = \frac{\partial\varphi}{\partial\beta_1}\,d\alpha_1 + \frac{\partial\varphi}{\partial\beta_2}\,d\alpha_2 + \frac{\partial\varphi}{\partial\beta_3}\,d\alpha_3. \tag{1.7}$$

LEMMA. *If $\varphi \in S(H)$, then $\kappa\varphi$ can be pushed down from E to the manifold $\widetilde{H}$ of oriented lines in $\mathbb{R}^3$.*

Indeed, it follows from (1.2) and (1.3) that the form $\kappa\varphi$ is invariant with respect to the transformations $\beta \mapsto \beta + \alpha t_0$ and $\alpha \mapsto \lambda\alpha$, $\lambda > 0$.

Note that the form $\kappa\varphi$ changes the sign if the orientation of the line alternates $(\alpha \mapsto -\alpha)$.

For any point $x \in \mathbb{R}^3$, denote by H^x the submanifold of lines passing through x, and by $\kappa_x\varphi$ the restriction of the form $\kappa_x\varphi$ to H^x.

PROPOSITION 1.1. *Condition (1.5) is equivalent to the condition that the differential form $\kappa_x\varphi$ on the submanifold H^x is closed for any $x \in \mathbb{R}^3$, i.e., to the condition $d\kappa_x\varphi = 0$.*

PROOF. We have

$$\kappa_x\varphi = \sum_{i=1}^{3} \frac{\partial\varphi(\alpha,\beta)}{\partial\beta_i}\bigg|_{\beta=x} d\alpha_i.$$

Hence, the exterior differential of the form $\kappa_x\varphi$ is given by

$$\begin{aligned} d\kappa_x\varphi &= \sum_{i,j} \frac{\partial^2\varphi(\alpha,\beta)}{\partial\alpha_i\,\partial\beta_j}\bigg|_{\beta=x} d\alpha_i \wedge d\alpha_j \\ &= \sum_{i<j}\Big(\frac{\partial^2\varphi(\alpha,\beta)}{\partial\alpha_i\,\partial\beta_j} - \frac{\partial^2\varphi(\alpha,\beta)}{\partial\alpha_j\,\partial\beta_i}\Big)\bigg|_{\beta=x} d\alpha_i \wedge d\alpha_j. \end{aligned}$$

This immediately implies the assertion.

By Proposition 1.1, Theorem 1.1 admits the following equivalent formulation.

THEOREM 1.2. *A function $\varphi \in S(H)$ belongs to the image of the space $S(\mathbb{R}^3)$ under the John transform $\mathcal{J}$ if and only if the differential form $\kappa_x\varphi$ on the submanifold $H^x \subset H$ is closed for any point $x \in \mathbb{R}^3$.*

REMARK. If the John transform is given in the local coordinates α_1, α_2, β_1, β_2 on H by (1.5), then (1.6) is equivalent to a single equation for the function ψ,

$$\frac{\partial^2\psi}{\partial\alpha_1\,\partial\beta_2} = \frac{\partial^2\psi}{\partial\alpha_2\,\partial\beta_1}, \tag{1.8}$$

and the operator κ is given in the local coordinates by the following formula:

$$\kappa\psi = \frac{\partial\psi}{\partial\beta_1}d\alpha_1 + \frac{\partial\psi}{\partial\beta_2}d\alpha_2. \tag{1.9}$$

1.4. Space $S(H')$. This subsection, together with the next two, is devoted to the proof of the main theorem concerning the image of the space $S(\mathbb{R}^3)$. The reader may omit them and immediately proceed to 1.7. To describe the image of the space $S(\mathbb{R}^3)$ under the John transform, we first pass (by using the Fourier transform) from the functions $\varphi \in S(H)$ to functions on the bundle $H' \to G$ dual to the bundle $H \to G$.

Note that the fiber of the bundle $H' \to G$ over $a \in G$, i.e., the space H'_a dual to $\mathbb{R}^3/a$, is canonically embedded in $(\mathbb{R}^3)'$ as the subspace of vectors orthogonal to a. Thus, the elements of H' are the pairs (a, ξ), where $a \in G$ and ξ is a vector in $(\mathbb{R}^3)'$ orthogonal to a.

DEFINITION. Denote by $S(H')$ the space of Schwartz functions on H', where $H' \to G$ is the bundle dual to the bundle $H \to G$.

Introduce the Fourier transform $\mathcal{F}_0$ of a function $\varphi(\alpha, \beta) \in S(H)$ along the fibers of the bundle $H \to G$,

$$(\mathcal{F}_0\varphi)(\alpha, \xi) = \frac{1}{2\pi} \int_{\mathbb{R}^3/\{\alpha\}} \varphi(\alpha, \beta)\, e^{i\langle \beta, \xi \rangle}\, d_\alpha \beta, \qquad \xi \in (\mathbb{R}^3/\{\alpha\})',$$

where $\{\alpha\}$ is the one-dimensional subspace spanned by the vector α, and the volume element $d_\alpha\beta$ on $\mathbb{R}^3/\{\alpha\}$ is determined by the volume element $dx = dx_1\, dx_2\, dx_3$ on $\mathbb{R}^3$ and the vector α, i.e.,

$$\int_{\mathbb{R}^3/\{\alpha\}} \int_{-\infty}^{+\infty} f(\alpha t + \beta)\, dt\, d_\alpha \beta = \int_{\mathbb{R}^3} f(x)\, dx$$

for any function $f \in S(\mathbb{R}^3)$.

It follows from the definition of $\mathcal{F}_0$ and from (1.3) that $\mathcal{F}_0\varphi(\lambda\alpha, \xi) = \mathcal{F}_0\varphi(\alpha, \xi)$ for any $\lambda \neq 0$, and hence $\mathcal{F}_0\varphi$ is a function on H'.

Recall that the Fourier transform $\mathcal{F}$ (of the functions on $\mathbb{R}^3$) given by

$$(\mathcal{F}f)(\xi) = (2\pi)^{-3/2} \int_{\mathbb{R}^3} f(x) e^{i\langle \xi, x \rangle}\, dx$$

defines an isomorphism of the Schwartz spaces

$$\mathcal{F} : S(\mathbb{R}^3) \to S((\mathbb{R}^3)').$$

Similarly, the next assertion readily follows from the properties of the ordinary Fourier transform.

PROPOSITION 1.2. *If $\varphi \in S(H)$, then $\mathcal{F}_0\varphi \in S(H')$, and the mapping $\varphi \mapsto \mathcal{F}_0\varphi$ defines an isomorphism of the Schwartz spaces*

$$\mathcal{F}_0 : S(H) \to S(H').$$

We state two important properties of the space $S(H')$ in the form of separate propositions. The first property is obvious, whereas the second needs a proof, which is omitted here.

PROPOSITION 1.3. *If $F \in S(\mathbb{R}^3)$, then the function φ on H' given by the relation*

$$\varphi(a, \xi) = F(\xi) \quad \textit{for any} \quad (a, \xi) \in H'$$

belongs to $S(H')$. Thus, we have a natural embedding

$$S((\mathbb{R}^3)') \hookrightarrow S(H').$$

PROPOSITION 1.4. *Let a function $\varphi \in S(H')$ be such that*

$$\varphi(a, \xi) = F(\xi) \quad \textit{for any} \quad \xi \neq 0,$$

for a function $F \in S(\mathbb{R}^3)$. Then the function $\varphi(a, 0)$ does not depend on the subspace a, and if we set $F(0) = \varphi(a, 0)$, then the resulting function F on the entire space $(\mathbb{R}^3)'$ belongs to $S((\mathbb{R}^3)')$.

1.5. Description of the image of $S(\mathbb{R}^3)$ in the space $S(H')$. We now find the image of $S(\mathbb{R}^3)$ in $S(H')$ under the composition mapping

$$\mathcal{F}_0\mathcal{J} : S(\mathbb{R}^3) \xrightarrow{\mathcal{J}} S(H) \xrightarrow{\mathcal{F}_0} S(H').$$

PROPOSITION 1.5. *For any function $f \in S(\mathbb{R}^3)$ we have*

$$\mathcal{F}_0\mathcal{J}f = (2\pi)^{1/2}\,\mathcal{F}f, \tag{1.10}$$

where $\mathcal{F}f$ is the Fourier transform of the function $f \in S(\mathbb{R}^3)$, and thus $\mathcal{F}f$ is an element of the subspace $S((\mathbb{R}^3)') \subset S(H')$.

PROOF. Let $(a, \xi) \in H'$ and let $\alpha \neq 0$ be an arbitrary vector in the subspace $a \in G$. Then

$$\begin{aligned}(\mathcal{F}_0\mathcal{J}f)(\alpha,\xi) &= \frac{1}{2\pi}\int_{\mathbb{R}^3/a} \mathcal{J}f(\alpha,\beta)\, e^{i\langle\beta,\,\xi\rangle}\, d_\alpha\beta \\ &= \frac{1}{2\pi}\int_{\mathbb{R}^3/\{\alpha\}}\int_{-\infty}^{+\infty} f(\alpha t+\beta)\, e^{i\langle\beta,\,\xi\rangle}\, dt\, d_\alpha\beta \\ &= \frac{1}{2\pi}\int_{\mathbb{R}^3} f(x)\, e^{i\langle x,\,\xi\rangle}\, dx = (2\pi)^{1/2}(\mathcal{F}f)(\xi).\end{aligned}$$

COROLLARY. *The image of $S(\mathbb{R}^3)$ under the mapping $\mathcal{F}_0\mathcal{J}$ is the subspace $S((\mathbb{R}^3)') \subset S(H')$.*

Indeed, it follows from Proposition 1.5 that the image of $S(\mathbb{R}^3)$ under the mapping $\mathcal{F}_0\mathcal{J}$ is contained in $S((\mathbb{R}^3)')$. Conversely, if $\widetilde{f} \in S((\mathbb{R}^3)')$, then $\widetilde{f} = \mathcal{F}f$, where $f \in S(\mathbb{R}^3)$. Therefore, $\mathcal{F}_0\mathcal{J}f = (2\pi)^{1/2}\mathcal{F}f = (2\pi)^{1/2}\widetilde{f}$.

REMARK. Relation (1.10) implies an inversion formula for the John transform. Namely, to reconstruct f, one must apply the inverse Fourier transform on $\mathbb{R}^3$ to the function $\mathcal{F}_0\mathcal{J}f$, i.e.,

$$f = (\mathcal{F}^{-1}\mathcal{F}_0)\mathcal{J}f.$$

1.6. Proof of Theorem 1.1 on the image of the John transform. Let $\varphi \in S(H)$ and $\tilde{\varphi} = \mathcal{F}_0\varphi$. To prove Theorem 1.1, is suffices to establish the following assertion.

PROPOSITION 1.6. *If a function φ satisfies (1.6), then $\tilde{\varphi}(a,\xi) = F(\xi)$ for any vector $\xi \neq 0$, i.e., the function $\tilde{\varphi}(a,\xi)$ does not depend on the subspace a orthogonal to ξ if $\xi \neq 0$.*

Indeed, it follows from Proposition 1.4 that, in this case, the function $\tilde{\varphi} = \mathcal{J}_0\varphi$ belongs to the subspace $S((\mathbb{R}^3)') \subset S(H')$. Then $\tilde{\varphi} = \mathcal{F}_0\mathcal{J}f$ for some function $f \in S(\mathbb{R}^3)$. Thus, $\varphi = \mathcal{J}f$.

PROOF OF PROPOSITION 1.6. Let a be a one-dimensional subspace and let $\alpha = (\alpha_1, \alpha_2, \alpha_3)$ be a nonzero vector in a. Denote by V_i, $i = 1, 2, 3$, the family of subspaces a for which $\alpha_i \neq 0$. Since the V_i form a covering of G, it suffices to prove the assertion for each V_i separately. To be definite, consider V_3.

We present every subspace $a \in V_3$ by a vector of the form $\alpha = (\alpha_1, \alpha_2, 1)$ and take the coordinates α_1, α_2 of this vector as the coordinates on V_3. Moreover, it is clear that on every line in $\mathbb{R}^3$ parallel to a there is a point $\beta = (\beta_1, \beta_2, \beta_3)$ such that $\beta_3 = 0$. Take the coordinates β_1, β_2 of this point as the coordinates on $\mathbb{R}^3/a$.

In this system of coordinates the measure on $\mathbb{R}^3/a$ becomes $d_\alpha\beta = d\beta_1\, d\beta_2$, and therefore

$$\tilde{\varphi}(a,\xi) = \int_{-\infty}^{+\infty}\int_{-\infty}^{+\infty} \psi(\alpha_1,\alpha_2,\beta_1,\beta_2)\, e^{i(\beta_1\xi_1+\beta_2\xi_2)}\, d\beta_1\, d\beta_2,$$

where $\psi(\alpha_1,\alpha_2,\beta_1,\beta_2)$ stands for $\varphi(\alpha_1,\alpha_2,1,\beta_1,\beta_2,0)$ and the pair α_1, α_2 and the triple $\xi = (\xi_1,\xi_2,\xi_3)$ are related as follows:

$$\alpha_1\xi_1 + \alpha_2\xi_2 + \xi_3 = 0. \tag{1.11}$$

By virtue of (1.11) we can take ξ_1, ξ_2 for the coordinates on $H'_a = (\mathbb{R}^3/a)'$. Moreover, if $\xi \neq 0$, then $(\xi_1,\xi_2) \neq 0$.

By the assumption of the theorem, the function ψ satisfies the equation

$$\frac{\partial^2\psi}{\partial\alpha_1\partial\beta_2} = \frac{\partial^2\psi}{\partial\alpha_2\partial\beta_1}.$$

Hence, the Fourier transform of ψ with respect to β_1, β_2 satisfies the equation

$$\xi_2\frac{\partial\tilde{\varphi}}{\partial\alpha_1} = \xi_1\frac{\partial\tilde{\varphi}}{\partial\alpha_2}.$$

For $(\xi_1,\xi_2) \neq 0$, the general solution of this equation is

$$\tilde{\varphi}(\alpha_1,\alpha_2,\xi_1,\xi_2) = F(\xi_1,\xi_2,-(\alpha_1\xi_1+\alpha_2\xi_2)),$$

or, by (1.11),

$$\tilde{\varphi}(\alpha_1,\alpha_2,\xi) = F(\xi_1,\xi_2,\xi_3).$$

Thus, for any chosen ξ, $\xi \neq 0$, the function $\tilde{\varphi}(a,\xi)$ does not depend on a.

1.7. Analogs of the operator κ. In 1.3 we have introduced a linear operator κ from the space $S(H)$ to the space $\Omega^1(H)$ of differential 1-forms on the manifold of lines H, and this operator has the following important property.

CONDITION A. If $\varphi = \mathcal{J}f$, then the differential form $\kappa\varphi$ is closed on the submanifolds H^x of the lines passing through an arbitrarily chosen point $x \in \mathbb{R}^3$.

We pose the following problem: Describe all linear operators κ having this property. In the statement of Proposition 1.7 we refine the setting of the problem by restricting the class of operators under consideration.

Suppose that the lines in $\mathbb{R}^3$ are given by the equations

$$x_1 = \alpha_1 x_3 + \beta_1, \qquad x_2 = \alpha_2 x_3 + \beta_2,$$

and the John transform $\psi = \mathcal{J}f$ is defined in the local coordinates α_1, α_2, β_1, β_2 on H by formula (1.5). By 1.3, the function ψ satisfies the equation

$$\frac{\partial^2\psi}{\partial\alpha_1\,\partial\beta_2} = \frac{\partial^2\psi}{\partial\alpha_2\,\partial\beta_1},$$

and the operator κ is given by the formula

$$\kappa\psi = \frac{\partial\psi}{\partial\beta_1}d\alpha_1 + \frac{\partial\psi}{\partial\beta_2}d\alpha_2. \tag{1.12}$$

One can readily see that, along with $\kappa = \kappa_1$, the following three operators also satisfy condition A:

$$\kappa_2\psi = \frac{\partial\psi}{\partial\beta_1}\,d\beta_1 + \frac{\partial\psi}{\partial\beta_2}\,d\beta_2,$$
$$\kappa_3\psi = \frac{\partial\psi}{\partial\alpha_1}\,d\alpha_1 + \frac{\partial\psi}{\partial\alpha_2}\,d\alpha_2,$$
$$\kappa_4\psi = \frac{\partial\psi}{\partial\alpha_1}\,d\beta_1 + \frac{\partial\psi}{\partial\alpha_2}\,d\beta_2.$$

Indeed, since $d\beta_i \big|_{H^x} = -x_3\,d\alpha_i$, $i = 1, 2$, on H^x, it follows that the forms $\kappa_2\psi$ and $\kappa_4\psi$ are proportional on H^x to $\kappa_1\psi$ and $\kappa_3\psi$, respectively. Moreover, $\kappa_2\psi + \kappa_3\psi = d\psi$. Hence, each of the forms $\kappa_i\psi$ is cohomology equivalent on H^x (up to a factor) to the form $\kappa_1\psi$.

Consider the operators κ of the form

$$\kappa\psi = \frac{\partial\psi}{\partial\alpha_1}\,\omega_1 + \frac{\partial\psi}{\partial\alpha_2}\,\omega_2 + \frac{\partial\psi}{\partial\beta_1}\,\omega_3 + \frac{\partial\psi}{\partial\beta_1}\,\omega_4, \tag{1.13}$$

where ω_i is an arbitrary 1-form on H.

PROPOSITION 1.7. *Every operator κ of the form* (1.13) *satisfying condition A is a linear combination with constant coefficients of the operators κ_i, $i = 1, 2, 3, 4$.*

PROOF. For convenience, we shall write α_3 and α_4 instead of β_1 and β_2, respectively. Then (1.13) becomes

$$\kappa\psi = \sum_{i=1}^{4} \frac{\partial\psi}{\partial\alpha_i}\,\omega_i, \quad \text{where} \quad \omega_i = \sum_{j=1}^{4} a_{ij}(\alpha)\,d\alpha_j, \qquad i = 1, 2, 3, 4.$$

It follows from the condition $d(\kappa\psi) \big|_{H^x} = 0$ that on each submanifold H^x the differential forms ω_i satisfy the relations

$$\omega_i \wedge d\alpha_i = 0, \qquad i = 1, 2, 3, 4, \tag{1.14}$$
$$\omega_1 \wedge d\alpha_2 + \omega_2 \wedge d\alpha_1 = 0, \qquad \omega_3 \wedge d\alpha_4 + \omega_4 \wedge d\alpha_3 = 0, \tag{1.15}$$
$$d\omega_i = 0, \qquad i = 1, 2, 3, 4. \tag{1.16}$$

It follows from (1.14) that

$$\omega_1 = a_{11}\,d\alpha_1 + a_{13}\,d\alpha_3, \qquad \omega_2 = a_{22}\,d\alpha_2 + a_{24}\,d\alpha_4,$$
$$\omega_3 = a_{31}\,d\alpha_1 + a_{33}\,d\alpha_3, \qquad \omega_4 = a_{42}\,d\alpha_2 + a_{44}\,d\alpha_4.$$

Substituting the resulting expressions for ω_i in (1.15), we obtain

$$a_{11} = a_{22}, \qquad a_{13} = a_{24}, \qquad a_{31} = a_{42}, \qquad a_{33} = a_{44}.$$

This implies that

$$\kappa = a_{11}\,\kappa_4 + a_{33}\,\kappa_2 + a_{31}\,\kappa_1 + a_{13}\,\kappa_3.$$

Finally, substituting the last expression into (1.16), we conclude that all coefficients a_{ij} in this expression are constant.

2. John transform of differential forms on $\mathbb{R}^3$

2.1. Definition of the John transform of differential forms. Here we consider integral transforms that take a differential form on $\mathbb{R}^3$ to the integrals of this form over the lines in $\mathbb{R}^3$. These transforms are defined not only for differential forms of degree 1 but also for differential forms of degrees 2 and 3. In this case we obtain differential forms (on the manifold of lines) whose degree is less by one. Similar transforms of differential forms on $\mathbb{R}^3$ related to planes in $\mathbb{R}^3$ were treated in § 7 of Chapter 1.

Although the definition of these transforms can be given in an invariant form, here it is more convenient to connect them with the choice of a specific local system of coordinates on the manifold of lines in $\mathbb{R}^3$.

Let us present lines in $\mathbb{R}^3$ by the equations

$$x_i = \alpha_i x_3 + \beta_i, \quad i = 1, 2, \tag{2.1}$$

and take the coefficients α_i, β_i of these equations for the local coordinates on the manifold of lines.

Introduce the four-dimensional flag manifold A whose elements are pairs (h, x), where h is an arbitrary line in $\mathbb{R}^3$ and x is an arbitrary point of $\mathbb{R}^3$ belonging to the line h. We equip the manifold A with two systems of coordinates. The first system is $(x_1, x_2, x_3, \alpha_1, \alpha_2)$, where (x_1, x_2, x_3) are coordinates of a point in $\mathbb{R}^3$ and (α_1, α_2) define the direction of a line passing through this point. The other system is $(\alpha_1, \alpha_2, \beta_1, \beta_2, x_3)$, where $(\alpha_1, \alpha_2, \beta_1, \beta_2)$ are the local coordinates of a line and x_3 is the coordinate of a point on this line. These two systems of coordinates are related by (2.1).

We now define the John transform of a differential form ω on $\mathbb{R}^3$ of degree $r \geq 1$. We regard ω as a differential form on A given in the coordinate system $(x_1,\ x_2,\ x_3,\ \alpha_1,\ \alpha_2)$, then pass to the other system $(\alpha_1, \alpha_2, \beta_1, \beta_2, x_3)$ and omit the terms not containing dx_3. The resulting expression is a differential 1-form on the line with coordinates α_i, β_i, $i = 1, 2$, for any chosen values α_i, β_i, $d\alpha_i$, $d\beta_i$. Integrating this 1-form along the line, we obtain a differential $(r-1)$-form on the manifold of lines. We denote this form by $\mathcal{J}\omega$ and call it the *John transform* of the original differential form ω.

The above definition is convenient for calculations, but it is related to a specific coordinate system in $\mathbb{R}^3$ and in the manifold of lines H. There is an invariant definition of the John transform of differential forms in terms of operations of pullback and pushdown. Namely, two operations on differential forms are defined for any smooth fibration $\pi : A \to B$, the *pullback* $\pi^\star$ and the *pushdown* $\pi_\star$. The former takes the differential forms on B to differential forms of the same degree on A; the latter (the integration over the fibers of the bundle) takes the differential forms of degree r on A to forms of degree $r - l$ on B, where l is the dimension of the fiber of the bundle $A \to B$ (it is assumed that $r \geq l$).

If a pair of fibrations $A \to B$ and $A \to C$ (double fibration) is given, then one can assign to this pair an integral transform

$$\mathcal{J} : \Omega(B) \to \Omega(C)$$

from the space of differential forms on B to the space of differential forms on C. This form is defined as the composition $\mathcal{J} = \pi_\star \pi^\star$ of the pullback $\pi^\star$ taking the forms on B to those on A and the pushdown $\pi_\star$ taking the forms on A to those on

C. It turns out that the operator $\mathcal{J}$ thus defined commutes with the operator of exterior differentiation d:

$$d\mathcal{J} = \mathcal{J}d.$$

The John transform $\mathcal{J} : \Omega(\mathbb{R}^3) \to \Omega(H)$ is an example of such an operation related to the double fibration $A \to \mathbb{R}^3$ and $A \to H$, where A is the flag space. The property $d\mathcal{J} = \mathcal{J}\, d$ of this operation can be verified by the immediate calculation.

2.2. John transform of 3-forms on $\mathbb{R}^3$. Let a differential 3-form ω be given,

$$\omega = f(x)\, dx_1 \wedge dx_2 \wedge dx_3, \qquad f \in S(\mathbb{R}^3).$$

In order to construct the John transform of ω, one must first represent this form in the coordinates $(\alpha_1, \alpha_2, \beta_1, \beta_2, x_3)$ on A:

$$\omega = f(\alpha_1\, x_3 + \beta_1,\, \alpha_2\, x_3 + \beta_2,\, x_3)\, (x_3\, d\alpha_1 + d\beta_1) \wedge (x_3\, d\alpha_2 + d\beta_2) \wedge dx_3.$$

According to the definition in 2.1, the John transform of the form ω is defined as follows:

$$\begin{aligned}(2.2)\quad \mathcal{J}\omega = {} & \varphi_2(\alpha, \beta)\, d\alpha_1 \wedge d\alpha_2 \\ & + \varphi_1(\alpha, \beta)\, (d\alpha_1 \wedge d\beta_2 - d\alpha_2 \wedge d\beta_1) + \varphi_0(\alpha, \beta)\, d\beta_1 \wedge d\beta_2,\end{aligned}$$

where

$$(2.3)\qquad \varphi_i(\alpha, \beta) = \int_{-\infty}^{+\infty} f(\alpha_1\, x_3 + \beta_1,\, \alpha_2\, x_3 + \beta_2,\, x_3)\, (x_3)^i\, dx_3, \qquad i = 0, 1, 2.$$

We study the properties of the resulting form. First, since $f \in S(\mathbb{R}^3)$, it follows that the coefficients φ_i of $\mathcal{J}\omega$ are rapidly decreasing together with all their derivatives.

PROPOSITION 2.1. *The differential form $\mathcal{J}\omega$ is closed.*

Indeed, it follows from the expressions (2.3) for the coefficients φ_i of the form $\mathcal{J}\omega$ that

$$(2.4)\qquad \frac{\partial \varphi_2}{\partial \beta_i} = \frac{\partial \varphi_1}{\partial \alpha_i}, \quad \frac{\partial \varphi_1}{\partial \beta_i} = \frac{\partial \varphi_0}{\partial \alpha_i}, \qquad i = 1, 2.$$

These relations are equivalent to the condition that the form $\mathcal{J}\omega$ is closed.

REMARK. There is another way to prove that the form $\mathcal{J}\omega$ is closed. Namely, the original form ω is closed as a form of maximum degree on $\mathbb{R}^3$, and the transform $\mathcal{J}$ commutes with the operator of exterior differentiation.

PROPOSITION 2.2. *The restriction of the differential form $\mathcal{J}\omega$ to the submanifold of lines belonging to an arbitrarily fixed plane vanishes.*

PROOF. The submanifold P of lines belonging to a given plane is defined by the following system of equations in the local coordinates (α, β):

$$\lambda_1\, (\alpha_1 - \alpha_1^0) + \lambda_2\, (\alpha_2 - \alpha_2^0) = 0, \qquad \lambda_1\, (\beta_1 - \beta_1^0) + \lambda_2\, (\beta_2 - \beta_2^0) = 0.$$

Here the pair (α^0, β^0) stands for the coordinates of an arbitrarily chosen line on the plane and λ_1 and λ_2 are constant. Hence, on this submanifold P we have $d\alpha_2 = \lambda\, d\alpha_1$ and $d\beta_2 = \lambda\, d\beta_1$, where $\lambda = -\lambda_1/\lambda_2$, and thus it follows from (2.2) that $\mathcal{J}\omega\, |_P = 0$.

THEOREM 2.1. *A differential* 2*-form* Ω *with rapidly decreasing coefficients on the manifold of lines* H *is the John transform of a* 3*-form* $\omega = f(x)\,dx_1 \wedge dx_2 \wedge dx_3$, $f \in S(\mathbb{R}^3)$, *if and only if the form* Ω *satisfies the conditions of Propositions* 2.1 *and* 2.2.

PROOF. The necessity of the conditions is already established. We prove the sufficiency. Suppose a differential 2-form Ω satisfies the conditions of Propositions 2.1 and 2.2. We first prove that this form can be represented as in (2.2), i.e., that Ω contains no terms with $d\alpha_i \wedge d\beta_i$, $i = 1, 2$, and the coefficients of $d\alpha_1 \wedge d\beta_2$ and $d\alpha_2 \wedge d\beta_1$ differ in sign only.

Indeed, let

$$\Omega = a_1\, d\alpha_1 \wedge d\beta_1 + a_2\, d\alpha_2 \wedge d\beta_2 + b_1\, d\alpha_1 \wedge d\beta_2 + b_2\, d\alpha_2 \wedge d\beta_1 + \cdots,$$

where the dots stand for the remaining terms. If h^0 is an arbitrary line and P is the manifold of lines in an arbitrary plane passing through h^0, then

$$\Omega\,|_P = (a_1 + \lambda^2\, a_2 + \lambda(b_1 + b_2))\, d\alpha_1 \wedge d\beta_1$$

(see the proof of Proposition 2.2). Since $\Omega\,|_P = 0$ by assumption and λ is arbitrary, it follows that $a_1 = a_2 = 0$ and $b_1 = -b_2$.

Thus, we have proved that the form Ω can be represented as

$$\begin{aligned}\Omega &= \psi_2(\alpha, \beta)\, d\alpha_1 \wedge d\alpha_2 \\ &\quad + \psi_1(\alpha, \beta)\,(d\alpha_1 \wedge d\beta_2 - d\alpha_2 \wedge d\beta_1) + \psi_0(\alpha, \beta)\, d\beta_1 \wedge d\beta_2.\end{aligned}$$

Since Ω is closed, the functions ψ_i are related as follows:

$$\frac{\partial \psi_2}{\partial \beta_i} = \frac{\partial \psi_1}{\partial \alpha_i}, \qquad \frac{\partial \psi_1}{\partial \beta_i} = \frac{\partial \psi_0}{\partial \alpha_i}, \qquad i = 1, 2. \tag{2.5}$$

Therefore, the function ψ_0 satisfies the equation

$$\frac{\partial^2 \psi_0}{\partial \alpha_1 \beta_2} = \frac{\partial^2 \psi_0}{\partial \alpha_2\, \partial \beta_1}.$$

Hence, by the theorem on the image of the John transform (see § 1), this function is the John transform of some function $f \in S(\mathbb{R}^3)$:

$$\psi_0(\alpha, \beta) = \int_{-\infty}^{+\infty} f(\alpha_1 x_3 + \beta_1,\ \alpha_2 x_3 + \beta_2,\ x_3)\, dx_3.$$

We now prove that $\Omega = \mathcal{J}\omega$, where $\omega = f(x)\, dx_1 \wedge dx_2 \wedge dx_3$. Indeed, it immediately follows from formulas (2.3) defining the coefficients φ_i of $\mathcal{J}\omega$ that $\psi_0 = \varphi_0$. Further, these formulas, together with (2.4) and (2.5), imply that $\frac{\partial \psi_1}{\partial \beta_i} = \frac{\partial \varphi_1}{\partial \beta_i}$, $i = 1, 2$. Hence, since ψ_1 and φ_1 are rapidly decreasing, it follows that $\psi_1 = \varphi_1$. We can similarly see that $\psi_2 = \varphi_2$, and hence $\Omega = \mathcal{J}\omega$.

Note that the form $\omega = f(x)\, dx_1 \wedge dx_2 \wedge dx_3$ can be reconstructed from only one coefficient φ_0 of the form $\mathcal{J}\omega$ because the function φ_0 is the John transform of the function f.

2.3. John transform of 2-forms on $\mathbb{R}^3$. Consider an arbitrary differential 2-form

$$\omega = f_1(x)\,dx_2 \wedge dx_3 + f_2(x)\,dx_3 \wedge dx_1 + f_3(x)\,dx_1 \wedge dx_2$$

with coefficients in $S(\mathbb{R}^3)$. In the coordinates $\alpha_1, \alpha_2, \beta_1, \beta_2, x_3$ on the flag space this form can be represented as

$$\begin{aligned}\omega = f_1\,(x_3\,d\alpha_2 + d\beta_2) \wedge dx_3 \ + \ f_2\,dx_3 \wedge (x_3\,d\alpha_1 + d\beta_1)\\ + f_3\,(\alpha_1\,dx_3 + x_3\,d\alpha_1 + d\beta_1) \wedge (\alpha_2\,dx_3 + x_3\,d\alpha_2 + d\beta_2).\end{aligned}$$

According to the definition in 2.1, the John transform of the 2-form ω is the 1-form

$$\mathcal{J}\omega = a_1\,d\alpha_1 + a_2\,d\alpha_2 + b_1\,d\beta_1 + b_2\,d\beta_2 \tag{2.6}$$

with the coefficients

$$\begin{aligned} a_1 &= \alpha_2\,\psi_3 - \psi_2, \qquad & a_2 &= -\alpha_1\,\psi_3 + \psi_1,\\ b_1 &= \alpha_2\,\varphi_3 - \varphi_2, \qquad & b_2 &= -\alpha_1\,\varphi_3 + \varphi_1,\end{aligned} \tag{2.7}$$

where

$$\varphi_i(\alpha, \beta) = \int_{-\infty}^{+\infty} f_i(\alpha_1\,x_3 + \beta_1,\ \alpha_2\,x_3 + \beta_2,\ x_3)\,dx_3, \qquad i = 1, 2, 3, \tag{2.8}$$

$$\psi_i(\alpha, \beta) = \int_{-\infty}^{+\infty} x_3\,f_i(\alpha_1\,x_3 + \beta_1,\ \alpha_2\,x_3 + \beta_2,\ x_3)\,dx_3, \qquad i = 1, 2, 3. \tag{2.9}$$

Consider first the problem of reconstructing the form ω from its John transform $\mathcal{J}\omega$. This problem can be reduced to the corresponding problem for functions by using the following assertion.

PROPOSITION 2.3. *The functions φ_i are expressed in terms of the coefficients of the form $\mathcal{J}\omega$ as follows*:

$$\varphi_3 = \frac{\partial b_1}{\partial \alpha_2} - \frac{\partial a_1}{\partial \beta_2}, \qquad \varphi_1 = \alpha_1\,\varphi_3 + b_2, \qquad \varphi_2 = \alpha_2\,\varphi_3 - b_1. \tag{2.10}$$

Relations (2.10) immediately follow from (2.7)–(2.9).

Since the functions φ_i are the John transforms of the coefficients f_i of ω, it suffices to apply the inversion formula for the John transform of functions to the φ_i's to reconstruct the coefficients f_i.

We now proceed with the description of the image of the John transform $\mathcal{J}$. To this end, we first establish relations for the coefficients of the differential form $\mathcal{J}\omega$.

PROPOSITION 2.4. *The coefficients a_i, b_i of the differential form $\mathcal{J}\omega$ satisfy the following relations*:

$$\frac{\partial a_1}{\partial \beta_1} = \frac{\partial b_1}{\partial \alpha_1}, \qquad \frac{\partial a_2}{\partial \beta_2} = \frac{\partial b_2}{\partial \alpha_2}, \qquad \frac{\partial a_1}{\partial \beta_2} - \frac{\partial b_1}{\partial \alpha_2} = -\Big(\frac{\partial a_2}{\partial \beta_1} - \frac{\partial b_2}{\partial \alpha_1}\Big) \tag{2.11}$$

and

$$\Delta\Big(\frac{\partial a_1}{\partial \beta_2} - \frac{\partial b_1}{\partial \alpha_2}\Big) = 0, \quad \textit{where} \quad \Delta = \frac{\partial^2}{\partial \alpha_1\,\partial \beta_2} - \frac{\partial^2}{\partial \alpha_2 \partial \beta_1}. \tag{2.12}$$

Indeed, relations (2.11) immediately follow from (2.7), (2.8), and (2.9). Relation (2.12) follows from the above equation $\varphi_3 = \frac{\partial b_1}{\partial \alpha_2} - \frac{\partial a_1}{\partial \beta_2}$ because $\Delta\varphi_3 = 0$.

PROPOSITION 2.5. *Relations* (2.11) *are equivalent to the condition that the restriction* $\mathcal{J}\omega\mid_P$ *of the differential form* $\mathcal{J}\omega$ *to the submanifold of lines that all belong to a plane* P *is a closed form on* P, *i.e.,* $d\mathcal{J}\omega\mid_P = 0$.

PROOF. We have

$$d\mathcal{J}\omega = \Big(\frac{\partial b_1}{\partial \alpha_1} - \frac{\partial a_1}{\partial \beta_1}\Big)\, d\alpha_1 \wedge d\beta_1 + \Big(\frac{\partial b_2}{\partial \alpha_1} - \frac{\partial a_1}{\partial \beta_2}\Big)\, d\alpha_1 \wedge d\beta_2$$
$$+ \Big(\frac{\partial b_1}{\partial \alpha_2} - \frac{\partial a_2}{\partial \beta_1}\Big)\, d\alpha_2 \wedge d\beta_1 + \Big(\frac{\partial b_2}{\partial \alpha_2} - \frac{\partial a_2}{\partial \beta_2}\Big)\, d\alpha_2 \wedge d\beta_2 + \cdots .$$

Therefore, relations (2.11) are equivalent to the condition that the form $d\mathcal{J}\omega$ contains no terms with $d\alpha_i \wedge d\beta_i$, $i = 1, 2$, and the coefficients of $d\alpha_1 \wedge d\beta_2$ and $d\alpha_2 \wedge d\beta_1$ differ in the sign only. In turn, according to 2.2, this condition is equivalent to the one formulated in the proposition.

THEOREM 2.2. *Suppose* Ω *is a differential* 1*-form with infinitely differentiable rapidly decreasing coefficients on the manifold of lines in* $\mathbb{R}^3$,

$$\Omega = a_1\, d\alpha_1 + a_2\, d\alpha_2 + b_1\, d\beta_1 + b_2\, d\beta_2.$$

For the form Ω *to be the John transform of a differential* 2*-form* ω *on* $\mathbb{R}^3$ *with coefficients in* $S(\mathbb{R}^3)$, *it is necessary and sufficient that its restriction* $\Omega\mid_P$ *to submanifolds* P *of lines that belong to the same plane be closed forms for any plane and that the coefficients of* Ω *satisfy the additional condition* (2.12).

PROOF. We have already proved the necessity. Let us prove the sufficiency. Suppose a differential form Ω satisfies the hypothesis of the theorem. Then the coefficients of this form satisfy conditions (2.11) and (2.12).

Define functions φ_i, $i = 1, 2, 3$, by (2.10). It follows from (2.12) that $\Delta\varphi_3 = 0$. We prove that also $\Delta\varphi_1 = 0$ and $\Delta\varphi_2 = 0$. Indeed,

$$\Delta\varphi_2 = \alpha_2\, \Delta\varphi_3 - \frac{\partial \varphi_3}{\partial \beta_1} - \Delta b_1$$
$$= -\frac{\partial}{\partial \beta_1}\Big(\frac{\partial b_1}{\partial \alpha_2} - \frac{\partial a_1}{\partial \beta_2}\Big) - \frac{\partial^2 b_1}{\partial \alpha_1 \partial \beta_2} + \frac{\partial^2 b_1}{\partial \alpha_2 \partial \beta_1}$$
$$= \frac{\partial}{\partial \beta_2}\Big(\frac{\partial a_1}{\partial \beta_1} - \frac{\partial b_1}{\partial \alpha_1}\Big) = 0.$$

One can similarly prove that $\Delta\varphi_1 = 0$.

This implies (see § 1) that the functions φ_i are the John transforms of some functions $f_i \in S(\mathbb{R}^3)$, i.e., the functions φ_i are given by (2.8). Introduce the differential form

$$\omega = f_1(x)\, dx_2 \wedge dx_3 + f_2(x)\, dx_3 \wedge dx_1 + f_3(x)\, dx_1 \wedge dx_2.$$

We prove that $\Omega = \mathcal{J}\omega$. Indeed,

$$\mathcal{J}\omega = \tilde{a}_1\, d\alpha_1 + \tilde{a}_2\, d\alpha_2 + \tilde{b}_1\, d\beta_1 + \tilde{b}_2\, d\beta_2,$$

where

$$\tilde{a}_1 = \alpha_2\, \psi_3 - \psi_2, \qquad \tilde{a}_2 = -\alpha_1\, \psi_3 + \psi_1,$$
$$\tilde{b}_1 = \alpha_2\, \varphi_3 - \varphi_2, \qquad \tilde{b}_2 = -\alpha_1\, \varphi_3 + \varphi_1,$$

and the functions φ_i and ψ_i are expressed in terms of the coefficients of the form ω by (2.8) and (2.9).

By Proposition 2.3,

$$\varphi_3 = \frac{\partial \tilde{b}_1}{\partial \alpha_2} - \frac{\partial \tilde{a}_1}{\partial \beta_2}, \qquad \varphi_1 = \alpha_1 \varphi_3 + \tilde{b}_2, \qquad \varphi_2 = \alpha_2 \varphi_3 - \tilde{b}_1. \tag{2.13}$$

Comparing (2.8) and (2.13), we conclude that $\tilde{b}_i = b_i$, $i = 1, 2$. Hence, comparing relation (2.11) for a_i and b_i with the similar relations for $\tilde{a}_i$ and $\tilde{b}_i$, we see that $\frac{\partial \tilde{a}_i}{\partial \beta_j} = \frac{\partial a_i}{\partial \beta_j}$, $i, j = 1, 2$. Since the functions a_i and $\tilde{a}_i$ are rapidly decreasing, we have $\tilde{a}_i = a_i$, $i = 1, 2$, and therefore $\Omega = \mathcal{J}\omega$.

2.4. John transform of 1-forms on $\mathbb{R}^3$. Consider a differential 1-form on $\mathbb{R}^3$,

$$\omega = f_1(x)\, dx_1 + f_2(x)\, dx_2 + f_3(x)\, dx_3, \qquad f_i \in S(\mathbb{R}^3).$$

By definition, the John transform $\varphi = \mathcal{J}\omega$ of ω is the following function on the manifold of lines in $\mathbb{R}^3$:

$$\begin{aligned} \varphi(\alpha, \beta) = \int_{-\infty}^{+\infty} [f_1(\alpha_1 x_3 + \beta_1,\ &\alpha_2 x_3 + \beta_2,\ x_3)\, \alpha_1 \\ &+ f_2(\alpha_1 x_3 + \beta_1,\ \alpha_2 x_3 + \beta_2,\ x_3)\, \alpha_2 \\ &\quad + f_3(\alpha_1 x_3 + \beta_1,\ \alpha_2 x_3 + \beta_2,\ x_3)]\, dx_3. \end{aligned}$$

We find the kernel and the image of the transform $\mathcal{J}$.

PROPOSITION 2.6. *The kernel* Ker $\mathcal{J}$ *of the John transform coincides with the subspace of all exact differential* 1*-forms on* $\mathbb{R}^3$ *with coefficients in* $S(\mathbb{R}^3)$.

PROOF. Obviously, each exact 1-form ω on $\mathbb{R}^3$ belongs to Ker $\mathcal{J}$. Conversely, let $\omega \in \operatorname{Ker} \mathcal{J}$, i.e., $\mathcal{J}\omega = 0$. Then $\mathcal{J}\, d\omega = d\mathcal{J}\omega = 0$. Since the John transform $\mathcal{J}$ has zero kernel on the differential 2-forms, it follows from the relation $\mathcal{J}\, d\omega = 0$ that $d\omega = 0$, i.e., the form ω is closed. Since the form ω is closed, it is exact.

PROPOSITION 2.7. *A function* φ *in the Schwartz space on the manifold of lines* H *belongs to the image* Im $\mathcal{J}$ *of the John transform of* 1*-forms if and only if the differential* $d\varphi$ *belongs to the image of the John transform of differential* 2*-forms on* H.

PROOF. Let $\omega \in \operatorname{Im} \mathcal{J}$, i.e., $\varphi = \mathcal{J}\omega_1$, where ω_1 is a 1-form with coefficients in $S(\mathbb{R}^3)$. Then $d\varphi = d\mathcal{J}\omega_1 = \mathcal{J} d\omega_1$, i.e., $d\varphi \in \operatorname{Im} \mathcal{J}$. Conversely, let $d\varphi \in \operatorname{Im} \mathcal{J}$, i.e., $d\varphi = \mathcal{J}\omega_2$, where ω_2 is a 2-form on $\mathbb{R}^3$ with coefficients in $S(\mathbb{R}^3)$. Then $\mathcal{J} d\omega_2 = d\mathcal{J}\omega_2 = 0$, and since the John transform has zero kernel on the 3-forms in $\mathbb{R}^3$, it follows that $d\omega_2 = 0$, i.e., the form ω_2 is closed. Since each closed 2-form on $\mathbb{R}^3$ with coefficients in $S(\mathbb{R}^3)$ is exact, it follows that $\omega_2 = d\omega_1$, where ω_1 is a 1-form on $\mathbb{R}^3$ with coefficients in $S(\mathbb{R}^3)$. Thus, $d\varphi = \mathcal{J} d\omega_1 = d\mathcal{J}\omega_1$, and hence $d\,(\varphi - \mathcal{J}\omega_1) = 0$. Therefore, $\varphi = \mathcal{J}\omega_1$.

THEOREM 2.3. *A function* φ *in the Schwartz space on* H *is the John transform of a* 1*-form on* $\mathbb{R}^3$ *if and only if this function satisfies the equation*

$$\Delta^2 \varphi = 0, \quad \text{where} \quad \Delta = \frac{\partial^2}{\partial \alpha_1 \beta_2} - \frac{\partial^2}{\partial \alpha_2 \beta_1}. \tag{2.14}$$

PROOF. It follows from Proposition 2.7 and the description of the image for the John transform of differential 2-forms (see 2.3) that $\varphi \in \operatorname{Im} \mathcal{J}$ if and only if the coefficients $a_i = \frac{\partial \varphi}{\partial \alpha_i}$ and $b_i = \frac{\partial \varphi}{\partial \beta_i}$, $i = 1, 2$, of the differential form $d\varphi$

satisfy relations (2.11) and (2.12). Obviously, relations (2.11) hold identically for the differential form $d\varphi$, and then (2.12) coincides with (2.14).

3. John transform in the three-dimensional real projective space

In this section, the main definitions and results of § 1 are extended to the case of projective space $\mathbb{P}^3$. The passage from the affine space to the projective space has two important advantages. First, the main definitions and results in the projective case have a simpler and more invariant form as compared with those in the affine case. Second, since the projective space $\mathbb{P}^3$ itself and the manifold of lines in $\mathbb{P}^3$ are compact, the function spaces related to these manifolds are more natural as compared with those in the affine case.

We also give a group-theoretic interpretation of the John transform.

3.1. Manifold of lines in $\mathbb{P}^3$. Let us define points in $\mathbb{P}^3$ by homogeneous coordinates x_1, x_2, x_3, x_4, i.e., by vectors $x \in \mathbb{R}^4 \setminus O$ defined up to a nonzero factor. In other words, the points of the space $\mathbb{P}^3$ are interpreted as one-dimensional subspaces of $\mathbb{R}^4$. To the set of points on a line there corresponds the set of one-dimensional subspaces in $\mathbb{R}^4$ belonging to some two-dimensional subspace. Thus, the lines in $\mathbb{P}^3$ are interpreted as two-dimensional subspaces of $\mathbb{R}^4$, and hence the manifold of lines in $\mathbb{P}^3$ is identified with the Grassmannian $G_{2,4}$ of two-dimensional subspaces in $\mathbb{R}^4$.

We introduce a coordinate system on the Grassmannian $G_{2,4}$. Let $h \in G_{2,4}$ and let $x = (x_1, x_2, x_3, x_4)$, $y = (y_1, y_2, y_3, y_4)$ be an arbitrary basis in the subspace h.

DEFINITION. By the *Plücker coordinates* of a two-dimensional subspace h in $\mathbb{R}^4$ one means the coordinates of the bivector $x \wedge y$, i.e., the following number set:

$$p_{ij} = x_i\, y_j - x_j\, y_i, \qquad i, j = 1, 2, 3, 4, \quad i \neq j.$$

Under any change of a basis in h, in which case the vectors x and y are replaced by some linear combinations of them, all Plücker coordinates are multiplied by the same number. Thus, the Plücker coordinates of a subspace h are defined uniquely up to a factor. Conversely, if the Plücker coordinates of two subspaces are proportional, then these subspaces coincide. Thus, each two-dimensional subspace of $\mathbb{R}^4$ is uniquely determined by its Plücker coordinates, and they can be taken as homogeneous coordinates on the Grassmannian $G_{2,4}$.

Since $p_{ij} = -p_{ji}$, it follows that we can restrict ourselves to the six coordinates p_{ij}, $i < j$.

The Plücker coordinates p_{ij} satisfy the quadratic relation

$$p_{12}\, p_{34} - p_{13}\, p_{24} + p_{14}\, p_{23} = 0. \tag{3.1}$$

This relation can be obtained, for instance, by expanding the following determinant of order four with respect to the first two rows:

$$\begin{vmatrix} x_1 & x_2 & x_3 & x_4 \\ y_1 & y_2 & y_3 & y_4 \\ x_1 & x_2 & x_3 & x_4 \\ y_1 & y_2 & y_3 & y_4 \end{vmatrix} = 0.$$

We prove that, conversely, if some numbers p_{ij}, $i \neq j$, are not simultaneously zero and satisfy (3.1), then these numbers are the Plücker coordinates of a subspace $h \in G_{2,4}$. Indeed, to be definite, let $p_{12} \neq 0$; then one can assume that $p_{12} = 1$.

Consider the subspace spanned by the vectors $x = (1, 0, -p_{23}, -p_{24})$ and $y = (0, 1, p_{13}, p_{14})$. Using relation (3.1), one can readily see that the Plücker coordinates of this subspace are equal to p_{ij}.

Thus, there is an embedding

$$G_{2,4} \hookrightarrow \mathbb{P}^5$$

that assigns, to each two-dimensional subspace with Plücker coordinates p_{ij}, $i < j$, a point of the projective space $\mathbb{P}^5$ with the homogeneous coordinates p_{ij}. The image of $G_{2,4}$ under this embedding is the quadratic surface (3.1) in $\mathbb{P}^5$.

3.2. John transform in $\mathbb{P}^3$. The first problem that arises when passing from the affine space to the projective one is to choose a function space on which the John transform $\mathcal{J}$ must be constructed. As in the case of Radon transform, spaces of functions on $\mathbb{P}^3$ are not convenient for this purpose. The natural object on which the John transform will be defined is the space $\mathcal{F}(-2)$ of infinitely differentiable functions on $\mathbb{R}^4 \setminus O$, i.e., on a one-dimensional bundle over $\mathbb{P}^3$, that satisfy the following homogeneity condition:

$$f(\lambda x) = \lambda^{-2} f(x) \quad \text{for each} \quad \lambda \neq 0. \tag{3.2}$$

Let us define the John transform of an arbitrary function $f \in \mathcal{F}(-2)$. Take $h \in G_{2,4}$. Choose an arbitrary basis u, v in h and denote by s and t the coordinates of points on h with respect to this basis,

$$x = s\,u + t\,v.$$

To a function $f \in \mathcal{F}(-2)$ we assign the following differential 1-form on h:

$$f(su + tv)\,(s\,dt - t\,ds).$$

It follows from the homogeneity condition (3.2) that this form is homogeneous of degree zero, and the integral of the form over a contour S in h enclosing the point O does not depend on the choice of this contour.

DEFINITION. By the John transform of a function $f \in \mathcal{F}$ we mean the following function of a pair of linearly independent vectors u, $v \in \mathbb{R}^4$:

$$(\mathcal{J}f)\,(u,\, v) = \frac{1}{2} \int_S f(su + tv)\,(s\,dt - t\,ds); \tag{3.3}$$

the integral is taken over an arbitrary contour S surrounding the point O in the subspace h spanned by u and v.

It has already been mentioned that the integral does not depend on the choice of the contour S.

The function $\varphi = \mathcal{J}f$ thus defined depends not only on the subspace $h \in G_{2,4}$ but also on the choice of the basis in h. It follows from (3.3) that, when passing from one basis u, v to any other basis $u' = \alpha\,u + \beta\,v$, $v' = \gamma\,u + \delta\,v$, the function φ is transformed by the following law:

$$\varphi(\alpha\,u + \beta\,v,\ \gamma\,u + \delta\,v) = |\alpha\delta - \beta\gamma|^{-1}\,\varphi(u,\, v).$$

In particular, if the area elements in the bases u, v and u', v' coincide, i.e., if $|\alpha\delta - \beta\gamma| = 1$, then $\varphi(\alpha\,u + \beta\,v,\ \gamma\,u + \delta\,v) = \varphi(u,\, v)$.

Thus, the John transform of $\varphi(u, v)$ depends only on the Plücker coordinates $p_{ij} = u_i v_j - u_j v_i$, i.e., this transform is a function on the quadratic cone $K \subset \mathbb{R}^6$

given by (3.1). Obviously, $K\setminus O$ is a one-dimensional bundle over the Grassmannian $G_{2,4}$.

In what follows, we often write $\varphi(p)$ instead of $\varphi(u,v)$ having in mind that p is a point of the cone K, i.e., a point of $\mathbb{R}^6$ with coordinates p_{ij}, $i<j$, satisfying (3.1).

DEFINITION. Denote by $\mathcal{H}(-1)$ the space of C^∞ functions φ on $K\setminus O$ satisfying the homogeneity condition

$$\varphi(\lambda p) = |\lambda|^{-1}\varphi(p). \tag{3.4}$$

It follows from the definition of the John transform $\mathcal{J}$ that $\mathcal{J}\varphi \in \mathcal{H}(-1)$ for $\varphi \in \mathcal{F}(-2)$. Thus, the John transform is a mapping

$$\mathcal{J} : \mathcal{F}(-2) \to \mathcal{H}(-1).$$

3.3. Relation to the John transform in the affine space. For convenience, in this subsection we denote the John transforms in the affine and projective spaces by $\mathcal{J}_a$ and $\mathcal{J}_p$, respectively. Recall that the John transform $\mathcal{J}_a$ of a function F on $\mathbb{R}^3$ is given in local coordinates $(\alpha_1,\alpha_2,\beta_1,\beta_2)$ on the manifold of lines in $\mathbb{R}^3$ as follows:

$$(\mathcal{J}_aF)(\alpha_1,\alpha_2,\beta_1,\beta_2) = \int_{-\infty}^{+\infty} F(\alpha_1 x_3+\beta_1,\ \alpha_2 x_3+\beta_2,\ x_3)\,dx_3. \tag{3.5}$$

Suppose $\mathbb{R}^3$ is embedded in $\mathbb{R}^4$ as the hyperplane $x_4=1$, and assign to every function $f\in\mathcal{F}(-2)$ the restriction F of f to this hyperplane, i.e., put

$$F(x_1,x_2,x_3) = f(x_1,x_2,x_3,1).$$

By the homogeneity condition (3.2), the function f can be reconstructed uniquely from the function F, and conversely, any C^∞ function F on $\mathbb{R}^3$ which appropriately decreases at infinity is a restriction to $\mathbb{R}^3$ of a certain function $f\in\mathcal{F}(-2)$.

It follows from (3.5) that

$$(\mathcal{J}_aF)(\alpha_1,\alpha_2,\beta_1,\beta_2) = \int_{-\infty}^{+\infty} f(\beta_1+\alpha_1 t,\ \beta_2+\alpha_2 t,\ t,\ 1)\,dt = (\mathcal{J}_pf)(u,v),$$

where $u=(\beta_1,\beta_2,0,1)$ and $v=(\alpha_1,\alpha_2,1,0)$. (For the contour S used in the definition of the function $\mathcal{J}_pf$ in (3.3), one takes the pair of lines $s=\pm1$.) Thus, passing from the vectors u, v to the Plücker coordinates, we obtain the following assertion.

PROPOSITION 3.1. *The following relation holds between the John transform $\mathcal{J}_pf$ of a function $f\in\mathcal{F}(-2)$ and the affine John transform $\mathcal{J}_aF$, where F stands for the restriction of f to the hyperplane $x_4=1$:*

$$(\mathcal{J}_aF)(\alpha_1,\alpha_2,\beta_1,\beta_2) = (\mathcal{J}_pf)(p), \tag{3.6}$$

where $p_{12}=\beta_1\alpha_2-\beta_2\alpha_1$, $p_{13}=\beta_1$, $p_{14}=-\alpha_1$, $p_{23}=\beta_2$, $p_{24}=-\alpha_2$, and $p_{34}=-1$.

In turn, it follows from (3.6) and the homogeneity condition (3.2) that

$$(\mathcal{J}_pf)(p) = |p_{34}|^{-1}(\mathcal{J}_aF)(p_{14}/p_{34},\ p_{24}/p_{34},\ -p_{13}/p_{34},\ -p_{23}/p_{34}).$$

3.4. Description of the image of the John transform. As in the affine case, any function $\varphi(u,v)$ in the image of the John transform $\mathcal{J}$ satisfies additional relations. Namely, it immediately follows from the definition (3.3) of the John transform that

$$\frac{\partial^2\varphi}{\partial u_i\partial v_j} = \frac{\partial^2\varphi}{\partial u_j\partial v_i}, \qquad i,j=1,2,3,4 \quad (i\neq j), \tag{3.7}$$

where u_i, v_i are the coordinates of the vectors u and v.

We represent these conditions on the functions $\varphi = \mathcal{J}f$ in an invariant form by viewing φ as a function of the Plücker coordinates p_{ij}, i.e., as a homogeneous function on the cone K given by the equation $p_{12}\,p_{34} - p_{13}\,p_{24} + p_{14}\,p_{23} = 0$ in $\mathbb{R}^6$.

Let $H(-1)$ be the space of C^∞ functions on $\mathbb{R}^6 \setminus O$ satisfying the homogeneity condition (3.4). Introduce the following differential operator on $H(-1)$:

$$\Delta = \frac{\partial^2}{\partial p_{12}\,\partial p_{34}} - \frac{\partial^2}{\partial p_{13}\,\partial p_{24}} + \frac{\partial^2}{\partial p_{14}\,\partial p_{23}}. \tag{3.8}$$

PROPOSITION 3.2. *The operator Δ can be pushed down from $H(-1)$ to $\mathcal{H}(-1)$ under the restriction operation $H(-1) \to \mathcal{H}(-1)$. In other words, if $\varphi\,|_K = 0$, then $\Delta\varphi\,|_K = 0$.*

PROOF. Write

$$r = p_{12}\,p_{34} - p_{13}\,p_{24} + p_{14}\,p_{23}$$

and pass to new coordinates on $\mathbb{R}^6$ in a neighborhood of an arbitrary point $p^0 \in K$ by replacing one of the coordinates p_{ij} $(i<j)$, say, p_{34}, with r (this change of coordinates is admissible if $p_{12} \neq 0$). We must show that the expression $\Delta\varphi\,|_K \equiv \Delta\varphi\,|_{r=0}$ represented in the new coordinates contains no derivatives with respect to r.

Denote by ψ the function φ in the new coordinates, i.e.,

$$\varphi(p) = \psi(p_{12}, p_{13}, p_{14}, p_{23}, p_{24}, r).$$

Immediate calculations show that

$$\Delta\varphi = -\frac{\partial^2\psi}{\partial p_{13}\partial p_{24}} + \frac{\partial^2\psi}{\partial p_{14}\partial p_{23}} + \sum{}' p_{ij}\,\frac{\partial^2\psi}{\partial p_{ij}\partial r} + 3\,\frac{\partial\psi}{\partial r} + r\,\frac{\partial^2\psi}{\partial r^2}, \tag{3.9}$$

where the prime means that the sum is taken over all indices i and j $(i<j)$ except for $(i,j) = (3,4)$. Further, it follows from the homogeneity condition for φ that $\sum' p_{ij}\,\frac{\partial\psi}{\partial p_{ij}} + 2\,r\,\frac{\partial\psi}{\partial r} = -\psi$, and hence

$$\sum{}' p_{ij}\,\frac{\partial^2\psi}{\partial p_{ij}\partial r} = -3\,\frac{\partial\psi}{\partial r} - 2\,r\,\frac{\partial^2\psi}{\partial r^2}.$$

Substituting this expression into (3.9), we obtain

$$\Delta\varphi = -\frac{\partial^2\psi}{\partial p_{13}\partial p_{24}} + \frac{\partial^2\psi}{\partial p_{14}\partial p_{23}} - r\,\frac{\partial^2\psi}{\partial r^2}. \tag{3.10}$$

The assertion follows from (3.10).

COROLLARY. *The operator Δ is well defined on the function space $\mathcal{H}(-1)$.*

PROPOSITION 3.3. *Each of the equations* (3.7) *for the function* $\varphi \in \mathcal{H}(-1)$ *is equivalent to the equation*

$$\Delta\varphi \equiv \frac{\partial^2\varphi}{\partial p_{12}\partial p_{34}} - \frac{\partial^2\varphi}{\partial p_{13}\partial p_{24}} + \frac{\partial^2\varphi}{\partial p_{14}\partial p_{23}} = 0. \tag{3.11}$$

PROOF. Let us prove, for instance, that the relation $\frac{\partial^2\varphi}{\partial u_1\partial v_2} = \frac{\partial^2\varphi}{\partial u_2\partial v_1}$ is equivalent to (3.11).

For the local coordinates on the cone K we take the Plücker coordinates p_{ij}, $i < j$, $(i,j) \neq (1,2)$. By (3.10), in these coordinates we have

$$\Delta\varphi = -\frac{\partial^2\varphi}{\partial p_{13}\partial p_{24}} + \frac{\partial^2\varphi}{\partial p_{14}\partial p_{23}}.$$

Since $p_{ij} = u_i v_j - u_j v_i$, it follows that

$$\frac{\partial^2\varphi}{\partial u_1\partial v_2} = -\sum_{i,j=3,4} v_i u_j \frac{\partial^2\varphi}{\partial p_{1i}\partial p_{2j}}, \qquad \frac{\partial^2\varphi}{\partial u_2\partial v_1} = -\sum_{i,j=3,4} v_i u_j \frac{\partial^2\varphi}{\partial p_{2i}\partial p_{1j}}.$$

This implies that

$$\frac{\partial^2\varphi}{\partial u_1\partial v_2} - \frac{\partial^2\varphi}{\partial u_2\partial v_1} = (v_3u_4 - v_4u_3)\,\Delta\varphi,$$

which proves our assertion.

THEOREM 3.1. *A function* $\varphi \in \mathcal{H}(-1)$ *is representable in the form* $\varphi = \mathcal{J}f$, *where* $f \in \mathcal{F}(-2)$, *if and only if this function satisfies the equation* $\Delta\varphi = 0$.

The "only if" part is already proven. For the proof of the "if" part, see 3.6.

3.5. Another way to define the John transform. Let us define two-dimensional subspaces of $\mathbb{R}^4$ by systems of independent equations

$$\langle \xi,\, x\rangle = 0, \qquad \langle \eta,\, x\rangle = 0, \tag{3.12}$$

i.e., by pairs ξ, η of linearly independent vectors of the dual space $(\mathbb{R}^4)'$ rather than by pairs u, v of linearly independent vectors in $\mathbb{R}^4$.

Denote by $h_{\xi\eta}$ the subspace of $\mathbb{R}^4$ given by (3.12) and denote by $Ph_{\xi\eta}$ the projectivization of this subspace. Introduce a differential 1-form $\sigma_{\xi\eta}(x)$ on $h_{\xi\eta}$ as follows:

$$\langle \xi, dx\rangle \wedge \langle \eta, dx\rangle \wedge \sigma_{\xi\eta}(x) = \omega(x), \tag{3.13}$$

where $\omega(x) = x_1\, dx_2 \wedge dx_3 \wedge dx_4 + \cdots$ (cyclic permutations).

If $f \in \mathcal{F}(-2)$, then the differential 1-form on $h_{\xi\eta}$ given by

$$f(x)\,\sigma_{\xi\eta}(x)$$

is homogeneous of degree zero, and this form is orthogonal to the fibers of the bundle $h_{\xi,\eta} \setminus O \to Ph_{\xi,\eta}$ (i.e., it vanishes on vectors tangent to the fibers).

DEFINITION. By the *John transform* of a function $f \in \mathcal{F}$ we mean the following function of a pair of linearly independent vectors ξ, $\eta \in (\mathbb{R}^4)'$:

$$(\mathcal{J}'f)(\xi,\, \eta) = \int_{\gamma_{\xi\eta}} f(x)\,\sigma_{\xi\eta}(x), \tag{3.14}$$

where $\gamma_{\xi\eta}$ is any section of the bundle $h_{\xi,\eta} \setminus O \to Ph_{\xi,\eta}$; the integral does not depend on the choice of $\gamma_{\xi\eta}$.

Using the symbolism of delta functions, one can represent the integral (3.14) as follows:

$$(\mathcal{J}'f)(\xi,\eta)=\int_\Gamma f(x)\,\delta(\langle\xi,x\rangle)\,\delta(\langle\eta,x\rangle)\,\omega(x), \tag{3.15}$$

where Γ is an arbitrary section of the bundle $\mathbb{R}^4\setminus O\to\mathbb{P}^3$.

Similarly to $\varphi=\mathcal{J}f$, the function $\psi=\mathcal{J}'f$ satisfies the following conditions.

1^0. It depends only on the bivector $q=\xi\wedge\eta$, i.e., on the minors $q^{ij}=\xi^i\eta^j-\xi^j\eta^i$.

2^0. When viewed as a function of q, the function ψ is a C^∞ function on the cone $q^{12}q^{34}-q^{13}q^{24}+q^{14}q^{23}=0$ in $(\mathbb{R}^6)'$, and ψ satisfies the homogeneity condition

$$\psi(\lambda q)=|\lambda|^{-1}\psi(q). \tag{3.16}$$

3^0. The following relations hold:

$$\frac{\partial^2\psi}{\partial\xi^i\partial\eta^j}=\frac{\partial^2\psi}{\partial\xi^j\partial\eta^i},\qquad i,j=1,2,3,4\quad(i\neq j), \tag{3.17}$$

where ξ^i and η^i are the coordinates of the vectors ξ and η.

As in the case of functions, $\varphi=\mathcal{J}f$, one can show that each of relations (3.17) is equivalent to the relation

$$\Delta\psi\equiv\frac{\partial^2\psi}{\partial q^{12}\partial q^{34}}-\frac{\partial^2\psi}{\partial q^{13}\partial q^{24}}+\frac{\partial^2\psi}{\partial q^{14}\partial q^{23}}=0. \tag{3.18}$$

Let us find a relation between the operators $\mathcal{J}$ and $\mathcal{J}'$. We say that bivectors $p=\{p_{ij}\}$ in $\mathbb{R}^4$ and $q=\{q^{ij}\}$ in $(\mathbb{R}^4)'$ are *dual* and write $p\sim q$ if

$$p_{ij}=\operatorname{sgn}(i,j,k,l)\,q^{kl}$$

for any permutation (i,j,k,l) of the indices $(1,2,3,4)$.

One can readily see that the same subspace $h\subset\mathbb{R}^4$ corresponds to dual bivectors p and q, and the differential 1-forms on h related to these bivectors coincide. This implies the following assertion.

PROPOSITION 3.4. *If $q\sim p$, then $\mathcal{J}f(p)=\mathcal{J}'f(q)$.*

Denote by $\mathcal{H}'(-1)$ the space of C^∞ functions on $K'\setminus O$ satisfying the homogeneity condition (3.16), where K' stands for the cone $q^{12}q^{34}-q^{13}q^{24}+q^{14}q^{23}=0$ in $(\mathbb{R}^6)'$. By Proposition 3.4, Theorem 3.1 can be restated in terms of functions in $H'(-1)$.

THEOREM 3.2. *A function $\psi\in\mathcal{H}'(-1)$ is representable in the form $\psi=\mathcal{J}'f$, where $f\in\mathcal{F}(-2)$, if and only if this function satisfies the equation $\Delta\psi=0$.*

3.6. Proof of the theorem on the image of the John transform. The "only if" part of Theorems 3.1 and 3.2 was already established. We prove the "if" part in Theorem 3.2.

Let $\psi(\xi,\eta)$ be a C^∞ function of the bivector $\xi\wedge\eta$, and let ψ satisfy the homogeneity conditions (3.16) and equations (3.17). We prove that $\psi=\mathcal{J}'f$ for some $f\in\mathcal{F}(-2)$.

Introduce the manifold A of pairs (x,η), where $x\in\mathbb{R}^4\setminus O$, $\eta\in(\mathbb{R}^4)'\setminus O$, and $\langle\eta,x\rangle=0$. Let $(x,\eta)\in A$. Denote by $L_\eta\subset\mathbb{R}^4$ the subspace orthogonal to η. Note that $x\in L_\eta$. For any $\xi\in(\mathbb{R}^4)'\setminus\{\eta\}$, where $\{\eta\}$ is the subspace spanned by η, we set

$$\psi_\eta(\xi)=\psi(\xi,\eta).$$

Since ψ depends on $\xi \wedge \eta$ only, it follows that $\psi_\eta(\xi + \lambda\eta) = \psi_\eta(\xi)$ for each $\lambda \in \mathbb{R}$, i.e., ψ_η is a function on $((\mathbb{R}^4)'/\{\eta\}) \setminus O$. It is clear that this function is infinitely differentiable on $((\mathbb{R}^4)'/\{\eta\}) \setminus O$ and satisfies the homogeneity condition $\psi_\eta(\lambda\xi) = |\lambda|^{-1}\,\psi_\eta(\xi)$. Hence, by the theorem on the image of the Radon transform for the projective plane, the function ψ_η is the Radon transform of a C^∞ function $f_\eta(x)$ defined on the subspace $L_\eta \subset \mathbb{R}^4$ dual to $(\mathbb{R}^4)'/\{\eta\}$, and f_η satisfies the homogeneity condition $f_\eta(\lambda x) = \lambda^{-2}\,f_\eta(x)$.

Thus, the following function F is well defined on the manifold A:

$$F(x, \eta) = f_\eta(x).$$

LEMMA. *For any $x \in \mathbb{R}^4 \setminus O$, the function F does not depend on η.*

PROOF. It suffices to prove the assertion for some special point x, say, for $x = x_0 = (1, 0, 0, 0)$. In this case, the condition that the vectors η and x are orthogonal becomes $\eta^1 = 0$, i.e., $\eta = (0, \eta^2, \eta^3, \eta^4)$. We must show that $\frac{\partial F(x_0,\eta)}{\partial \eta_i} = 0$, $i = 2, 3, 4$. For instance, let us prove that $\frac{\partial F(x_0,\eta)}{\partial \eta_2} = 0$.

We write out an explicit expression for $F(x, \eta)$ in terms of $f_\eta(x)$ by using the inversion formula for the Radon transform on the projective plane.

Let L be the coordinate subspace $\xi^4 = 0$ in $(\mathbb{R}^4)'$. If $\eta^4 \neq 0$, then L is transversal to the vector η, and we can assume that ψ_η is a function on the subspace L.

Since $\langle \xi, x_0 \rangle = \xi^1$, the inversion formula for the Radon transform implies that

$$\begin{aligned} F(x_0, \eta) &= c \int \psi_\eta(\xi) \Big|_{\xi^3 = 1} (\xi^1)^{-2}\, d\xi^1 \wedge d\xi^2 \\ &= c \int \frac{\partial \psi(\eta, \xi)}{\partial \xi^1} \Big|_{\xi^3 = 1} (\xi^1)^{-1}\, d\xi^1 \wedge d\xi^2. \end{aligned}$$

Differentiating the resulting relation with respect to η^2 and applying the formula $\frac{\partial^2 \psi}{\partial \xi^1 \partial \eta^2} = \frac{\partial^2 \psi}{\partial \xi^2 \partial \eta^1}$, we obtain

$$\frac{\partial F(x_0, \eta)}{\partial \eta^2} = c \int \frac{\partial^2 \psi(\eta, \xi)}{\partial \xi^2 \partial \eta^1} \Big|_{\xi^3 = 1} (\xi^1)^{-1}\, d\xi^1 \wedge d\xi^2 = 0.$$

We now return to the proof of the theorem. Using the above lemma, we define a function $f(x)$ on $\mathbb{R}^4$ by the formula

$$f(x) = f_\eta(x),$$

where η is an arbitrary vector in $(\mathbb{R}^4)'$ orthogonal to x. The function f is homogeneous, $f(\lambda x) = \lambda^{-2}\, f(x)$, and infinitely differentiable on $L_\eta \setminus O$, where L_η is a 3-dimensional subspace of $\mathbb{R}^4$, and hence f is C^∞-differentiable on $\mathbb{R}^4 \setminus O$. Therefore, $f \in \mathcal{F}(-2)$. Moreover, it is clear that $\mathcal{J}' f = \psi$ because this relation holds for the restriction of ψ to the set of subspaces belonging to an arbitrary three-dimensional space.

3.7. John transform as an intertwining operator. The John transform has a simple group-theoretic interpretation. Consider the group $G = GL(4, \mathbb{R})$ of nondegenerate matrices of order four. Any element $g \in G$ defines a linear transform $x \mapsto xg$ of $\mathbb{R}^4$ sending a point $x = (x_1, x_2, x_3, x_4)$ to the point x' with coordinates $x'_j = \sum_{i=1}^4 x_i\, g_{ij}$.

To any element $g \in G$ we assign linear operators $T_g^{(1)}$ and $T_g^{(2)}$ on the spaces $\mathcal{F}(-2)$ and $\mathcal{H}(-1)$, respectively,

$$(T_g^{(1)}\, f)\,(x) = f(xg), \qquad (T_g^{(2)}\, \varphi)\,(u,v) = \varphi(ug, vg).$$

Obviously,

$$T_{g_1}^{(i)}\, T_{g_2}^{(i)} = T_{g_1 g_2}^{(i)}, \qquad i = 1, 2,$$

for any g_1, $g_2 \in G$, i.e., the operators $T_g^{(1)}$ and $T_g^{(2)}$ form representations of the group G in the corresponding function spaces.

The next assertion immediately follows from the definition of the operator $\mathcal{J}$.

PROPOSITION 3.5. *The operator $\mathcal{J}$ of John transform satisfies the relation*

$$T_g^{(2)}\, \mathcal{J} = \mathcal{J}\, T_g^{(1)} \quad \textit{for each} \quad g \in G. \tag{3.19}$$

In terms of representation theory, relation (3.19) means that the operator $\mathcal{J} : \mathcal{F}(-2) \to \mathcal{H}(-1)$ is an intertwining operator for the representations $T^{(1)}$ and $T^{(2)}$ of the group G.

One can prove that an intertwining operator $\mathcal{F}(-2) \to \mathcal{H}(-1)$ is unique up to a factor. Thus, the condition that $\mathcal{J}$ is an intertwining operator uniquely determines the John transform up to a factor.

We now establish the group-theoretic meaning of the additional relations (3.7) for a function $\varphi = \mathcal{J} f$.

Let $\mathfrak{g}$ be the Lie algebra of the group G and let $\{e_{ij}\}$ be the standard basis in $\mathfrak{g}$ (i.e., e_{ij} is the matrix with only one nonzero entry which is equal to 1 and stands at the (ij)th place).

The representations $T^{(1)}$ and $T^{(2)}$ of the group G induce the corresponding representations of the Lie algebra $\mathfrak{g}$ of G. Denote by $E_{ij}^{(1)}$ and $E_{ij}^{(2)}$ the operators of these representations corresponding to $e_{ij} \in \mathfrak{g}$. The formulas for the operators $T_g^{(1)}$ and $T_g^{(2)}$ immediately imply the following expressions for $E_{ij}^{(1)}$ and $E_{ij}^{(2)}$:

$$E_{ij}^{(1)}\, f = x_i\, \frac{\partial f}{\partial x_j}, \tag{3.20}$$

$$E_{ij}^{(2)}\, \varphi = u_i\, \frac{\partial \varphi}{\partial u_j} + v_i\, \frac{\partial \varphi}{\partial v_j}. \tag{3.21}$$

The next assertion immediately follows from (3.20).

PROPOSITION 3.6. *Each function $f \in \mathcal{F}(-2)$ satisfies the following relations:*

$$(E_{ij}^{(1)}\, E_{kl}^{(1)} - \delta_{jk}\, E_{il}^{(1)})\, f = (E_{il}^{(1)} E_{kj}^{(1)} - \delta_{lk}\, E_{ij}^{(1)})\, f, \tag{3.22}$$

$i, j, k, l = 1, 2, 3, 4$, where δ_{ij} is the Kronecker delta.

Since $\mathcal{J}$ is an intertwining operator, this implies the following corollary.

COROLLARY. *The function $\varphi = \mathcal{J} f$ satisfies the relations*

$$(E_{ij}^{(2)}\, E_{kl}^{(2)} - \delta_{jk}\, E_{il}^{(2)})\, \varphi = (E_{il}^{(2)} E_{kj}^{(2)} - \delta_{lk}\, E_{ij}^{(2)})\, \varphi, \tag{3.23}$$

where the operators $E_{ij}^{(2)}$ are given by (3.21).

Relations (3.23) obtained from group-theoretic considerations are equivalent to (3.7). Indeed, if we substitute the explicit expressions for the operators $E_{ij}^{(2)}$ into (3.23), then (3.23) becomes

$$u_i\, v_k\, \frac{\partial^2 \varphi}{\partial u_j \partial v_l} + v_i\, u_k\, \frac{\partial^2 \varphi}{\partial v_j \partial u_l} = u_i\, v_k\, \frac{\partial^2 \varphi}{\partial u_l \partial v_j} + v_i\, u_k\, \frac{\partial^2 \varphi}{\partial v_l \partial u_j},$$

i.e.,

$$(u_i\, v_k - u_k\, v_i)\left(\frac{\partial^2 \varphi}{\partial u_j \partial v_l} - \frac{\partial^2 \varphi}{\partial u_l \partial v_j}\right) = 0.$$

4. John transform in the complex affine space

The definition of John transform for the complex affine space and the description of the image of this transform are similar to those given in § 1 for the real case. We briefly present this material in 4.1 and 4.2. The main point at which the complex John transform differs from the real one is that the former admits local inversion formulas. Subsections 4.3 and 4.4 are devoted to the construction of these formulas.

4.1. John transform in $\mathbb{C}^3$. Similarly to the case of real affine space, we present complex lines in $\mathbb{C}^3$ by equations of the form

$$x = \alpha t + \beta, \qquad \alpha \in \mathbb{C}^3 \setminus O, \quad \beta \in \mathbb{C}^3,$$

and the integral of a function f on $\mathbb{C}^3$ over a line $x = \alpha t + \beta$ is defined by the relation

$$\varphi(\alpha, \beta) = \frac{i}{2} \int_{\mathbb{C}} f(\alpha t + \beta)\, dt \wedge d\bar{t}. \tag{4.1}$$

The function $\varphi(\alpha, \beta)$ thus defined on the manifold E of pairs (α, β) is denoted by $\mathcal{J} f$ and referred to as the *John transform of the function* f. In what follows, it is assumed that all functions f belong to the Schwartz space $S(\mathbb{C}^3)$ of infinitely differentiable functions rapidly decreasing, together with all their derivatives, on $\mathbb{C}^3$.

Similarly to the real case, the functions $\varphi = \mathcal{J} f$ satisfy the relations

$$\varphi(\alpha, \beta + \alpha t_0) = \varphi(\alpha, \beta) \quad \text{for each} \quad t_0 \in \mathbb{C}, \tag{4.2}$$

i.e., the function φ does not depend on the choice of a point β on the line, and

$$\varphi(\lambda \alpha, \beta) = |\lambda|^{-2}\, \varphi(\alpha, \beta) \quad \text{for each} \quad \lambda \neq 0. \tag{4.3}$$

Thus, φ depends only on the line itself and on the choice of the direction vector α.

If the space $\mathbb{C}^3$ is equipped with Euclidean metric, then in the definition of the John transform in (4.1) one can assume that $|\alpha| = 1$. In this case, condition (4.3) is reduced to a simpler one: $\varphi(\lambda\alpha, \beta) = \varphi(\alpha, \beta)$ and $|\lambda| = 1$, and φ can be regarded as a function on the manifold of complex lines H.

By analogy with the real case, denote by $S(H)$ the space of all infinitely differentiable functions $\varphi(\alpha, \beta)$ satisfying condition (4.2), (4.3) and such that, for $|\alpha| = 1$, these functions (regarded as functions on H) rapidly decrease together with all their derivatives. One can readily see that the image of any function $f \in S(\mathbb{C}^3)$ under the John transform $\mathcal{J}$ belongs to the space $S(H)$.

Sometimes it is convenient to present lines on $\mathbb{C}^3$ by equations of the form

$$x_1 = \alpha_1 x_3 + \beta_1, \qquad x_2 = \alpha_2 x_3 + \beta_2,$$

and define the *John transform* of $\psi = \mathcal{J}f$ in the local coordinates α_1, α_2, β_1, β_2 on H by the formula

$$\psi(\alpha_1\alpha_2, \beta_1, \beta_2) = \frac{i}{2}\int_{\mathbb{C}} f(\alpha_1 x_3 + \beta_1,\, \alpha_2 x_3 + \beta_2,\, x_3)\, dx_3 \wedge d\bar{x}_3. \tag{4.4}$$

4.2. Differential form $\kappa\varphi$ and the theorem on the image of the John transform. The complex analog of the differential form $\kappa\varphi$ introduced in § 1 is the following differential $(1,1)$-form on the manifold E of pairs (α, β):

$$\kappa\varphi = \sum_{i,j=1}^{3} \frac{\partial^2\varphi}{\partial\beta_i\,\partial\bar{\beta}_j}\, d\alpha_i \wedge d\bar{\alpha}_j. \tag{4.5}$$

Here φ stands for an arbitrary function on E. If φ satisfies (4.2) and (4.3), then the differential form $\kappa\varphi$ can be pushed down from the manifold E to the manifold of lines H. Correspondingly, in the local coordinates $\alpha_1\alpha_2, \beta_1, \beta_2$ the form $\kappa\psi$ becomes

$$\kappa\psi = \sum_{i,j=1}^{2} \frac{\partial^2\psi}{\partial\beta_i\,\partial\bar{\beta}_j}\, d\alpha_i \wedge d\bar{\alpha}_j. \tag{4.6}$$

Similarly to the real case, one can describe the image of the space $S(\mathbb{C}^3)$ under the John transform by using the form $\kappa\varphi$. Namely, denote by H^x the submanifold of lines in $\mathbb{C}^3$ passing through an arbitrary point $x \in \mathbb{C}^3$ and by $\kappa_x\varphi$ the restriction of the form $\kappa\varphi$ to H^x.

THEOREM 4.1. *The function $\varphi \in S(H)$ belongs to the image of the Schwartz space $S(\mathbb{C}^3)$ under the John transform $\mathcal{J}$ if and only if the differential form $\kappa_x\varphi$ on the submanifold H^x is closed for any point $x \in \mathbb{C}^3$.*

The condition that the differential form $\kappa_x\varphi$ is closed for any point $x \in \mathbb{C}^3$ is equivalent to the following conditions on the function φ:

$$\frac{\partial^2\varphi}{\partial\alpha_i\,\partial\beta_j} = \frac{\partial^2\varphi}{\partial\alpha_j\,\partial\beta_i}, \qquad \frac{\partial^2\varphi}{\partial\bar{\alpha}_i\,\partial\bar{\beta}_j} = \frac{\partial^2\varphi}{\partial\bar{\alpha}_j\,\partial\bar{\beta}_i}.$$

The proof of Theorem 4.1 is the same as in the real case.

4.3. Inversion formula. In the complex case, in contrast to the real case, the operator κ can be used not only to describe the image of the John transform $\mathcal{J}$ but also to obtain an inversion formula reconstructing a function $f \in S(\mathbb{C}^3)$ from its John transform $\varphi = \mathcal{J}f$.

THEOREM 4.2. *If $\varphi = \mathcal{J}f$, $f \in S(\mathbb{C}^3)$ and $\kappa_x\varphi$ is the restriction of the differential $(1,1)$-form $\kappa\varphi$ to the submanifold $H^x \subset H$ of lines passing through a point $x \in \mathbb{C}^3$, then*

$$\int_\gamma \kappa_x\varphi = c(\gamma)\, f(x) \tag{4.7}$$

for each cycle $\gamma \subset H^x$ of real dimension 2, where $c(\gamma)$ depends on the homology class of the cycle γ. Moreover, if the cycle γ is not homologous to zero, then $c(\gamma) \neq 0$.

PROOF. Since the differential form $\kappa_x\varphi$ is closed, it follows that, for a fixed f, the integral (4.7) depends only on the homology class of the cycle γ. It is known that the second homology group of a two-dimensional manifold H^x is a group with a single generator $\tilde{\gamma}$, and for a representative of the homology class $\tilde{\gamma}$ one can take an Euler cycle, i.e., the manifold of lines passing through the point x and belonging to a chosen plane. Therefore, it suffices to evaluate the integral (4.7) over some Euler cycle. We show that, if γ is an Euler cycle, then the inversion formula (4.7) coincides with the inversion formula obtained in Chapter 1 for the Radon transform in $\mathbb{C}^2$.

We express the John transform in the local coordinates by (4.4).

Consider the Euler cycle γ formed by the lines passing through a point $x^0 = (x_1^0, x_2^0, x_3^0)$ and belonging to the plane h given by the equation $x_1 = x_1^0$. Let f_0 be the restriction of the function f to the plane h and let ψ_0 be the restriction of the function $\psi = \mathcal{J}f$ to the submanifold of lines lying in the plane h, i.e.,

$$f_0(x_2, x_3) = f(x_1^0, x_2, x_3),$$

$$\psi_0(\alpha, \beta) = \psi(0, \alpha; x_1^0, \beta) = \frac{i}{2} \int_{\mathbb{C}} f_0(\alpha x_3 + \beta, x_3)\, dx_3 \wedge d\bar{x}_3.$$

We have

$$\int_\gamma \kappa_x \psi = \int_{\mathbb{C}} \frac{\partial^2 \psi_0(\alpha, \beta)}{\partial \beta\, \partial \bar{\beta}} \Big|_{\beta = x_2^0 - \alpha x_3^0} d\alpha \wedge d\bar{\alpha}.$$

On the other hand, the function ψ_0 is the Radon transform of the function f_0 on the plane h. Hence, by the inversion formula for the Radon transform on the complex plane (see (9.11) in § 9 of Chapter 1) we obtain

$$\int_{\mathbb{C}} \frac{\partial^2 \psi_0(\alpha, \beta)}{\partial \beta\, \partial \bar{\beta}} \Big|_{\beta = x_2^0 - \alpha x_3^0} d\alpha \wedge d\bar{\alpha} = 2\pi^2 i\, f_0(x_2^0, x_3^0).$$

Thus, it is proved that

$$\int_\gamma \kappa_{x^0} \varphi = 2\pi^2 i\, f(x^0)$$

for an Euler cycle $\gamma \subset H^{x^0}$.

Obviously, the inversion formula (4.7) is local; namely, to reconstruct a function f at an arbitrary point of $x \in \mathbb{C}^3$ by this formula, it suffices to know only the integrals of the function f over the lines infinitely close to x.

We emphasize that there is no analog of this inversion formula for the John transform in the real space because in the real case we have $\int_\gamma \kappa_x \varphi = 0$ for each one-dimensional cycle $\gamma \subset H^x$.

4.4. Analogs of the operator κ. In this subsection we construct a family of linear operators κ from the space $S(H)$ to the space $\Omega^{(1,1)}(H)$ of differential $(1,1)$-forms on H satisfying the following condition similar to that satisfied by the above operator κ.

CONDITION A. If $\varphi = \mathcal{J}f$, where $f \in S(\mathbb{C}^3)$, then the differential form $\kappa\varphi$ is closed on the submanifold of lines passing through an arbitrarily chosen point $x \in \mathbb{C}^3$.

Each of these operators can be used to construct an inversion formula in the same way in which we used the original operator κ above.

Consider the John transform in local coordinates (4.4). In these coordinates, the conditions for a function ψ to belong to the image of the John transform become

$$\frac{\partial^2 \psi}{\partial \alpha_1\, \partial \beta_2} = \frac{\partial^2 \psi}{\partial \alpha_2\, \partial \beta_1}, \qquad \frac{\partial^2 \psi}{\partial \bar{\alpha}_1\, \partial \bar{\beta}_2} = \frac{\partial^2 \psi}{\partial \bar{\alpha}_2\, \partial \bar{\beta}_1}. \tag{4.8}$$

We introduce the following four linear mappings $\Omega^{(p,q)} \to \Omega^{(p+1,q)}$, where $\Omega^{(p,q)}$ denotes the space of differential (p,q)-forms on H:

$$\kappa^1 = \frac{\partial}{\partial\beta_1}\,d\alpha_1 + \frac{\partial}{\partial\beta_2}\,d\alpha_2,$$
$$\kappa^2 = \frac{\partial}{\partial\beta_1}\,d\beta_1 + \frac{\partial}{\partial\beta_2}\,d\beta_2,$$
$$\kappa^3 = \frac{\partial}{\partial\alpha_1}\,d\alpha_1 + \frac{\partial}{\partial\alpha_2}\,d\alpha_2,$$
$$\kappa^4 = \frac{\partial}{\partial\alpha_1}\,d\beta_1 + \frac{\partial}{\partial\alpha_2}\,d\beta_2.$$

Further, denote by $\bar\kappa^i$, $i = 1,2,3,4$, the linear mapping $\Omega^{(p,q)} \to \Omega^{(p,q+1)}$ obtained from κ^i by replacing $\frac{\partial}{\partial\alpha_j}$, $\frac{\partial}{\partial\beta_j}$, $d\alpha_j$, and $d\beta_j$ with $\frac{\partial}{\partial\bar\alpha_j}$, $\frac{\partial}{\partial\bar\beta_j}$, $d\bar\alpha_j$, and $d\bar\beta_j$, respectively. Write

$$\kappa^{ij} = \kappa^i\,\bar\kappa^j, \quad i,j = 1,2,3,4.$$

According to this notation, if ψ is a function on H, then $\kappa^{ij}\psi$ is a differential (1,1)-form on H. In particular, $\kappa^{11}\psi$ coincides with the above form $\kappa\psi$. Denote by $\kappa_x^{ij}\psi$ the restriction of the form $\kappa^{ij}\psi$ to the submanifold $H^x \subset H$ of lines passing through a given point $x \in \mathbb{C}^3$.

PROPOSITION 4.1. *For any function $\psi = \mathcal{J}f$ and any point $x \in \mathbb{C}^3$, the differential forms $\kappa_x^{ij}\psi$ are related as follows*:

$$\begin{aligned}\kappa_x^{11}\psi &= -x_3^{-1}\,\kappa_x^{21}\psi = -\bar x_3^{-1}\,\kappa_x^{12}\psi = |x_3|^{-2}\,\kappa_x^{22}\psi,\\ \kappa_x^{13}\psi &= -x_3^{-1}\,\kappa_x^{23}\psi = -\bar x_3^{-1}\,\kappa_x^{14}\psi = |x_3|^{-2}\,\kappa_x^{24}\psi,\\ \kappa_x^{31}\psi &= -x_3^{-1}\,\kappa_x^{41}\psi = -\bar x_3^{-1}\,\kappa_x^{32}\psi = |x_3|^{-2}\,\kappa_x^{42}\psi,\\ \kappa_x^{33}\psi &= -x_3^{-1}\,\kappa_x^{43}\psi = -\bar x_3^{-1}\,\kappa_x^{34}\psi = |x_3|^{-2}\,\kappa_x^{44}\psi,\end{aligned} \tag{4.9}$$

$$\kappa_x^{2i}\psi + \kappa_x^{3i}\psi = d(\bar\kappa^i\psi), \quad \kappa_x^{i2}\psi + \kappa_x^{i3}\psi = -d(\bar\kappa^i\psi), \quad i = 1,2,3,4. \tag{4.10}$$

PROOF. Relations (4.9) are immediate because $d\beta_i = -x_3\,d\alpha_i$ and $d\bar\beta_i = -\bar x_3\,d\bar\alpha_i$, $i = 1,2$, on the submanifold H^x. Relations (4.10) follow from (4.8).

COROLLARY 1. *The differential forms $\kappa_x^{ij}\varphi$, $i,j = 1,2,3,4$, are pairwise cohomology equivalent up to factors.*

COROLLARY 2. *If $\psi = \mathcal{J}f$, then all differential forms $\kappa^{ij}\varphi$ satisfy Condition A.*

By Proposition 4.1, to construct an inversion formula, one can use not only the original operator $\kappa = \kappa^{11}$ but also any other operator of the form κ^{ij} (and any nonzero linear combination of these operators). For instance, for the operators κ^{44} and κ^{33}, the inversion formulas are

$$\begin{aligned}\int_\gamma \kappa_x^{44}\psi &= c(\gamma)\,|x_3|^4\,f(x),\\ \int_\gamma \kappa_x^{33}\psi &= c(\gamma)\,|x_3|^2\,f(x).\end{aligned} \tag{4.11}$$

5. Problems of integral geometry for line complexes in $\mathbb{C}^3$

It was already noted above that the problem of reconstructing a function f on $\mathbb{R}^3$ or $\mathbb{C}^3$ from its John transform $\varphi = \mathcal{J}f$ is overdetermined; to reconstruct f, it suffices to know the function φ on some three-dimensional submanifold K of the manifold of lines H rather than on the entire manifold H. We refer to three-dimensional algebraic subvarieties of lines $K \subset H$ as *line complexes*.

We present examples of complexes for which the problem of reconstructing a function f has an elementary solution. These are the line complexes having the following property: For almost every point $x \in \mathbb{C}^3$, there is a plane h_x containing x and such that almost all lines on this plane belong to the complex. In particular, this property holds for the complex formed by the lines that intersect a given line and for the complex of lines parallel to a given plane (the latter can be regarded as the complex of lines that intersect a line at infinity). It is clear that the problem of reconstructing the function at a point x from the integrals of this function over the lines of this complex is reduced to the inversion formula for the Radon transform on the plane h_x.

In this section, the problem of reconstructing a function f is solved for various line complexes in $\mathbb{C}^3$. A specific feature of the resulting inversion formulas is that they are local, i.e., the function at a point $x \in \mathbb{C}^3$ is reconstructed from the integrals of this function over the lines of the complex that are infinitely close to x.

We begin with the complex of lines in $\mathbb{C}^3$ that intersect a chosen algebraic curve. We claim that the inversion formula for this complex is a simple consequence of the inversion formula obtained in § 4 for the John transform. After this we shall consider a broader class of line complexes ("admissible complexes") admitting an explicit local inversion formula.

5.1. Problem of integral geometry for a complex of lines in $\mathbb{C}^3$ intersecting a curve. Consider the complex K of lines in $\mathbb{C}^3$ intersecting a chosen algebraic curve $\Lambda \subset \mathbb{C}^3$. We formulate the following problem: Reconstruct a function $f \in S(\mathbb{C}^3)$ from the integrals of f over the lines of the complex, i.e., from the function

$$\Phi(\alpha, \lambda) = \frac{i}{2} \int_{\mathbb{C}} f(\alpha t + \lambda)\, dt\, d\bar{t}, \qquad (\alpha, \lambda) \in (\mathbb{C}^3 \setminus O) \times \Lambda. \tag{5.1}$$

Denote by ∂_Λ and $\bar{\partial}_\Lambda$ the $(1,0)$- and $(0,1)$-components, respectively, of the operator d of exterior differentiation on the manifold Λ.

Formula (5.1) defines the integral transform $\Phi(\alpha, \lambda)$ of a function $f \in S(\mathbb{C}^3)$. To any point $x^0 \in \mathbb{C}^3$ we assign the following differential $(1,1)$-form on Λ:

$$(\partial_\Lambda \wedge \bar{\partial}_\Lambda)\, \Phi(\alpha, \lambda) \mid_{\alpha = x^0 - \lambda} .$$

THEOREM 5.1. *The following inversion formula for the integral transform* (5.1) *holds for almost every point* $x^0 \in \mathbb{C}^3$:

$$f(x^0) = (2\, k\, \pi^2\, i)^{-1} \int_\Lambda (\partial_\Lambda \wedge \bar{\partial}_\Lambda)\, \Phi(\alpha, \lambda) \mid_{\alpha = x^0 - \lambda}, \tag{5.2}$$

where k is the degree of the algebraic curve Λ.

PROOF. We apply the inversion formula (4.7) for the John transform $\varphi = \mathcal{J}f$ (see § 4), where for γ^{x^0} we take the submanifold of lines passing through the point

x^0 and intersecting the curve Λ, i.e., the manifold of lines

$$x = (x^0 - \lambda)\,t + \lambda, \qquad \lambda \in \Lambda.$$

By the definition of the operator κ,

$$\kappa\varphi \,\Big|_{\gamma^{x^0}} = \sum_{i,j=1}^{3} \frac{\partial^2 \varphi(\alpha,\beta)}{\partial\beta_i\,\partial\bar{\beta}_j} \,\Big|_{\substack{\alpha = x^0 - \lambda \\ \beta=\lambda}} \; d\lambda_i \wedge d\bar{\lambda}_j. \tag{5.3}$$

Since the restriction $\varphi(\alpha,\beta)\,\big|_{\beta=\lambda}$ is equal to $\Phi(\alpha,\lambda)$, the right-hand side of (5.3) is equal to $(\partial_\Lambda \wedge \bar{\partial}_\Lambda)\,\Phi(\alpha,\lambda)\,|_{\alpha=x^0-\lambda}$. Hence, by the inversion formula (4.7) for the John transform,

$$\int_\Lambda (\partial_\Lambda \wedge \bar{\partial}_\Lambda)\,\Phi(\alpha,\lambda)\,\big|_{\alpha=x^0-\lambda} = c(\gamma^{x^0})\,f(x^0).$$

Let us find the coefficient $c(\gamma^{x^0})$. If the order of the algebraic curve Λ is equal to k, then the cycle γ^{x^0} is homology equivalent to $k\,\gamma$, where γ is an Euler cycle. Since $c(\gamma) = 2\,\pi^2\, i$ by § 4, it follows that $c(\gamma^{x^0}) = 2\,k\,\pi^2\, i$.

REMARK. If the curve Λ is given parametrically, $\lambda = u(s)$, then we set

$$\Psi(\alpha, s) = \Phi(\alpha, u(s)).$$

In this case, the inversion formula takes the form

$$f(x) = (2\,k\,\pi^2\, i)^{-1} \int_{\mathbb{C}} \Psi''_{s,\bar{s}}(x - u(s), s)\, ds \wedge d\bar{s}.$$

REMARK. Real analogs of the inversion formula (5.2) play an important role in tomography. The real case is more complicated, and the reason is not only that real inversion formulas for line complexes are nonlocal. As a rule, it is impossible to reconstruct a functioin in $\mathbb{R}^3$ in terms of integrals along the lines intersecting a curve. The reconstruction is possible only for functions with strong restrictions on their supports. We do not consider such problems in this book.

5.2. Definition of admissible line complexes in $\mathbb{C}^3$. Consider an arbitrary line complex K in $\mathbb{C}^3$. As above, the problem is to reconstruct a function $f \in S(\mathbb{C}^3)$ from the integrals of this function over the lines of the complex. We need to find conditions on K under which the reconstruction problem can be solved similarly to the case of a complex of lines intersecting a curve, using the inversion formula of § 4 for the John transform.

Let $H^x \subset H$ be a submanifold of lines passing through a chosen point $x \in \mathbb{C}^3$ and let $\gamma^x = K \cap H^x$. The manifold γ^x is a cycle of real dimension two for almost every point x. Therefore, the following inversion formula holds for the John transform $\mathcal{J}$: if $\varphi = \mathcal{J} f$, then

$$\int_{\gamma^x} \kappa\varphi = c\, f(x), \tag{5.4}$$

where κ is the operator introduced in § 4.

By the assumption, we know only the restriction $\Phi = \varphi\,|_K$ of the function φ to the manifold of lines of the complex K. Thus, formula (5.4) gives a solution of the problem of integral geometry for the complex K if and only if the differential form $\kappa\varphi\big|_{\gamma^x}$ can be expressed in terms of the function Φ and its derivatives along K only. An algebraic 3-dimensional subvariety of lines for which this condition holds is called an *admissible line complex.* One can prove that this definition does not depend on the specific choice of the form $\kappa\varphi$. An example of an admissible line complex is the complex of lines intersecting an algebraic curve in $\mathbb{C}^3$, which was treated in 5.1.

If the complex K is admissible, then, substituting the explicit expression for $\kappa\psi\big|_{\gamma^x}$ in terms of the function Φ into (5.4), we obtain an inversion formula for the integral transform related to this complex.

REMARK. Our condition of admissibility is a purely local condition on a complex K. The requirement that K is an algebraic variety implies that γ^x is a cycle for almost all x and the number of lines in K passing through a generic point x is constant. Combining these global geometric conditions with the local admissibility condition, we can write down the inversion formula for smooth complexes.

The next three subsections are devoted to the study of the analytic and geometric structure of admissible complexes. In particular, we claim that each admissible complex is either the variety of lines intersecting an algebraic curve in $\mathbb{C}^3$ or a complex of lines tangent to some algebraic surface in $\mathbb{C}^3$.

5.3. Necessary and sufficient conditions for a complex K to be admissible. Consider the John transform $\mathcal{J}$ in the local coordinates α_1, α_2, β_1, β_2 on the manifold of lines H (see 4.1).

If $\psi = \mathcal{J}f$ is the John transform of a function $f \in S(\mathbb{C}^3)$, then, according to § 4, the following inversion formula holds:

$$\int_{\gamma^x} \sum_{i,j=1,2} \frac{\partial^2 \psi}{\partial \alpha_i \bar{\alpha}_j}\, d\alpha_i \wedge d\bar{\alpha}_j = c\,|x_3|^2 f(x), \tag{5.5}$$

where $\gamma^x \subset H^x$ is an arbitrary cycle of real dimension two.

Let K be an arbitrary complex. We set $\gamma^x = H^x \cap K$ and, for each $h \in \gamma^x$, denote by T_hH, T_hK, and $T_h\gamma^x$ the tangent spaces at the point h to the manifold of lines H, the complex $K \subset H$, and the cycle γ^x, respectively.

DEFINITION. Introduce the endomorphism $A_h : T_hH \to T_hH$ by setting

$$A_h(d\alpha_1, d\alpha_2, d\beta_1, d\beta_2) = (d\alpha_1, d\alpha_2, 0, 0).$$

LEMMA. *A complex K is admissible if and only if*

$$A_h\big(T_h\gamma^x\big) \subset T_hK$$

for almost every point $x \in \mathbb{C}^3$ and almost every line $h \in \gamma^x$.

Indeed, the integrand in (5.5) is the derivative of ψ taken in the complex direction $A_h\big(T_h\gamma^x\big)$ for any $h \in \gamma^x$. Hence, the form in the integrand in (5.5) can be expressed in terms of $\Phi = \psi\,|_K$ if and only if this direction belongs to the subspace T_hK.

Let a complex K be given in the local coordinates (α, β) by the equation

$$F(\alpha, \beta) = 0.$$

PROPOSITION 5.1. *The complex K is admissible if and only if*

$$\frac{\partial F}{\partial \alpha_1}\frac{\partial F}{\partial \beta_2} - \frac{\partial F}{\partial \alpha_2}\frac{\partial F}{\partial \beta_1} = 0 \tag{5.6}$$

at any point $h = (\alpha, \beta)$ of K.

PROOF. Let $\xi = (d\alpha_1, d\alpha_2, d\beta_1, d\beta_2)$ be a vector in T_hH. Then the condition that $\xi \in T_hK$ becomes

$$\frac{\partial F}{\partial \alpha_1}\,d\alpha_1 + \frac{\partial F}{\partial \alpha_2}\,d\alpha_2 + \frac{\partial F}{\partial \beta_1}\,d\beta_1 + \frac{\partial F}{\partial \beta_2}\,d\beta_2 = 0,$$

and the condition that $\xi \in T_hH^x$ is

$$x_3\,d\alpha_1 + d\beta_1 = 0, \qquad x_3\,d\alpha_2 + d\beta_2 = 0.$$

Hence, the vector $\xi \in T_h\gamma^x = T_h(H^x \cap K)$ is of the form

$$\xi = (a_1, a_2, -x_3a_1, -x_3a_2),$$

where $a_1 = \frac{\partial F}{\partial \alpha_2} - x_3\frac{\partial F}{\partial \beta_2}$ and $a_2 = -\big(\frac{\partial F}{\partial \alpha_1} - x_3\frac{\partial F}{\partial \beta_1}\big)$. It follows that $A_h(\xi) = (a_1, a_2, 0, 0)$. By the lemma, for the complex to be admissible, it is necessary and sufficient that $a_1\frac{\partial F}{\partial \alpha_1} + a_2\frac{\partial F}{\partial \alpha_2} = 0$, which is equivalent to (5.6).

We now present another proof of Proposition 5.1; this proof uses direct computations. First we assume that the complex K is given by an equation solved with respect to β_2:

$$\beta_2 = u(\alpha_1, \alpha_2, \beta_1).$$

On the manifold of lines H we introduce new coordinates α_1, α_2, β_1, $s = \beta_2 - u(\alpha_1, \alpha_2, \beta_1)$ instead of the coordinates α_1, α_2, β_1, β_2 and denote by $\chi(\alpha_1, \alpha_2, \beta_1, s)$ the function ψ in these new coordinates, i.e.,

$$\chi(\alpha_1, \alpha_2, \beta_1, \beta_2 - u(\alpha_1, \alpha_2, \beta_1)) = \psi(\alpha_1, \alpha_2, \beta_1, \beta_2).$$

Note that in the new coordinates the complex K is given by the equation $s = 0$.

Let us represent the form $\kappa\psi$ in the new system of coordinates. We have

$$\frac{\partial \psi}{\partial \alpha_i} = \frac{\partial \chi}{\partial \alpha_i} - \frac{\partial \chi}{\partial s}\frac{\partial \chi}{\partial \alpha_i}, \qquad i = 1, 2.$$

Hence,

$$\sum \frac{\partial \psi}{\partial \alpha_i}\,d\alpha_i = \sum \frac{\partial \chi}{\partial \alpha_i}\,d\alpha_i - \frac{\partial \chi}{\partial s}\sum \frac{\partial u}{\partial \alpha_i}\,d\alpha_i.$$

For the complex K to be admissible, it is necessary and sufficient that the restriction of this form to γ^x contain no term with $\partial\chi/\partial s$, i.e., that

$$\text{(5.7)} \qquad \frac{\partial u}{\partial \alpha_1}\,d\alpha_1 + \frac{\partial u}{\partial \alpha_2}\,d\alpha_2\,\Big|_{\gamma^x} = 0.$$

On the other hand, on γ^x we have $x_3\,d\alpha_1 + d\beta_1 = 0$ and $x_3\,d\alpha_2 + d\beta_2 = 0$, and hence

$$\text{(5.8)} \qquad \left(-x_3\frac{\partial u}{\partial \beta_1} + \frac{\partial u}{\partial \alpha_1}\right)d\alpha_1 + \left(x_3 + \frac{\partial u}{\partial \alpha_2}\right)d\alpha_2 = 0.$$

It follows from (5.7) and (5.8) that the function $u(\alpha_1, \alpha_2, \beta_1)$ satisfies the equation

$$\text{(5.9)} \qquad \frac{\partial u}{\partial \alpha_1} + \frac{\partial u}{\partial \alpha_2}\frac{\partial u}{\partial \beta_1} = 0.$$

If the complex K is given by an equation of the form $F(\alpha, \beta) = 0$, then $\partial u/\partial\alpha_i = -F'_{\alpha_i}/F'_{\beta_2}$, $i = 1, 2$, and $\partial u/\partial\beta_1 = -F'_{\beta_1}/F'_{\beta_2}$. Substituting these expressions in (5.9), we obtain (5.6).

5.4. Geometric structure of admissible complexes. The admissibility condition obtained in 5.3 for a line complex in $\mathbb{C}^3$ has an interesting geometric interpretation. Let us first present some geometric notions and facts related to the manifold of lines H.

To any line $h \in H$ we assign a three-dimensional manifold H_h of all lines intersecting h. The family of one-dimensional subspaces of $T_h H$ tangent to H_h forms a three-dimensional cone $\Gamma_h \subset T_h H$.

One can readily see that the manifold H_h, where $h = (\alpha^0, \beta^0)$, is given in the local coordinates (α, β) by the equation

$$(\alpha_1 - \alpha_1^0)(\beta_2 - \beta_2^0) - (\alpha_2 - \alpha_2^0)(\beta_1 - \beta_1^0) = 0.$$

Hence, the equation of the cone Γ_h is

$$d\alpha_1\, d\beta_2 - d\alpha_2\, d\beta_1 = 0.$$

The generatrices of the cone Γ_h can be interpreted as isotropic directions with respect to the holomorphic metric

$$ds^2 = d\alpha_1\, d\beta_2 - d\alpha_2\, d\beta_1.$$

For this reason, it is natural to call $\Gamma_h \subset T_h H$ the *isotropic cone* at the point h.

We state several simple facts concerning the isotropic cone Γ_h.

Note first that the planes tangent to Γ_h are given by the equations

$$a_1\, d\alpha_1 + a_2\, d\alpha_2 + b_1\, d\beta_1 + b_2\, d\beta_2 = 0$$

whose coefficients satisfy the relation $a_1 b_2 - a_2 b_1 = 0$.

Further, it is clear from the equation of Γ_h that this cone contains two one-dimensional families Π_1, Π_2 of two-dimensional planes,

$$\Pi_1 : d\beta_1 = \lambda\, d\alpha_1, \quad d\beta_2 = \lambda\, d\alpha_2;$$
$$\Pi_2 : d\alpha_2 = \lambda\, d\alpha_1, \quad d\beta_2 = \lambda\, d\beta_1.$$

Moreover, exactly one plane of each of the families passes through any given point (of the cone) that differs from the vertex. We refer to these planes as *α-planes* and *β-planes*, respectively.

These families of planes in $T_h H$ have a simple geometric meaning; namely, the α-planes are tangent to the submanifolds of lines intersecting h at a single point and the β-planes are tangent to the submanifolds of lines belonging to some chosen plane of $\mathbb{C}^3$ passing through h.

Any two planes in the same family intersect only at zero, and any two planes in different families intersect along a line. Conversely, every generatrix of the isotropic cone Γ_h belongs to one of the planes of each of these families.

If a three-dimensional subspace of $T_h H$ is tangent to the isotropic cone Γ_h, then the intersection of this subspace with Γ_h is the union of a single α-plane and a single β-plane. Conversely, if a three-dimensional subspace of $T_h H$ contains a plane of the isotropic cone Γ_h, then this subspace is tangent to the cone.

PROPOSITION 5.2. *The admissible line complexes in $\mathbb{C}^3$ are exactly the line complexes K that are tangent to the corresponding isotropic cone Γ_h at any smooth point $h \in K$.*

Indeed, the admissibility condition proved in Proposition 5.1 is equivalent to the condition that the tangent subspace $T_h K$ to the complex K at any point $h \in K$ is tangent to the isotropic cone Γ_h.

As before, let γ^x be the manifold of lines of the complex K that pass through a point x, i.e., $\gamma^x = H^x \cap K$.

PROPOSITION 5.3. *A line complex K is admissible if and only if, for any smooth point $h \in K$, all subspaces $T_h\gamma^x \subset T_hK \cap \Gamma_h$, $x \in h$, belong to the same β-plane.*

Indeed, if the complex is admissible, then, for any smooth point $h \in K$, the subspace T_hK is tangent to the cone Γ_h, and hence the space $T_hK \cap \Gamma_h$ is the union of an α-plane and a β-plane. In this case, all subspaces $T_h\gamma^x$, where x ranges over the line h, belong to the same β-plane.

Conversely, let a complex K be such that, for any smooth point $h \in K$, the subspaces $T_h\gamma^x$ form a two-dimensional plane if x ranges over the line h. Then the space $T_hK \cap \Gamma_h$ contains this two-dimensional plane, and therefore the space T_hK is tangent to the isotropic cone Γ_h. Hence, the complex K is admissible by Proposition 5.2.

Proposition 5.3 admits another (equivalent) formulation. For any complex of lines K, denote by $\Gamma^x \subset \mathbb{C}^3$ the cone formed by the lines of K that pass through a point $x \in \mathbb{C}^3$. Further, if h is a smooth point of the complex K, then, for any point x of the line h, denote by $\pi_{h,x}$ the plane in $\mathbb{C}^3$ tangent to Γ^x along the generatrix h.

PROPOSITION 5.4. *A line complex K is admissible if and only if the planes $\pi_{h,x}$, where x ranges over a line h, coincide for any smooth point $h \in K$.*

5.5. Description of admissible complexes. It follows from Proposition 5.4 that, along with complexes of lines in $\mathbb{C}^3$ intersecting a chosen curve, the complexes of lines tangent to a chosen (algebraic) surface in $\mathbb{C}^3$ are also admissible. Indeed, let K be a line complex tangent to a surface P. For any smooth point $h \in K$, denote by π_h the plane in $\mathbb{C}^3$ which is tangent to the surface P at the tangency point of h and P. Obviously, if x is an arbitrary point of the line h and Γ^x is the cone formed by the lines of K passing through x, then the plane $\pi_{h,x}$ tangent to Γ^x along the generatrix h coincides with π_h. Hence, the complex K satisfies the conditions of Proposition 5.4, and thus it is admissible.

THEOREM 5.2. *The admissible line complexes in $\mathbb{C}^3$ are precisely those complex three-dimensional submanifolds of the manifold of lines H that consist either of the lines tangent to some complex surface in $\mathbb{C}^3$ or of the lines intersecting some complex curve $\mathbb{C}^3$ (possibly at infinity).*

To prove the theorem, we introduce the notion of a critical point on a line of an arbitrary line complex K. Consider a smooth point h of K (i.e., a line in $\mathbb{C}^3$). A point $x \in h$ is said to be *critical* if the α-plane (of the isotropic cone Γ^x) corresponding to this point belongs to T_hK. Similarly, a two-dimensional plane $\pi \supset h$ in $\mathbb{C}^3$ is said to be *critical* if the corresponding β-plane belongs to T_hK.

PROPOSITION 5.5. *If K is an admissible line complex in $\mathbb{C}^3$, then, for any smooth point $h \in K$, there is a unique critical point $x \in h$ (possibly at infinity) and a unique critical plane $\pi \supset h$.*

Indeed, if the complex K is admissible, then $T_hK \cap \Gamma_h$, where Γ_h is the isotropic cone, is the union of a single α-plane and a single β-plane. Corresponding to these planes are the critical point $x \in h$ and the critical plane $\pi \supset h$, respectively.

COROLLARY. *The manifold K' of smooth points of K admits a standard embedding in the manifold of complete flags on $\mathbb{C}P^3$.*

Namely, to any $h \in K'$ we assign a triple, a flag $(x,\, h,\, \pi)$, where x is the critical point and π is the critical plane for h.

Let K be an admissible line complex given by an equation $F(\alpha, \beta) = 0$. We find the coordinates of the critical point $x = (x_1, x_2, x_3)$ of a line $(\alpha, \beta) \in K$ assuming that this point is not at infinity.

The equations of the α-plane in Γ_h corresponding to x are

$$x_3\, d\alpha_i + d\beta_i = 0, \qquad i = 1, 2.$$

The conditions that this plane belongs to the hyperplane $F'_{\alpha_1}\, d\alpha_1 + F'_{\alpha_2}\, d\alpha_2 + F'_{\beta_1}\, d\beta_1 + F'_{\beta_2}\, d\beta_2 = 0$ tangent to K are

$$F'_{\alpha_1} - x_3\, F'_{\beta_1} = F'_{\alpha_2} - x_3\, F'_{\beta_2} = 0.$$

This implies the following assertion.

PROPOSITION 5.6. *The coordinates of the critical point of an admissible complex K on a line $(\alpha, \beta) \in K$ are given by the relations*

$$x_3 = \frac{F'_{\alpha_1}}{F'_{\beta_1}} = \frac{F'_{\alpha_2}}{F'_{\beta_2}}, \qquad x_1 = \alpha_1 x_3 + \beta_1, \quad x_2 = \alpha_2 x_3 + \beta_2.$$

PROPOSITION 5.7. *The dimension of the manifold of critical points on the lines of an admissible complex K does not exceed two.*

PROOF. Let $K'' \subset K$ be the subset of smooth points $h \in K$ whose critical points x are finite. To prove the proposition, it suffices to show that the mapping $K'' \to \mathbb{C}^3$ taking every line $(\alpha, \beta) \in K''$ to its critical point $x = (x_1, x_2, x_3)$ is everywhere degenerate.

It follows from the equations of the lines that

$$dx_i - \alpha_i\, dx_3 = x_3\, d\alpha_i + d\beta_i, \qquad i = 1, 2.$$

Hence, it follows from the relations $x_3 = \frac{F'_{\alpha_1}}{F'_{\beta_1}} = \frac{F'_{\alpha_2}}{F'_{\beta_2}}$ and $F'_{\alpha_1}\, d\alpha_1 + F'_{\alpha_2}\, d\alpha_2 + F'_{\beta_1}\, d\beta_1 + F'_{\beta_2}\, d\beta_2 = 0$ that the differentials of the coordinates of the critical points are related as follows:

$$F'_{\beta_1}\,(dx_1 - \alpha_1\, dx_3) + F'_{\beta_2}\,(dx_2 - \alpha_2\, dx_3) = 0. \tag{5.10}$$

This proves the assertion.

It follows from Proposition 5.6 that, for an admissible complex K, the manifold of critical points is either a complex curve Λ or a complex surface S in $\mathbb{C}^3$. In the first case the complex K consists of the lines intersecting the curve Λ. In the other case it follows from (5.10) that the lines (α, β) of the complex K are tangent to the surface at their critical points. Thus, the proof of the theorem is complete.

REMARK. There is a simple relation between the problem of reconstructing a function $f \in S(\mathbb{C}^3)$ from the integrals of f over the lines of an admissible line complex K in $\mathbb{C}^3$ and the following problem for functions $\psi(\alpha_1, \alpha_2, \beta_1, \beta_2)$ on $\mathbb{C}^4$: Find a solution ψ of the system of equations

$$\frac{\partial^2 \psi}{\partial \alpha_1\, \partial \beta_2} = \frac{\partial^2 \psi}{\partial \alpha_2\, \partial \beta_1}, \qquad \frac{\partial^2 \psi}{\partial \bar{\alpha}_1\, \partial \bar{\beta}_2} = \frac{\partial^2 \psi}{\partial \bar{\alpha}_2\, \partial \bar{\beta}_1}, \tag{5.11}$$

satisfying the boundary condition $\psi|_K = \Phi$. Here K stands for a complex submanifold of $\mathbb{C}^4$ given by an equation $F(\alpha, \beta) = 0$. It is assumed that F satisfies (5.6), i.e., that K is a characteristic submanifold of system (5.6). In this way we obtain a Goursat problem if derivatives do not enter the boundary data. Correspondingly, such a problem can be well defined only on characteristic submaniflds.

It follows from (5.11) that ψ is the John transform of a function f on $\mathbb{C}^3$, i.e., $\psi = \mathcal{J} f$. Therefore, the solution of the boundary value problem is reduced to the application of the inversion formula for admissible complexes. Namely, using this formula we first find the function f from the function $\Phi = \psi|_K$. Then the original function ψ is obtained as the John transform of the function f, i.e., $\psi = \mathcal{J} f$.

Conversely, the problem of integral geometry to reconstruct a function f from the function $\Phi = \psi|_K$ is reduced to the solution of a boundary value problem. Namely, if the boundary value problem for the system of equations (5.11) is solved, i.e., the function ψ is reconstructed from the function $\Phi = \psi\ |_K$, then the desired function f is determined by the inversion formula for the John transform.

CHAPTER 3

Integral Geometry and Harmonic Analysis on the Hyperbolic Plane and in the Hyperbolic Space

The specific feature of transforms in integral geometry as compared with other integral transforms is that the former often admit an immediate generalization from the flat case to the curved one. We illustrate this fact here by considering analogs of the Radon transform for the hyperbolic plane and hyperbolic space. We shall see that the scheme of constructing these new transforms is just like that for the Radon transform in the affine space, and even the structure of inversion formulas is similar. However, the nonzero curvature of the hyperbolic space leads to a richer geometry. In particular, on the hyperbolic plane there are two analogs of lines, namely, lines themselves (i.e., geodesics) and horocycles (circles of infinitely large radius). Two different versions of the Radon transform are related to these two analogs of lines.

For the hyperbolic plane (and space), there is an analog of the Fourier transform using unitary representations of their motion groups. Since these groups are noncommutative, the theory of Fourier integral for the Lobachevsky plane (space) is much more complicated than that of its Euclidean analog. Seemingly, the simplest way to construct a non-Euclidean theory of the Fourier integral (representation theory) is to use the Radon transform for the horocycles (horospheres). The point is that in the case of hyperbolic space, as in the affine case, the analog of the Radon transform for horocycles (horospheres) and the analog of the Fourier transform are related by the one-dimensional (affine) Fourier transform. For this reason, the horocycle (horospherical) Radon transform is quite important when constructing harmonic analysis on the hyperbolic plane and space. This transform admits generalizations to other important classes of homogeneous manifolds. By the example of hyperbolic plane and space we simulate quite general constructions of harmonic analysis on homogeneous manifolds. When reading this chapter, it is very useful to keep in mind (and compare with) the corresponding constructions of §§ 1 and 2 of Chapter 1.

1. Elements of the hyperbolic planimetry[1]

1.1. Models of the hyperbolic plane. The hyperbolic plane $\mathcal{L}^2$ is the homogeneous space of the group $SL(2,\mathbb{R})$ with the isotropy subgroup $SO(2)$ (the subgroup of orthogonal matrices). In the Poincaré modelthis plane is represented as the upper half-plane $\operatorname{Im} z > 0$ of the plane of complex variable $z = x + iy$, where the motions corresponding to matrices $g = \left(\begin{smallmatrix} \alpha & \beta \\ \gamma & \delta \end{smallmatrix}\right) \in SL(2,\mathbb{R})$ are given as follows:

$$z \mapsto z \circ g = \frac{\alpha z + \gamma}{\beta z + \delta}.$$

[1]The main facts about hyperbolic geometry are presented without proofs.

Obviously, the stabilizer of the point $z = i$ is the subgroup $SO(2)$.[2]

The Riemannian metric on $\mathcal{L}^2$ is

$$ds^2 = y^{-2}(dx^2 + dy^2). \tag{1.1}$$

The invariant distance d between points $z_1 = x_1 + iy_1$ and $z_2 = x_2 + iy_2$ is defined by the relation

$$4\sinh^2\frac{d}{2} = \frac{|z_1 - z_2|^2}{y_1 y_2}. \tag{1.2}$$

The area element in the metric (1.1) is

$$dv = y^{-2}\,dx\,dy. \tag{1.3}$$

Another way to define the hyperbolic plane is to consider the manifold of real symmetric matrices

$$u = \begin{pmatrix} a & b \\ b & c \end{pmatrix}, \quad \text{where} \quad a > 0 \quad \text{and} \quad \det u = 1, \tag{1.4}$$

with the motions

$$u \mapsto g'ug, \qquad g \in SL(2, \mathbb{R}),$$

where g' stands for the transposed matrix. Note that the manifold of matrices of the form (1.4) can be represented as a sheet of a hyperboloid of two sheets in $\mathbb{R}^3$:

$$ac - b^2 = 1, \qquad a, c > 0. \tag{1.5}$$

The correspondence between the points z of the upper half-plane and the matrices $u = u(z)$ is described as follows. The point $z = i$ corresponds to the identity matrix e because the point i and the matrix e have the same isotropy subgroup. Hence, for each $g = \begin{pmatrix} \alpha & \beta \\ \gamma & \delta \end{pmatrix} \in SL(2, \mathbb{R})$, the matrix $u = g'g$ corresponds to the point $i \circ g = \frac{\alpha i + \gamma}{\beta i + \delta}$. In particular, $i \circ g = x + iy$ for $g = \begin{pmatrix} y^{1/2} & 0 \\ xy^{1/2} & y^{-1/2} \end{pmatrix}$, and we see that to a point $z = x + iy$ corresponds the matrix $\begin{pmatrix} y^{-1}(x^2+y^2) & y^{-1}x \\ y^{-1}x & y^{-1} \end{pmatrix}$; in other words, to a matrix $\begin{pmatrix} a & b \\ b & c \end{pmatrix}$ corresponds the point $z = \frac{b}{c} + \frac{i}{c}$. Geometrically, the passage from the sheet of the hyperboloid (of two sheets) to the half-plane is carried out as follows. We first project the hyperboloid (1.5) parallel to the axis a to the plane (b, c); this projection is the half-plane $c > 0$. After this we make a projective transformation in the plane (b, c) by setting $(b, c) \mapsto (\frac{b}{c}, \frac{1}{c})$, and this transformation sends the half-plane $c > 0$ into itself.

It is convenient to pass in (1.5) to the new coordinates $x_1 = \frac{a+c}{2}$, $x_2 = \frac{a-c}{2}$, $x_3 = b$. Then we obtain a realization of the hyperbolic plane $\mathcal{L}^2$ as the upper sheet of the hyperboloid of two sheets in $\mathbb{R}^3$ given by

$$[x, x] \equiv x_1^2 - x_2^2 - x_3^2 = 1, \qquad x_1 > 0. \tag{1.6}$$

In this realization, the motions become linear transformations in $\mathbb{R}^3$ that preserve the upper sheet of the hyperboloid, i.e., the elements of the group $SO_0(1, 2) \cong SL(2, \mathbb{R})/\{\pm e\}$.

The Riemannian metric in this model is

$$ds^2 = -dx_1^2 + dx_2^2 + dx_3^2. \tag{1.7}$$

---

[2]Note that the center of the group $SL(2, \mathbb{R})$, which consists of two elements $\pm e$, acts ineffectively (preserves any point). Therefore, it is convenient to view the quotient group $SL(2, \mathbb{R})/\{\pm e\}$ as the motion group.

The invariant distance d between points x and y is

$$\cosh d = [x, y]. \tag{1.8}$$

The area element in the coordinates x_2, x_3 on $\mathcal{L}^2$ is

$$dv = \frac{dx_2\, dx_3}{x_1}. \tag{1.9}$$

1.2. Horocycles. On the hyperbolic plane one can define circles of infinitely large radius (horocycles), which are the limits of non-Euclidean circles as the center and the radius of these circles consistently tend to infinity. In the Poincaré model, the horocycles are represented either as (Euclidean) circles tangent to the real axis (the absolute) or as lines parallel to the real axis. The latter objects can naturally be interpreted as circles of infinitely large radius tangent to the real axis at the point $x = \infty$. It follows from this description that every horocycle is given by an equation of the form

$$|\xi_2 z - \xi_1|^2 = y, \qquad \xi = (\xi_1, \xi_2) \in \mathbb{R}^2 \setminus O. \tag{1.10}$$

Thus, the manifold of horocycles is parametrized by the points $(\xi_1, \xi_2) \in \mathbb{R}^2 \setminus O$, and the horocycles corresponding to the points (ξ_1, ξ_2) and $(-\xi_1, -\xi_2)$ coincide. Denote by $h(\xi) = h(\xi_1, \xi_2)$ the horocycle given by (1.10).

REMARK. If a plane $\mathcal{L}^2$ is represented as the upper sheet of the hyperboloid given by (1.6), then the horocycles are the plane parabolic sections of this hyperboloid, namely, the plane sections of the form $[\xi, x] \equiv \xi_1 x_1 - \xi_2 x_2 - \xi_3 x_3 = 1$, where ξ is a point of the upper nappe of the cone $[\xi, \xi] = 0$, $\xi_1 > 0$, $\xi \in \mathbb{R}^3$.

Obviously, the motion group $SL(2, \mathbb{R})$ acts on the manifold of horocycles H, and this action is transitive; the motion corresponding to a matrix $g = \left(\begin{smallmatrix} \alpha & \beta \\ \gamma & \delta \end{smallmatrix}\right)$ sends a horocycle $h(\xi_1, \xi_2)$ to the horocycle $h(\alpha\xi_1 + \gamma\xi_2, \beta\xi_1 + \delta\xi_2)$.

In the language of motions, the horocycles are described as follows. They are the orbits[3] of the unipotent subgroup Z of the matrices $\left(\begin{smallmatrix} 1 & 0 \\ t & 1 \end{smallmatrix}\right)$ and all their translations. A group conjugate to Z acts on every horocycle, and the horocycle is a Euclidean line with respect to this action. Thus, the manifold H of horocycles is a homogeneous space of the group $SL(2, \mathbb{R})$, namely, the homogeneous space with the isotropy subgroup $\pm Z$. Therefore, when passing from the hyperbolic plane $\mathcal{L}^2$ to the manifold of horocycles H, we obtain a new homogeneous space with the same motion group but with a different stabilizer. Constructions of this type play an important role in geometry.

Every horocycle has a unique common point with the absolute, which is called the *center* of the horocycle. In the Poincaré realization, this is the tangency point with the real line. Horocycles with common center are said to be *parallel.* Note that a horocycle $h(\xi_1, \xi_2)$ is tangent to the real axis at the point $x = \frac{\xi_1}{\xi_2}$; hence, every pencil of parallel horocycles is of the form $\{h(\lambda\xi_1, \lambda\xi_2) \mid 0 < \lambda < \infty\}$ for some chosen pair (ξ_1, ξ_2).

All points of a horocycle $h(\lambda\xi_1, \lambda\xi_2)$ are placed at the same distance $2|\log \lambda|$ from the parallel horocycle $h(\xi_1, \xi_2)$. We refer to the expression $\rho(z; \xi) = -2 \log \lambda$

[3]In the Poincaré model, these are the lines parallel to the real axis.

as the *oriented distance from a point* $z \in h(\lambda\xi)$ *to a horocycle* $h(\xi)$. This distance is given by

$$\rho(z;\xi) = \log(y^{-1}|\xi_2 z - \xi_1|^2) \qquad (z = x + iy,\ \xi = (\xi_1, \xi_2)) \tag{1.11}$$

(because $\lambda = (y^{-1}|\xi_2 z - \xi_1|^2)^{-1/2}$).

Every horocycle $h(\xi)$ partitions the hyperbolic plane $\mathcal{L}^2$ into two parts $\mathcal{L}_+(\xi)$ and $\mathcal{L}_-(\xi)$, where $\mathcal{L}_+(\xi)$ abuts on the absolute and the boundary of $\mathcal{L}_-(\xi)$ has only one common point with the absolute, namely, the center of the horocycle $h(\xi)$. The domains $\mathcal{L}_+(\xi)$ and $\mathcal{L}_-(\xi)$ consist of the points $z \in \mathcal{L}^2$ for which $\rho(z;\xi) > 0$ and $\rho(z;\xi) < 0$, respectively.

REMARK. The homogeneous spaces $\mathcal{L}^2$ ($\mathcal{L}^2 = SO(2) \setminus SL(2,\mathbb{R})$) and H ($H = \pm Z \setminus SL(2,\mathbb{R})$) are dual to each other. By construction, to points $\xi \in H$ correspond curves $h(\xi)$ in $\mathcal{L}^2$ (horocycles). At the same time one can assign to a point $z \in \mathcal{L}^2$ the curve $l(z) \in H$ corresponding to the set of horocycles passing through z. Each of these curves is a trajectory of the subgroup $SO(2)$ or a subgroup conjugate to $SO(2)$. This trajectory partitions the manifold H into two parts, namely, the domain $H_+(z)$ in which $\rho(z;\xi) > 0$ and the domain $H_-(z)$ in which $\rho(z;\xi) < 0$. In the realization $H = \mathbb{R}^2 \setminus O$, the curves $l(z)$ are all possible ellipses on $\mathbb{R}^2$ centered at the origin O and bounding a domain of unit area, and $H_+(z)$ and $H_-(z)$ are the exterior and interior domains of the ellipse $l(z)$, respectively. Note that the manifold of all orbits of the subgroup $SO(2)$ and the subgroups conjugate to $SO(2)$ is not homogeneous; in the realization at hand, this manifold consists of all ellipses on $\mathbb{R}^2$ centered at the origin O.

1.3. Geodesics. Consider now the "lines" (geodesics) on the hyperbolic plane $\mathcal{L}^2$. In the Poincaré model, these are the semicircles on the half-plane $\operatorname{Im} z > 0$ with centers on the real line, and the half-lines orthogonal to the real line. Hence, every "line" is given by an equation of the form

$$(x - a)(x - b) + y^2 = 0, \qquad a, b \in \mathbb{R} \cup (\infty), \quad a \neq b. \tag{1.12}$$

In particular, for $a = \infty$ (or $b = \infty$) we obtain the family of (Euclidean) half-lines orthogonal to the real line. It follows from (1.12) that the manifold of "lines" can be parametrized by the pairs (a, b), $a \neq b$, where a, b are the points of the projective line $\mathbb{P}^1$, and the "lines" corresponding to the pairs (a, b) and (b, a) coincide.

In the language of motions, every "line" is an orbit of the subgroup D of diagonal matrices or of a subgroup conjugate to D, and all "lines" can be obtained from one another by motions. Thus, the manifold of "lines" is the homogeneous space $D \setminus SL(2,\mathbb{R})$. We thus obtain another homogeneous space with the same motion group $SL(2,\mathbb{R})$ and a different isotropy subgroup. Note that the manifold of all orbits of the subgroup D and subgroups conjugate to D is not homogeneous. In the Poincaré model, this manifold coincides with the manifold of all half-lines on the half-plane $\operatorname{Im} z > 0$ and of circular arcs having exactly two common points with the real line.

If the hyperbolic plane is represented as the upper sheet of the hyperboloid of two sheets given by (1.6), then, in this realization, the geodesics are the curves of intersection (one branch of a hyperbola) with planes in $\mathbb{R}^3$ passing through the origin O, i.e., with the planes

$$\xi_1 x_1 + \xi_2 x_2 + \xi_3 x_3 = 0. \tag{1.13}$$

Denote by $\gamma(\xi)$ the geodesic given by (1.13). Obviously, a plane (1.13) intersects the hyperboloid $[x,x]=1$ if and only if $[\xi,\xi]<0$. Moreover, since $\gamma(\lambda\xi)=\gamma(\xi)$ for each $\lambda\neq 0$, we can set

$$[\xi,\xi]=-1.$$

Thus, the manifold of geodesics on the hyperbolic plane can be parametrized by the points ξ of the hyperboloid $[\xi,\xi]=-1$ of two sheets in $\mathbb{R}^3$, and the geodesics corresponding to the points ξ and $-\xi$ coincide.

2. Horocycle transform

2.1. Definition of the operator $\mathcal{R}^h$. In this section we study an analog of the Radon transform related to horocycles on the hyperbolic plane.

DEFINITION. By the *horocycle transform* we mean the integral transform $\mathcal{R}^h$ on $\mathcal{L}^2$ sending each smooth compactly supported function f on $\mathcal{L}^2$ to the integrals of f over the horocycles,

$$\mathcal{R}^h f(\xi)=\int_{h(\xi)} f(z)\,ds, \qquad \xi=(\xi_1,\xi_2), \tag{2.1}$$

where $h(\xi)$ is the horocycle given by (1.10) and $ds=\frac{|dz|}{y}$ is the length element on the hyperbolic plane.

The expression for $\mathcal{R}^h$ can also be written as follows:

$$\mathcal{R}^h f(\xi)=\int_{\mathcal{L}^2} f(z)\,\delta(\rho(z;\xi))\frac{dx\,dy}{y^2}, \tag{2.2}$$

where $\rho(z;\xi)$ is the distance from the point z to the horocycle $h(\xi)$ (see (1.11)), and $\delta(\cdot)$ is the delta function on $\mathbb{R}$.

It follows from the definition that the operator $\mathcal{R}^h$ commutes with the action of the group $SL(2,\mathbb{R})$. Thus, we have a pair of homogeneous spaces of the same group $SL(2,\mathbb{R})$, namely, the hyperbolic plane and the manifold of horocycles, and an operator sending functions on one space to functions on the other and commuting with the motions. Note that the integration is over the orbits of the isotropy subgroup of the second space and over the shifts of these orbits.

2.2. Inversion formula. Let us find the inversion formula for the integral transform $\mathcal{R}^h$ following the same scheme as in the case of Euclidean plane (see § 1 of Chapter 1).

First, take a radially symmetric function f depending only on the non-Euclidean distance from a point z to the point i:

$$f(z)=F(y^{-1/2}|z-i|).$$

Since the operator $\mathcal{R}^h$ commutes with the motions of the plane $\mathcal{L}^2$, it follows that $\mathcal{R}^h f(\xi)$ depends only on the distance from the point i to the horocycle $h(\xi)$, i.e.,

$$\mathcal{R}^h f(\xi)=\widehat{F}(r), \quad \text{where} \quad r=\rho(i;\xi)=\log(\xi_1^2+\xi_2^2).$$

Thus, the transform $\widehat{F}(r)$ is equal to the integral of the function f over the horocycle $h(e^{r/2}, 0)$ for which $y = e^r$, i.e.,

$$\begin{aligned}\widehat{F}(r) &= \int_{-\infty}^{+\infty} F(e^{-r/2}\sqrt{x^2+(e^r-1)^2})e^{-r}\,dx \\ &= 2\int_{2|\sinh\frac{r}{2}|}^{+\infty} \frac{F(s)s\,ds}{e^{r/2}\sqrt{s^2-4\sinh^2\frac{r}{2}}}\end{aligned} \tag{2.3}$$

(cf. (1.2) in § 1 of Chapter 1). Hence, the passage from $F(s)$ to $e^{r/2}\widehat{F}(r)$ reduces to the Abel transform, and it follows from (2.3) that

$$F(s) = -\frac{1}{\pi}\int_{2\sinh\frac{r}{2}>0} \frac{(e^{r/2}\widehat{F}(r))'\,dr}{\sqrt{4\sinh^2\frac{r}{2}-s^2}}. \tag{2.4}$$

In particular,

$$f(i) = F(0) = -\frac{1}{2\pi}\int_0^{\infty} \frac{(e^{r/2}\widehat{F}(r))'\,dr}{\sinh\frac{r}{2}}. \tag{2.5}$$

Note that $e^{r/2}\widehat{F}(r)$ is a smooth even function by (2.3). Therefore, the integral (2.5) converges (because $(e^{r/2}\widehat{F}(r))'_{r=0} = 0$), and the expression for $f(i)$ can also be represented in the form

$$f(i) = -\frac{1}{4\pi}\int_{-\infty}^{+\infty} \frac{(e^{r/2}\widehat{F}(r))'\,dr}{\sinh\frac{r}{2}}. \tag{2.6}$$

Assume now that f is an arbitrary smooth compactly supported function on $\mathcal{L}^2$; we apply to f the averaging $\overline{f}(z) = F(y^{-1/2}|z-i|)$ over the non-Euclidean circles centered at the point i,

$$F(y^{-1/2}|z-i|) = \frac{1}{2\pi}\int_0^{2\pi} f\left(\frac{z\cos\theta+\sin\theta}{-z\sin\theta+\cos\theta}\right)d\theta;$$

the expression for F can also be represented in the form

$$F(s) = \frac{2}{\pi}\int_{\mathcal{L}^2} f(z)\,\delta(y^{-1}|z-i|^2-s^2)\frac{dx\,dy}{y^2}.$$

Note that $f(i) = F(0)$. Let $\widehat{F}(r) = \mathcal{R}^h\,\overline{f}(\xi)$, where $r = \log(\xi_1^2+\xi_2^2)$. Since the operator $\mathcal{R}^h$ commutes with the action of the group $SL(2,\mathbb{R})$, it follows that

$$\begin{aligned}\widehat{F}(r) &= \frac{1}{2\pi}\int_0^{2\pi} \mathcal{R}^h\,f(e^{r/2}\cos\theta, e^{-r/2}\sin\theta)\,d\theta \\ &= \frac{1}{2\pi}e^{-r}\int_{\rho(i;\xi)=r} \mathcal{R}^h\,f(\xi)(\xi_1\,d\xi_2-\xi_2\,d\xi_1) \\ &= \frac{1}{\pi}e^{-r}\int_{\mathbb{R}^2} \mathcal{R}^h\,f(\xi)\,\delta(\rho(i;\xi)-r)\,d\xi_1\,d\xi_2,\end{aligned} \tag{2.7}$$

where $\rho(i;\xi)$ is the distance from the point $z = i$ to the horocycle $h(\xi)$. Using (2.5) and (2.6), we obtain

$$f(i) = \overline{f}(i) = -\frac{1}{2\pi}\int_0^{\infty} \frac{(e^{r/2}\widehat{F}(r))'}{\sinh\frac{r}{2}}\,dr = -\frac{1}{4\pi}\int_{-\infty}^{+\infty} \frac{(e^{r/2}\widehat{F}(r))'}{\sinh\frac{r}{2}}\,dr \tag{2.8}$$

(cf. (1.4) in § 1 of Chapter 1). Substituting the expression in (2.7) for $\widehat{F}(r)$ into (2.8), after elementary manipulations we obtain

$$f(i) = -\frac{1}{4\pi^2}\int_{H_\pm(i)} \frac{e^{-\rho(i;\xi)/2}}{\sinh\frac{\rho(i;\xi)}{2}}\left(\xi_1\frac{\partial}{\partial\xi_1}+\xi_2\frac{\partial}{\partial\xi_2}+1\right)\mathcal{R}^h f(\xi)\,d\xi_1\,d\xi_2, \tag{2.9}$$

where $H_\pm(i)$ are the domains of H^2 in which $\rho(i;\xi)>0$ and $\rho(i;\xi)<0$, respectively.

To obtain an expression for f at an arbitrary point z, one must apply the group translation taking the point i to the point z. We thus obtain the following assertion.

THEOREM 2.1. *A smooth compactly supported function f on $\mathcal{L}^2$ can be expressed in terms of its horocycle transform $\mathcal{R}^h f$ by the following inversion formula:*

$$f(z) = -\frac{1}{2\pi}\int_0^\infty \frac{(e^{r/2}\widehat{F}(z,r))'r}{\sinh\frac{r}{2}}\,dr = -\frac{1}{4\pi}\int_{-\infty}^{+\infty}\frac{(e^{r/2}\widehat{F}(z,r))'r}{\sinh\frac{r}{2}}\,dr, \tag{2.10}$$

where

$$\begin{aligned}\widehat{F}(z,r) &= \frac{1}{2\pi}e^{-r}\int_{\rho(z,\xi)=r}\mathcal{R}^h f(\xi)(\xi_1\,d\xi_2-\xi_2\,d\xi_1)\\ &= \frac{1}{\pi}e^{-r}\int_{\mathbb{R}^2}\mathcal{R}^h f(\xi)\,\delta(\rho(z;\xi)-r)\,d\xi_1\,d\xi_2.\end{aligned} \tag{2.11}$$

It also follows from (2.9) *that*

$$f(z) = -\frac{1}{4\pi^2}\int_{H_\pm(z)} \frac{e^{-\rho(z;\xi)/2}}{\sinh\frac{\rho(r;\xi)}{2}}\left(\xi_1\frac{\partial}{\partial\xi_1}+\xi_2\frac{\partial}{\partial\xi_2}+1\right)\mathcal{R}^h f(\xi)\,d\xi_1\,d\xi_2 \tag{2.12}$$

(*cf.* (1.5) *and* (1.6) *of Chapter* 1), *where $H_\pm(z)$ are the domains of H in which $\rho(x;\xi)>0$ and $\rho(z;\xi)<0$, respectively.*

Note that $\xi_1\frac{\partial}{\partial\xi_1}+\xi_2\frac{\partial}{\partial\xi_2}$ is the operator of differentiation along a pencil of parallel horocycles.

2.3. Asgeirsson relations. To invert the transform $\mathcal{R}^h$, we used (2.4) for $s=0$ only. For $s\neq 0$, this formula contains an additional information on the relation between the averages of the function f and the horocycle transform $\mathcal{R}^h f$ of f.

THEOREM 2.2. *The following Asgeirsson relations hold for the hyperbolic plane:*

$$F(z;s) = -\frac{1}{\pi}\int_{2\sinh r/2>s}\frac{\left(e^{r/2}\widehat{F}(z,r)\right)'_r}{\sqrt{4\sinh^2\frac{r}{2}-s^2}}\,dr, \tag{2.13}$$

where $F(r;s)$ is the mean of the function f over the non-Euclidean circles centered at the point z,

$$F(z,s) = \frac{2}{\pi}\int_{\mathcal{L}^2} f(z')\,\delta(y^{-1}{y'}^{-1}|z'-z|^2-s^2)\,\frac{dx'\,dy'}{y'^2}, \tag{2.14}$$

and $\widehat{F}(z,r)$ is the mean of the function $\mathcal{R}^h f$ over the set of horocycles equidistant from the point z; see (2.11). *We also have*

$$\widehat{F}(z,r) = 2\int_{2|\sinh\frac{r}{2}|}^\infty \frac{F(z,s)s\,ds}{e^{r/2}\sqrt{s^2-4\sinh^2\frac{r}{2}}}. \tag{2.15}$$

Formulas (2.13) and (2.15) are obtained from (2.4) and (2.3) by a group translation; compare with the Asgeirsson relations (1.8) and (1.10) in § 1 of Chapter 1.

2.4. Symmetry relation. It follows from (2.15) that $r \mapsto e^{r/2}\widehat{F}(z,r)$ is an even function of r, i.e.,

$$e^{r/2}\widehat{F}(z,r) = e^{-r/2}\widehat{F}(z,-r)$$

for any z and r. Substituting the expression (2.11) for $\widehat{F}(z,r)$, we obtain the following symmetry relation for the function $\mathcal{R}^h f$:

$$e^{-r/2}\int \mathcal{R}^h f(\xi)\,\delta(\rho(z;\xi)-r)\,d\xi_1\,d\xi_2 = e^{r/2}\int \mathcal{R}^h f(\xi)\,\delta(\rho(z;\xi)+r)\,d\xi_1\,d\xi_2. \tag{2.16}$$

2.5. Inversion formula for the horocycle transform in another model of the hyperbolic plane. Let the hyperbolic plane $\mathcal{L}^2$ be represented as the upper sheet of the hyperboloid in $\mathbb{R}^3$ given by

$$[x,x] \equiv x_1^2 - x_2^2 - x_3^2 = 1, \qquad x_1 > 0.$$

In this model, the horocycles are the sections of the hyperboloid by the planes

$$[\xi,x] \equiv \xi_1\,x_1 - \xi_2\,x_2 - \xi_3\,x_3 = 1,$$

where ξ is a point of the upper nappe ($\xi_1 > 0$) of the cone K given by $[\xi,\xi] = 0$. The horocycle transform is then given by the following formula:

$$\mathcal{R}^h f(\xi) = \int_{\mathcal{L}^2} f(x)\,\delta([\xi,x]-1)\,dx, \qquad \xi \in K, \tag{2.17}$$

where $dx = x_1^{-1}\,dx_2\,dx_3$ is an invariant measure on $\mathcal{L}^2$.

THEOREM 2.3. *A smooth compactly supported function f on $\mathcal{L}^2$ can be expressed in terms of its horocycle transform $\mathcal{R}^h f$ by the following inversion formula*:

$$f(x) = \frac{1}{(2\pi)^2}\int_K \mathcal{R}^h f(\xi)\,([\xi,x]-1)^{-2}\,d\xi, \quad \textit{where} \quad d\xi = \xi_1^{-1}\,d\xi_2\,d\xi_3. \tag{2.18}$$

The integral must be understood in the sense of regularized value [**10**], namely, as the value at $\mu = -2$ of the analytic function of μ given for $\operatorname{Re}\mu > 0$ by the convergent integral

$$(2\pi)^{-2}\int_K \mathcal{R}^h f(\xi)\,|[\xi,x]-1|^{\mu}\,d\xi.$$

Note that formulas (2.17) and (2.18) are similar to the corresponding formulas for the Radon transform in Chapter 1. Similarly to the case of Radon transform, it suffices to prove the inversion formula (2.18) for functions on $\mathcal{L}^2$ depending only on the distance from the point $x^0 = (1,0,0)$. We can readily see that, for these functions f, the inversion formula (2.18) coincides with the one proven above.

3. Analog of the Fourier transform on the hyperbolic plane and the relation between this analog and the horocycle transform

3.1. Fourier transform on $\mathbb{R}^2$. At the naive level, the classical Fourier integral on $\mathbb{R}^2$ is the expansion of a function in $L^2(\mathbb{R}^2)$ in the eigenfunctions of the operators of parallel translation, i.e., in the exponential functions $e^{i(\xi_1 x+\xi_2 y)}$. However, this integral is also closely related to a larger group, namely, the full motion group of the Euclidean plane. Let us indicate this relation.

Consider the Laplace operator Δ on the Euclidean plane, i.e., the second-order differential operator $\Delta = \frac{\partial^2}{\partial x^2} + \frac{\partial^2}{\partial y^2}$ commuting with the motions. It is known that this operator on $L^2(\mathbb{R}^2)$ has continuous Lebesgue spectrum coinciding with the half-line $(-\infty, 0)$. The problem is to construct sufficiently many distributional eigenfunctions of this operator such that any function $f \in L^2(\mathbb{R}^2)$ can be expanded in these eigenfunctions. We shall argue at the naive level without specifying what object must be regarded as a distributional eigenfunction in the case of continuous spectrum and what is an expansion of a function $f \in L^2(\mathbb{R}^2)$ in these distributions.

Choose a one-parameter subgroup of translations in the motion group of the Euclidean plane; for instance, take the subgroup of translations parallel to the y axis. Each of the eigenspaces of the operator Δ, say, the eigenspace corresponding to the eigenvalue $-\lambda^2$ ($\lambda \in \mathbb{R} \setminus 0$), contains a pair of distributions that are invariant with respect to this subgroup, namely, $e^{\pm i\lambda x}$. Take one of these functions, for instance, $e^{i\lambda x}$, and apply to it all possible motions. Obviously, the function $e^{i\lambda x}$ is multiplied by a number under any translation and turns into $e^{i\lambda(x\cos\theta + y\sin\theta)}$ under the rotation through the angle θ about the origin. Thus, for every eigenvalue $-\lambda^2$ of the operator Δ we obtain the family of eigenfunctions $e^{i\lambda(x\cos\theta+y\sin\theta)}$ (in particular, the function $e^{-i\lambda x}$). There are no linear relations among these functions, and every eigenfunction f_λ of the operator Δ corresponding to the eigenvalue $-\lambda^2$ can be expanded in these functions,

$$f_\lambda(x, y) = \frac{1}{2\pi}\int_0^\infty u_\lambda(\theta) e^{i\lambda(x\cos\theta + y\sin\theta)}\, d\theta \qquad (\lambda > 0). \tag{3.1}$$

We introduce the following norm in the space of eigenfunctions of the operator Δ that correspond to the eigenvalue $-\lambda^2$:

$$\|f_\lambda\|^2 = \frac{1}{2\pi}\int_0^{2\pi} |u_\lambda(\theta)|^2\, d\theta \tag{3.2}$$

and consider the related Hilbert space H_λ. In H_λ we obtain a unitary representation T_λ of the motion group of the Euclidean plane; namely, in terms of the functions u_λ, to each translation $(x, y) \mapsto (x + x_0, y + y_0)$ corresponds the operator of multiplication by the function $e^{i\lambda(x_0\cos\theta + y_0\sin\theta)}$, and to the rotation through the angle α about the origin corresponds the operator of rotation $u_\lambda(\cdot) \mapsto u_\lambda(\cdot + \alpha)$. It can readily be seen that the representation T_λ is irreducible.

The representation of the motion group in the space $L^2(\mathbb{R}^2)$ can be decomposed into the representations T_λ. Let us describe this decomposition. Every function $f \in L^2(\mathbb{R}^2)$ can be expanded in the integral

$$f = \int_0^\infty \lambda f_\lambda\, d\lambda, \tag{3.3}$$

where f_λ is given by (3.1) with $u_\lambda(\theta) = \widetilde{f}(\lambda\cos\theta, \lambda\sin\theta)$ ($\widetilde{f}$ stands for the Fourier transform of f). Moreover, the following Plancherel formula holds:

$$\|f\|^2 = \int_0^\infty \lambda \|f_\lambda\|^2\, d\lambda, \tag{3.4}$$

where $\|f_\lambda\|$ is given by (3.2).

3.2. Fourier transform on the hyperbolic plane. We now carry out similar considerations (at the same nonrigorous level) in the case of hyperbolic plane. Consider the Laplace–Beltrami operator Δ on the hyperbolic plane $\mathcal{L}^2$, i.e., a second-order differential operator commuting with the motions of $\mathcal{L}^2$; it is of the form

$$\Delta = y^2\left(\frac{\partial^2}{\partial x^2} + \frac{\partial^2}{\partial y^2}\right).$$

Let $L^2(\mathcal{L}^2)$ be the space of measurable functions on $\mathcal{L}^2$ with square-integrable absolute value (with respect to the invariant measure $dv = \frac{dx\,dy}{y^2}$).

On $L^2(\mathcal{L}^2)$ the operator Δ has continuous Lebesgue spectrum coinciding with the interval $(-\infty, -1/4)$. Let Z be the unipotent subgroup of matrices of the form $\left(\begin{smallmatrix} 1 & 0 \\ t & 1 \end{smallmatrix}\right)$. By the *Fourier integral on the hyperbolic plane* we mean the expansion of a function $f \in L^2(\mathcal{L}^2)$ in some special distributional eigenfunctions of the operator Δ, namely, in the distributional eigenfunctions that are invariant with respect to the subgroup Z, and the shifts of these functions. Just these functions are analogs of the exponential functions $e^{i(\xi_1 x + \xi_2 y)}$.

Each eigenspace of the operator Δ corresponding to some eigenvalue of the form $-\frac{1+s^2}{4}$ ($s \in \mathbb{R}$) contains two functions invariant with respect to the subgroup Z, namely, the functions $y^{\frac{1\pm is}{2}}$. Let us take one of them, say $y^{\frac{1+is}{2}}$; under all possible translations, this function transforms to functions of the form

$$\Phi_s(z;\xi) = (y|\xi_2 z - \xi_1|^{-2})^{\frac{1+is}{2}} = \exp\left(-\frac{1+is}{2}\rho(z;\xi)\right), \tag{3.5}$$

where $\rho(z;\xi)$ is the distance from a point $z = x + iy$ to the horocycle $h(\xi)$. We call functions $y^{\frac{1\pm is}{2}}$ *zonal horocycle functions* and translations of the form (3.5) are simply referred to as *horocycle functions*. We see that if ξ and ξ' are proportional, then the corresponding functions Φ_s differ by a factor only. There are no other relations among the functions Φ_s for $s > 0$, and if we take a family of functions $\Phi_s(z;\xi)$ in which s ranges over the positive reals and ξ over a set having exactly one representative on each line in $\mathbb{R}^2$ passing through the origin O, then we obtain a complete function set in which every function in $L^2(\mathcal{L}^2)$ can be expanded.

Let us define the Fourier transform of a function on the hyperbolic plane $\mathcal{L}^2$ by analogy with the classical Fourier transform.

DEFINITION. By the *Fourier transform* of a smooth compactly supported function f on $\mathcal{L}^2$ we mean the inner product of f and the generalized horocycle function (3.5),

$$\mathcal{F}f(\xi, s) = \int_{\mathcal{L}^2} f(z)\Phi_s(z;\xi)\frac{dx\,dy}{y^2}. \tag{3.6}$$

Since the functions Φ_s are homogeneous with respect to ξ, it follows that the function $\mathcal{F}f$ is also homogeneous with respect to ξ,

$$\mathcal{F}f(\lambda\xi; s) = |\lambda|^{-is-1}\mathcal{F}f(\xi; s). \tag{3.7}$$

3.3. Relation to the horocycle transform and the inversion formula. he completeness of the family of horocycle functions Φ_s means that an inversion mula for the Fourier transform $\mathcal{F}$ must exist. Let us construct it.

First we establish a relation between the Fourier transform $\mathcal{F}f$ and the horo-e transform $\mathcal{R}^h f$. As in the Euclidean case, the passage from a function f to

$\mathcal{F} f$ can be carried out in two steps. We first integrate the function f over the horocycles $\rho(z; e^{-\lambda/2}\xi) \equiv \rho(z;\xi) - \lambda = 0$, and then integrate the resulting function multiplied by $e^{-\frac{is+1}{2}\lambda}$ with respect to λ.

The first integration gives

$$\int_{h(e^{-\lambda/2}\xi)} f(z)\, ds = \mathcal{R}^h f(e^{-\lambda/2}\xi).$$

Thus, the following assertion holds.

THEOREM 3.1. *The Fourier transform on $\mathcal{L}^2$ and the horocycle transform are related as follows*:

$$\mathcal{F} f(\xi; s) = \int_{-\infty}^{+\infty} \mathcal{R}^h f(e^{-\lambda/2}\xi)\, e^{-\frac{is+1}{2}\lambda}\, d\lambda. \tag{3.8}$$

By the inversion formula for the one-dimensional Fourier transform, this implies that

$$\mathcal{R}^h f(\xi) = (2\pi)^{-1} \int_{-\infty}^{+\infty} \mathcal{F} f(\xi; s)\, ds. \tag{3.9}$$

One can obtain the inversion formula for the Fourier transform $\mathcal{F}$ from the known inversion formula (2.12) for the horocycle transform $\mathcal{R}^h$. Note that in the Euclidean case we obtained the inversion formula for the Radon transform from the inversion formula for the Fourier transform; here we proceed in opposite direction. Substituting in (2.12) the expression for $\mathcal{R}^h f$ in terms of $\mathcal{F} f$ given in (3.9) we obtain

(3.10)

$$f(z) = -\frac{1}{8\pi^3} \int_{\mathbb{R}^2} \int_{-\infty}^{+\infty} \frac{e^{-\rho(z;\xi)/2}}{\sinh \frac{\rho(z;\xi)}{2}} \left(\xi_1 \frac{\partial}{\partial \xi_1} + \xi_2 \frac{\partial}{\partial \xi_2} + 1 \right) \mathcal{F} f(\xi; s)\, ds\, d\xi_1\, d\xi_2.$$

It remains to simplify the resulting expression. We first note that the homogeneity condition (3.7) for $\mathcal{F} f$ implies

$$\left(\xi_1 \frac{\partial}{\partial \xi_1} + \xi_2 \frac{\partial}{\partial \xi_2} + 1 \right) \mathcal{F} f(\xi; s) = -is\, \mathcal{F} f(\xi; s).$$

Further, we pass in (3.10) to the "polar" coordinates $\xi = e^r \xi'$, where $-\infty < r < +\infty$ and ξ' belongs to a contour $\Gamma \subset \mathbb{R}^2 \setminus O$ that intersects once every ray issuing from the point O. We obtain

$$f(z) = \frac{i}{8\pi^3} \int_\Gamma \int_{-\infty}^{+\infty} \int_{-\infty}^{+\infty} s \frac{e^{-\frac{1}{2}\rho(z;\xi)}}{\sinh\left(r + \frac{\rho(z;\xi)}{2}\right)} \times e^{-irs}\, \mathcal{F} f(\xi; s)\, ds\, dr (\xi_1\, d\xi_2 - \xi_2\, d\xi_2).$$

Integrating over r (the integral over r must be understood in the principal value sense), we obtain the following assertion.

THEOREM 3.2. *A smooth compactly supported function f on $\mathcal{L}^2$ can be expressed in terms of its Fourier transform $\mathcal{F} f$ by the following inversion formula:*[4]

$$
\begin{aligned}
f(z) &= \frac{1}{8\pi^2}\int_{-\infty}^{+\infty} s\tanh\frac{\pi s}{2} \\
&\quad \times \Big(\int_\Gamma \mathcal{F} f(\xi;s)\exp\Big(-\frac{1-is}{2}\rho(z;\xi)\Big)(\xi_1\,d\xi_2-\xi_2\,d\xi_1)\Big)\,ds \\
&= \frac{1}{8\pi^2}\int_{-\infty}^{+\infty} s\tanh\frac{\pi s}{2}\Big(\int_\Gamma \mathcal{F} f(\xi;s)\Phi_{-s}(z;\xi)(\xi_1\,d\xi_2-\xi_2\,d\xi_1)\Big)\,ds.
\end{aligned}
\tag{3.11}
$$

In the proof of (3.11) we changed the order of integration; however, we did not prove rigorously that this change is indeed possible.

3.4. Symmetry relation. We show that the function f can be expressed in terms of the values $\mathcal{F} f(\xi;s)$ for $s>0$ only. This is related to the following fact: the shifts of the functions $y^{\frac{1\pm is}{2}}$ belonging to the same eigenspace of the operator Δ can be expressed in terms of one another. Namely,

$$
y^{\frac{1-is}{2}} = c(s)\int_{-\infty}^{+\infty}(y|z-\xi_1|^{-2})^{\frac{1+is}{2}}\,d\xi_1,
$$

where $c(s)=\pi^{-1/2}\Gamma\left(\frac{1+is}{2}\right)/\Gamma\left(\frac{is}{2}\right)$.

This readily implies the following *symmetry relation* for $\mathcal{F} f$: *the integral*

$$
\int_\Gamma \mathcal{F} f(\xi;s)\Phi_{-s}(z;\xi)(\xi_1\,d\xi_2-\xi_2\,d\xi_1)
$$

is an even function of s. Thus, by the symmetry relation, the inversion formula for the Fourier transform $\mathcal{F}$ can be represented as

$$
f(z)=\frac{1}{4\pi^2}\int_0^\infty s\tanh\frac{\pi s}{2}\Big(\int_\Gamma \mathcal{F} f(\xi;s)\Phi_{-s}(z;\xi)(\xi_1\,d\xi_2-\xi_2\,d\xi_1)\Big)\,ds. \tag{3.12}
$$

REMARK. The symmetry relation for $\mathcal{F} f$ can also be obtained from the symmetry relation (2.16) for the horocycle transform $\mathcal{R}^h f$. Namely, substituting in (2.16) the expression for $\mathcal{R}^h f$ in terms of $\mathcal{F} f$ and passing to the "polar" coordinates $\xi=e^t\xi'$, $\xi'\in\Gamma$, we see that the integral

$$
e^{-r/2}\int_\Gamma\int_{-\infty}^{+\infty}\int_{-\infty}^{+\infty} e^{(1-is)t}\,\mathcal{F} f(\xi;s)\,\delta(2t+\rho(z;\xi)-r)\,ds\,dt(\xi_1\,d\xi_2-\xi_2\,d\xi_1) \tag{3.13}
$$

is an even function of r. Integrating over t, we prove that the integral

$$
\int_\Gamma\int_{-\infty}^{+\infty} e^{-isr}\,\mathcal{F} f(\xi;s)e^{-\frac{1-is}{2}\rho(z;\xi)}\,ds(\xi_1\,d\xi_2-\xi_2\,d\xi_1)
$$

is an even function of r. Obviously, this condition is equivalent to the symmetry condition for $\mathcal{F} f$.

---

[4]This follows from P.V. $\int_{-\infty}^{+\infty}\frac{e^{-irs}\,dr}{\sinh(z+\frac{\rho}{2})} = e^{\frac{i}{2}s\rho}\,\text{P.V.}\int_{-\infty}^{+\infty}\frac{e^{-irs}}{\sinh r}\,dr = ie^{\frac{i}{2}s\rho}\int_{-\infty}^{+\infty}\frac{\sin rs}{\sinh r}\,dr = -\pi i\tanh\frac{\pi s}{2}e^{-\frac{i}{2}s\rho}$.

3.5. Plancherel formula. The definition of the Fourier transform $\mathcal{F} f$ and the inversion formula (3.12) imply the following assertion.

THEOREM 3.3. *The following Plancherel formula holds for any smooth compactly supported function f on $\mathcal{L}^2$:*

$$\int_{\mathcal{L}^2} |f(z)|^2 \frac{dx\,dy}{y^2} = \frac{1}{4\pi^2} \int_0^\infty \int_\Gamma s \tanh \frac{\pi s}{2} |\mathcal{F} f(\xi; s)|^2 (\xi_1\, d\xi_2 - \xi_2\, d\xi_1)\, ds. \tag{3.14}$$

Indeed, replacing the function $\overline{f(z)}$ in the integrand of $\int f(z)\overline{f(z)}\frac{dx\,dy}{y^2}$ by the expression for $\overline{f(z)}$ in terms of $\mathcal{F} f$, we obtain

$$\begin{aligned}\int_{\mathcal{L}^2} |f(z)|^2 \frac{dx\,dy}{y^2} &= \frac{1}{4\pi^2} \int_{\mathcal{L}^2} \int_0^\infty \int_\Gamma s \tanh \frac{\pi s}{2} f(z) \overline{\mathcal{F} f(\xi; s)} \\ &\qquad \times \exp\left[-\frac{1+is}{2}\rho(z;\xi)\right] (\xi_1\, d\xi_2 - \xi_2\, d\xi_1)\, ds \frac{dx\,dy}{y^2} \\ &= \frac{1}{4\pi^2} \int_0^\infty \int_\Gamma s \tanh \frac{\pi s}{2} |\mathcal{F} f(\xi; s)|^2 (\xi_1\, d\xi_2 - \xi_2\, d\xi_1)\, ds.\end{aligned}$$

By the Plancherel formula, the Fourier transform can be extended from the smooth compactly supported functions to all functions $f \in L^2(\mathcal{L}^2)$ according to the classical scheme.

4. Relation to the representation theory of the group $SL(2,\mathbb{R})$

Let us show the relation between integral geometry and the representation theory of Lie groups by the example of the hyperbolic plane $\mathcal{L}^2$.

We have the Hilbert space $L^2(\mathcal{L}^2)$ and a unitary representation T of the group $SL(2,\mathbb{R})$ in $L^2(\mathcal{L}^2)$,

$$T_g f(z) = f(z \circ g), \qquad g \in SL(2,\mathbb{R}),$$

where $z \mapsto z \circ g$ is the motion (corresponding to g) of the hyperbolic plane. The problem of representation theory (harmonic analysis for the hyperbolic plane) is to decompose $L^2(\mathcal{L}^2)$ into irreducible invariant subspaces. This problem is similar to that of decomposing a function in $L^2(\mathbb{R}^2)$ in the Fourier integral, and irreducible invariant subspaces can be treated as analogs of exponential functions. We show that the solution of this problem is given by (3.12) and (3.14) (see § 3).

We first describe the irreducible invariant subspaces themselves. Consider the eigenspace corresponding to an eigenvalue $-\frac{1+s^2}{4}$ of the Laplace–Beltrami operator Δ. As was already said before, this space is spanned by the distributions $\Phi_s(z;\xi) = (y|\xi_2 z - \xi_2|^{-2})^{\frac{1+is}{2}}$, where ξ ranges over a set containing exactly one representative of every line in $\mathbb{R}^2$ passing through the origin O. Thus, the elements of this eigenspace can be presented in the form

$$f_s(z) = \int_\Gamma u_s(\xi)\Phi_s(z;\xi)(\xi_1\, d\xi_2 - \xi_2\, d\xi_1), \tag{4.1}$$

where the function u_s is homogeneous,

$$u_s(\lambda\xi) = |\lambda|^{is-1} u_s(\xi),$$

and the integral is taken over any contour $\Gamma \subset \mathbb{R}^2 \setminus O$ intersecting at one point every ray issuing from the origin O. (The integral does not depend on the choice of Γ because the differential form in the integrand is invariant with respect to the homothetic transformations $\xi \mapsto \lambda\xi$; cf. (1.14) in Chapter 1.) In the space of functions f_s we introduce the norm

$$\|f_s\|^2 = \int_\Gamma |u_s(\xi)|^2(\xi_1\,d\xi_2 - \xi_2\,d\xi_1) \tag{4.2}$$

and define the Hilbert space H_s as the L^2 space with respect to this norm. It follows from what was said in 3.4 that the spaces H_s and H_{-s} coincide.

The motions on $\mathcal{L}^2$ determine a unitary representation T_s of $SL(2,\mathbb{R})$ in H_s. In terms of the functions $u_s(\xi)$, this representation looks as follows:

$$(T_s(g)u_s)(\xi_1,\xi_2) = u_s(\alpha\xi_1 + \gamma\xi_2, \beta\xi_1 + \delta\xi_2), \qquad g = \begin{pmatrix} \alpha & \beta \\ \gamma & \delta \end{pmatrix}.$$

It is known (see, e.g., **[8]**) that this representation is irreducible.

Introduce the projection operator $P_s : L^2(\mathcal{L}^2) \to H_s$ by the rule

$$P_s f = f_s,$$

where f_s is given by (4.1) in which $u_s(\xi) = \mathcal{F} f(\xi, s)$. Obviously, these operators P_s commute with the action of the group $SL(2,\mathbb{R})$ on the spaces $L^2(\mathcal{L}^2)$ and H_s. Thus, the problem of decomposing the space $L^2(\mathcal{L}^2)$ into irreducible subspaces consists in expanding the functions $f \in L^2(\mathcal{L}^2)$ in their projections f_s.

The solution of this problem is given by (3.12) and (3.14), since these formulas can be written as follows:

$$f = \frac{1}{4\pi^2}\int_0^{+\infty} s\tanh\frac{\pi s}{2} f_s\,ds, \qquad f_s = P_s f; \tag{4.3}$$

$$\int_{\mathcal{L}^2} |f(z)|^2\frac{dx\,dy}{y^2} = \frac{1}{4\pi^2}\int_0^\infty s\tanh\frac{\pi s}{2}\|f_s\|^2\,ds, \tag{4.4}$$

where $\|f_s\|$ is given by (4.2).

Recall that the proof of the main formula, (3.12), used the inversion formula for the operator $\mathcal{R}^h$ of horocycle transform. We clarify the role of this operator from the group-theoretic point of view.

The operator $\mathcal{R}^h$ sends a function $f(z)$ on the homogeneous space $\mathcal{L}^2 = SO(2)\setminus SL(2,\mathbb{R})$ to the function $\mathcal{R}^h f(\xi)$ on the homogeneous space $H = \pm Z \setminus SL(2,\mathbb{R})$, and this operator commutes with the action of the group $SL(2,\mathbb{R})$ on these spaces. Moreover, the irreducible components transform into irreducible components; however, *a priori* this mapping may have a nonzero kernel. In fact, the kernel of the mapping is trivial because there is an inversion formula for $\mathcal{R}^h$, and the problem of decomposing the representation of $SL(2,\mathbb{R})$ in $L^2(\mathcal{L}^2)$ is reduced to the problem of decomposing the representation in a function space on $H = \pm Z \setminus SL(2,\mathbb{R})$. The solution of the latter problem is very simple for the following reason. The homogeneous space H is endowed with an action of the commutative group of homothetic transformations $\xi \mapsto \lambda\xi$ commuting with the action of the group $SL(2,\mathbb{R})$. Consider the corresponding commutative group of the operators A_λ:

$$(A_\lambda f)(\xi) = f(\lambda\xi).$$

The common eigenspaces of A_λ consist of homogeneous functions. Since the operators A_λ commute with the operators of the representation of $SL(2,\mathbb{R})$, it follows that their eigenspaces are invariant with respect to this group. It is known that the representations of $SL(2,\mathbb{R})$ acting on these subspaces are irreducible. Thus, the decomposition of the function space on H into irreducible components is reduced to the decomposition into the eigenspaces of the operators A_λ, and since the group of operators A_λ is commutative and isomorphic to the additive group $\mathbb{R}$, this decomposition is reduced to the one-dimensional Fourier transform. The composition of the horocycle transform $\mathcal{R}^h$ and the above one-dimensional Fourier transform is the Fourier transform $\mathcal{F}$ on the hyperbolic plane.

Thus, the role of the horocycle transform is that this transform reduces the problem of expanding functions on the homogeneous space $\mathcal{L}^2 = SO(2)\backslash SL(2,\mathbb{R})$ to the expansion problem for functions on another homogeneous space $\pm Z \setminus SL(2,\mathbb{R})$, where this problem admits an elementary solution. Similar transforms reducing decomposition problems for group representations related to homogeneous spaces to a problem of integral geometry can be defined not only for $\mathcal{L}^2$ but also for other homogeneous spaces. This tool, known as the *method of horospheres*, was first introduced in [**5**].

The construction used in this section to introduce irreducible representations of the group $SL(2,\mathbb{R})$ can be included in the general scheme of realization of irreducible representations in function spaces on homogeneous spaces; this construction originates from Frobenius. A finite-dimensional irreducible unitary representation T of a group G can be realized in functions on a homogeneous space $U \setminus G$ if and only if the representation space contains a nonzero vector invariant with respect to the subgroup U; the dimension of the space of these vectors is equal to the multiplicity of occurrence of the representation T in the space $L^2(U \setminus G)$. The Frobenius duality theorem explicitly defines an operator that embeds T in $L^2(U \setminus G)$. In the infinite-dimensional case one needs some analytic refinements.

These are the functions $y^{\frac{1+is}{2}}$ considered above; they are invariant with respect to the subgroup Z, and the construction of irreducible representations of the group $G = SL(2,\mathbb{R})$ that correspond to these functions follows the Frobenius scheme. Since every irreducible space contains two (distribution) vectors invariant with respect to Z, it follows that this space is embedded in $L^2(\pm Z \setminus G)$ with multiplicity two. However, every irreducible representation enters the decomposition of $L^2(\mathcal{L}^2)$ with multiplicity one, due to the symmetry relation for $\mathcal{F} f$.

5. Integral transform related to lines (geodesics) on the hyperbolic plane $\mathcal{L}^2$

5.1. Definition and the inversion formula in the Poincaré model. Consider another analog of the Radon transform for the hyperbolic plane, namely, the integral transform related to "lines" (geodesics) on $\mathcal{L}^2$.

According to § 1, every "line" on $\mathcal{L}^2$ is given in the Poincaré realization by an equation on the half-plane $\operatorname{Im} z > 0$ of the form

$$(x-a)(x-b)+y^2=0, \tag{5.1}$$

where $a \neq b$ and $a, b \in \mathbb{R} \cup (\infty)$; we denote this "line" by $G(a,b)$. It readily follows from (5.1) that under the motion $z \mapsto \frac{\alpha z+\gamma}{\beta z+\delta}$ a line $G(a,b)$ transforms into the line $G\left(\frac{\alpha a+\gamma}{\beta a+\delta}, \frac{\alpha b+\gamma}{\beta b+\delta}\right)$.

DEFINITION. Denote by $\mathcal{R}^g$ the integral transform taking each smooth compactly supported function f on the hyperbolic plane $\mathcal{L}^2$ to the integrals of this function over the lines in $\mathcal{L}^2$ (i.e., over the geodesics with respect to the metric in $\mathcal{L}^2$),

$$\mathcal{R}^g f(a,b) = \int_{G(a,b)} f(z)\,ds, \qquad ds = \frac{|dz|}{y}. \tag{5.2}$$

Obviously, this transform commutes with the action of $SL(2,\mathbb{R})$.

Let us find the inversion formula reconstructing f from $\mathcal{R}^g f$. In the proof of this formula we follow the scheme used in the case of Radon transform on the Euclidean plane and the horocycle transform on $\mathcal{L}^2$.

We first consider a function $f \in D(\mathcal{L}^2)$ that is constant on the non-Euclidean circles centered at the point i, i.e., on the orbits of the subgroup $SO(2)$ of matrices $g = \left(\begin{smallmatrix} \cos\theta & -\sin\theta \\ \sin\theta & \cos\theta \end{smallmatrix}\right)$; this condition means that f has the form

$$f(z) = F\left(\frac{|z-i|^2}{2y}\right).$$

Since the transform $\mathcal{R}^g$ commutes with the action of the group $SL(2,\mathbb{R})$, it follows that the function $\mathcal{R}^g f(a,b)$ is also invariant with respect to the transforms in $SO(2)$. One can readily see that the fraction $\frac{1+ab}{a-b}$ is a full invariant of a pair of points $a, b \in \mathbb{P}^1$ with respect to $SO(2)$. Thus, $\mathcal{R}^g f(a,b) = \widehat{F}\left(\frac{1+ab}{a-b}\right)$. It follows from (5.2) that

$$\begin{aligned}\widehat{F}(r) = \mathcal{R}^g f(\infty, r) &= \int_0^\infty F\left(\frac{r^2+(y-1)^2}{2y}\right)\frac{dy}{y} \\ &= \int_{(r^2+1)^{1/2}}^\infty F(s-1)\frac{ds}{\sqrt{s^2-(r^2+1)}} \\ &= \int_{(r^2+1)^{1/2}}^\infty (s^{-1}F(s-1))\frac{s\,ds}{\sqrt{s^2-(r^2+1)}}.\end{aligned}$$

Thus, $\widehat{F}(r)$ is an even function of r, and it follows from the inversion formula for the Abel transform that

$$s^{-1}F(s-1) = -\frac{2}{\pi}\int_{\sqrt{s^2-1}}^\infty \frac{\widehat{F}'(r)\,dr}{\sqrt{r^2+1-s^2}}; \qquad s \geq 1.$$

In particular, for $s = 1$ we obtain

$$f(i) = F(0) = -\frac{2}{\pi}\int_0^\infty \frac{\widehat{F}'(r)\,dr}{r} \tag{5.3}$$

(the integral converges because $\widehat{F}'(r) = o(r)$ as $r \to 0$).

We now assume that f is an arbitrary smooth compactly supported function on $\mathcal{L}^2$ and $\overline{f}$ is the result of averaging f over the non-Euclidean circles centered at the point i,

$$\overline{f}(z) = F\left(\frac{|z-i|^2}{2y}\right).$$

Since the transform $\mathcal{R}^g$ commutes with the motions of $\mathcal{L}^2$, it follows that

$$\mathcal{R}^g\,\overline{f}(a,b) = \widehat{F}\left(\frac{1+ab}{a-b}\right)$$
$$= \frac{1}{2\pi}\int_0^{2\pi} \mathcal{R}^g\, f\left(\frac{a\cos\theta+\sin\theta}{-a\sin\theta+\cos\theta}, \frac{b\cos\theta+\sin\theta}{-b\sin\theta+\cos\theta}\right) d\theta,$$

and hence

$$\widehat{F}(r) = \mathcal{R}^g\,\overline{f}(\infty, r) = \frac{1}{2\pi}\int_0^{2\pi} \mathcal{R}^g\, f\left(-\cot\theta, \frac{r\cos\theta+\sin\theta}{-r\sin\theta+\cos\theta}\right) d\theta$$
$$= \frac{1}{\pi}\int_{-\infty}^{+\infty} \mathcal{R}^g\, f\left(t, \frac{rt-1}{r+t}\right)\frac{dt}{1+t^2}. \tag{5.4}$$

In view of (5.3), we obtain

$$f(i) = \overline{f}(i) = -\frac{2}{\pi}\int_0^{\infty} \frac{\widehat{F}'(r)}{r}\, dr, \tag{5.5}$$

where $\widehat{F}(r)$ is given by (5.4).

To find an expression for f at an arbitrary point z, it suffices to apply a group translation g sending the point i to the point z, for instance,

$$g = \begin{pmatrix} y^{1/2} & 0 \\ xy^{-1/2} & y^{-1/2} \end{pmatrix}.$$

Under this translation, $f(i)$ goes to $f(z)$ and the transform $\mathcal{R}^g\, f(a,b)$ goes to $\mathcal{R}^g\, f(x+ay, x+by)$. We obtain the following result.

THEOREM 5.1. *A smooth compactly supported function f on $\mathcal{L}^2$ can be expressed in terms of its transform $\mathcal{R}^g\, f$ related to the lines in $\mathcal{L}^2$ by the following inversion formula*:

$$f(z) = -\frac{2}{\pi}\int_0^{\infty} \frac{\widehat{F}_s'(z;r)}{r}\, dr,$$

where

$$\widehat{F}(z;r) = \frac{1}{\pi}\int_{-\infty}^{+\infty} \mathcal{R}^g\, f\left(x+ty, x+\frac{rt-1}{r+t}\,y\right)\frac{dt}{1+t^2}.$$

5.2. Relation to the Radon transform on the projective plane. We now assume that the hyperbolic plane $\mathcal{L}^2$ is represented as the upper sheet of the hyperboloid given by $[x,x]=1$, where $[x,x]=x_1^2-x_2^2-x_3^2$. In this realization, the geodesics on $\mathcal{L}^2$ are the curves $\gamma(\xi)$ which are the intersections of the hyperboloid with the planes

$$\langle \xi,\, x\rangle \equiv \xi_1x_1+\xi_2x_2+\xi_3x_3 = 0, \qquad [\xi,\xi] = -1,$$

and the integral transform $\mathcal{R}^g$ is given by the formula

$$\mathcal{R}^g\, f(\xi) = \int_{\gamma(\xi)} f(x)\, ds, \qquad [\xi,\xi]=-1, \tag{5.6}$$

where ds is the length element on $\mathcal{L}^2$. One can readily see that the expression for $\mathcal{R}\, f$ can also be represented in the form

$$\mathcal{R}^g\, f(\xi) = \int_{\mathcal{L}^2} f(x)\,\delta(\langle \xi,\, x\rangle)\,\frac{dx_2\,dx_3}{x_1}, \tag{5.7}$$

where $\delta(\cdot)$ is the delta function on $\mathbb{R}$.

In this model, one can obtain the inversion formula for the integral transform $\mathcal{R}^g$ similarly to that for the Minkowski–Funk transform related to the unit two-dimensional sphere; in other words, the inversion formula can be found as a consequence of the inversion formula for the Radon transform $\mathcal{R}$ on the projective plane; see § 8 of Chapter 1. Namely, to any function $f \in \mathcal{L}^2 C^\infty$ we assign a function F in $\mathbb{R}^3 \setminus O$ satisfying the homogeneity condition $F(\lambda x) = \lambda^{-2} F(x)$ and such that $F\,\big|_{[x,x]=1} = f$ and $F = 0$ for $[x, x] < 0$.

It follows from the definition of the operator $\mathcal{R}$ that $\mathcal{R}F(\xi) = \mathcal{R}^g f(\xi)$ for $[\xi, \xi] = -1$ and $\mathcal{R}F(\xi) = 0$ for $[\xi, \xi] > 0$. Therefore, the following assertion is an immediate corollary of the inversion formula given in Chapter 1 for the integral transform $\mathcal{R}$ (see (8.10) on page 36).

THEOREM 5.2. *A function f on $\mathcal{L}^2$ can be expressed in terms of its transform $\mathcal{R}^g f$ related to the geodesics by the following inversion formula*:

$$f(x) = \frac{1}{4\pi} \int_{[\xi,\xi]=-1} \mathcal{R}^g f(\xi)\, \langle \xi, x \rangle^{-2}\, \omega(\xi). \tag{5.8}$$

For the meaning of this integral, see § 8 of Chapter 1.

6. Horospherical transform in the three-dimensional hyperbolic space $\mathcal{L}^3$

The definitions and results of §§ 1–5 for the hyperbolic plane can be extended to hyperbolic spaces of higher dimensions; however, the inversion formula has different forms for spaces of odd and even dimensions. For this reason, we give an independent exposition for the three-dimensional hyperbolic space $\mathcal{L}^3$. Here we put an emphasis on explicit formulas.

6.1. Models of the hyperbolic space. Let us briefly present the elements of geometry in $\mathcal{L}^3$. The hyperbolic space $\mathcal{L}^3$ is the homogeneous space of the group $SL(2, \mathbb{C})$ with the isotropy subgroup $SU(2)$ (the subgroup of unitary matrices). The following quaternion model of the space $\mathcal{L}^3$ is convenient for our purposes: a point of $\mathcal{L}^3$ is a quaternion with three components,

$$w = x + iy + zj, \quad z > 0,$$

where $x, y, z \in \mathbb{R}$, $i^2 = j^2 = -1$, and $ij = -ji$; the motion of $\mathcal{L}^3$ corresponding to a matrix $\left(\begin{smallmatrix} \alpha & \beta \\ \gamma & \delta \end{smallmatrix}\right) \in SL(2, \mathbb{C})$ is given as follows:

$$w \mapsto w' = (w\beta + \delta)^{-1}(w\alpha + \gamma).$$

Obviously, the isotropy subgroup of the point $w = j$ is $SU(2)$.

The Riemannian metric on $\mathcal{L}^3$ is

$$ds^2 = z^{-2}(dx^2 + dy^2 + dz^2). \tag{6.1}$$

The distance d between points $w_1 = x_1 + y_1 i + z_1 j$ and $w_2 = x_2 + y_2 i + z_2 j$ is given by the relation

$$4 \sinh^2 \frac{d}{2} = \rho^2(w_1, w_2), \tag{6.2}$$

where

$$\rho^2(w_1, w_2) = \frac{|w_1 - w_2|^2}{z_1 z_2}, \tag{6.3}$$

and $|\cdot|$ stands for the norm of a quaternion ($|w|^2 = x^2 + y^2 + z^2$). The volume element in the metric (6.1) is of the form

$$dv = z^{-3}\, dx\, dy\, dz. \tag{6.4}$$

In a different way, the hyperbolic space can be given by the manifold of Hermitian-symmetric matrices

$$u = \begin{pmatrix} a & b \\ \bar{b} & c \end{pmatrix}, \qquad a, c > 0, \quad \det u = 1, \tag{6.5}$$

with motions

$$u \mapsto g^* u g, \qquad g \in SL(2, \mathbb{C}),$$

where g^* stands for the Hermitian-conjugate matrix.

Note that the manifold of matrices (6.5) can be interpreted as a sheet of the following hyperboloid of two sheets in $\mathbb{R}^4$:

$$ac - b_1^2 - b_2^2 = 1, \qquad a, c > 0 \quad (b_1 + ib_2 = b). \tag{6.6}$$

The correspondence between the quaternions w and the matrices $u = u(w)$ is as follows:

$$x + yi + zj \mapsto \begin{pmatrix} z^{-1}(x^2 + y^2 + z^2) & z^{-1}(x + iy) \\ z^{-1}(x - iy) & z^{-1} \end{pmatrix};$$

in other words, to a matrix $\left(\begin{smallmatrix} a & b \\ \bar{b} & c \end{smallmatrix}\right)$ corresponds the quaternion $w = \frac{b}{c} + \frac{1}{c} j$ (cf. the case of the hyperbolic plane). In particular, the identity matrix corresponds to the point $w = j$.

It is convenient to pass in the equation of the hyperboloid given by (6.6) to the new coordinates $x_1 = \frac{a+c}{2}$, $x_2 = \frac{a-c}{2}$, $x_3 = b_1$, $x_4 = b_2$. Then we obtain a realization of the hyperbolic space as the upper sheet of the following hyperboloid:

$$[x, x] \equiv x_1^2 - x_2^2 - x_3^2 - x_4^2 = 1, \qquad x_1 > 0. \tag{6.7}$$

In this realization, the motions are given by linear transforms in $\mathbb{R}^4$ preserving the upper sheet of the hyperboloid, i.e., by elements of the group $SO_0(1,3) \cong SL(2, \mathbb{C})/\{\pm e\}$.

The Riemannian metric in this model is

$$ds^2 = -dx_1^2 + dx_2^2 + dx_3^2 + dx_4^2;$$

the invariant distance d between points x and y is

$$\cosh d = [x, y];$$

the volume element in the coordinates x_2, x_3 on $\mathcal{L}^2$ is

$$dv = \frac{dx_2\, dx_3\, dx_4}{x_1}.$$

6.2. Horospheres. In the hyperbolic space one can define spheres of infinitely large radius (horospheres), which are limits of non-Euclidean spheres as the center and the radius of these spheres simultaneously tend to infinity. A horosphere can be represented either as a (Euclidean) sphere (in the half-space $z > 0$ of $\mathbb{R}^3$) that is tangent to the plane $z = 0$ (the absolute), or as a plane (in this half-space) parallel to the plane $z = 0$. The latter can be naturally interpreted as a sphere of infinitely large radius tangent to the plane $z = 0$ at a point at infinity. In the quaternion model, every horosphere is given by an equation of the form

$$|w\xi_2 - \xi_1|^2 = z, \qquad w = x + ui + zj, \tag{6.8}$$

where $\xi = (\xi_1, \xi_2) \in \mathbb{C}^2 \setminus O$. Thus, the manifold of horospheres is parametrized by the points $\xi \in \mathbb{C}^2 \setminus O$, and the horospheres corresponding to the points ξ and $\lambda\xi$, $|\lambda| = 1$, coincide (cf. the equation of horocycles on the hyperbolic plane). Denote by $h(\xi)$ the horosphere given by (6.8).

REMARK. If the space $\mathcal{L}^3$ is represented as the upper sheet of the hyperboloid (6.7), then the horospheres are the plane parabolic sections of this hyperboloid, namely, the sections by the planes of the form $[\xi, x] = \xi_1 x_1 + \xi_2 x_2 + \xi_3 x_3 + \xi_4 x_4 = 1$, where ξ is a point of the upper nappe of the cone $[\xi, \xi] = 0$, and thus $\xi_1 > 0$ ($\xi \in \mathbb{R}^4 \setminus O$).

The motion group $SL(2, \mathbb{C})$ acts on the manifold of horospheres, and this action is transitive; the motion corresponding to a matrix $g = \left(\begin{smallmatrix} \alpha & \beta \\ \gamma & \delta \end{smallmatrix}\right)$ takes a horosphere $f(\xi_1, \xi_2)$ to the horosphere $f(\alpha\xi_1 + \gamma\xi_2, \beta\xi_1 + \delta\xi_2)$.

In the language of motions, the horospheres are either orbits of the subgroup Z of the matrices of the form $\left(\begin{smallmatrix} 1 & 0 \\ t & 1 \end{smallmatrix}\right)$, $t \in \mathbb{C}$ (these are the planes parallel to the plane $z = 0$) or shifts of these orbits. Thus, the manifold of horospheres is the homogeneous space of the group $SL(2, \mathbb{C})$ with the isotropy subgroup $\pm Z$.

Each horosphere has a unique point of tangency with the absolute; this point is called the *center* of the horosphere. In our interpretation, these are the points of tangency with the plane $z = 0$. Horospheres with a common center are said to be *parallel*. Note that the horosphere $h(\xi)$ is tangent to the plane $z = 0$ at the point $\frac{\xi_1}{\xi_2}$; hence, every pencil of parallel horospheres is of the form $\{h(\lambda\xi) \mid 0 < \lambda < \infty\}$, where ξ is fixed.

The (oriented) distance $\rho(w, \xi)$ from a point $w \in \mathcal{L}^3$ to a horosphere $h(\xi)$ is defined in the same way as the distance from a point to a horocycle on the hyperbolic plane, and is given by a similar formula

$$\rho(w; \xi) = \log(z^{-1}|w\xi_2 - \xi_1|^2), \qquad w = x + yi + zj. \tag{6.9}$$

6.3. Horospherical transform.

DEFINITION. By the *horospherical transform* in $\mathcal{L}^3$ we mean the integral transform $\mathcal{R}^h$ taking a smooth compactly supported function f on $\mathcal{L}^3$ to the integrals of this function over the horospheres,

$$\mathcal{R}^h f(\xi) = \int_{h(\xi)} f(w)\, d\sigma, \tag{6.10}$$

where $h(\xi)$ stands for the horosphere given by (6.8) and $d\sigma$ for the (non-Euclidean) area element on $h(\xi)$.

The expression for $\mathcal{R}^h$ can also be written as

$$\mathcal{R}^h f(\xi) = \int_{\mathcal{L}^3} f(w)\, \delta(\rho(w; \xi)) \frac{dx\, dy\, dz}{z^3}, \tag{6.11}$$

where $\rho(w; \xi)$ is the distance from the point $w = x + yi + zj$ to the horosphere $h(\xi)$; see (6.9).

It follows from the definition that the operator $\mathcal{R}^h$ commutes with the action of the group $SL(2, \mathbb{C})$.

6.4. Inversion formula. We now proceed with the construction of an inversion formula for the integral transform $\mathcal{R}^h$ by repeating the arguments used in the case of hyperbolic plane. We first assume that f is a radial function depending only on the distance from the point w to the point j,

$$f(w) = F(z^{-1/2}|w - j|).$$

In this case, $\mathcal{R}^h f(\xi)$ depends only on the distance from the point j to the horosphere $h(\xi)$, i.e.,

$$\mathcal{R}^h f(\xi) = \widehat{F}(r), \quad \text{where} \quad r = \log(|\xi_1|^2 + |\xi_2|^2).$$

Thus, the value $\widehat{F}(r)$ is equal to the integral of the function f over the horosphere $h(e^{r/2}, 0)$ given by $z = e^r$, i.e.,

$$\begin{aligned}\widehat{F}(r) &= \int_{\mathbb{R}^2} F\left(\frac{\sqrt{x^2+y^2+(e^r-1)^2}}{e^{r/2}}\right)\frac{dx\,dy}{e^{2r}} \\ &= 2\pi\int_0^\infty F\left(\frac{\sqrt{t^2+(e^r-1)^2}}{e^{r/2}}\right)\frac{t\,dt}{e^{2r}} = 2\pi\int_{2|\sinh\frac{r}{2}|}^\infty F(s)\frac{s\,ds}{e^r}.\end{aligned} \tag{6.12}$$

First, this implies that $e^r\widehat{F}(r)$ is an even function of r,

$$e^r\widehat{F}(r) = e^{-r}\widehat{F}(-r),$$

and second,

$$(e^r\widehat{F}(r))' = -2\pi\Big(\cosh r F\Big(2\sinh\frac{r}{2}\Big) + \sinh r\cosh\frac{r}{2}F'\Big(2\sinh\frac{r}{2}\Big)\Big). \tag{6.13}$$

It follows from (6.13) for $r = 0$ that

$$f(j) = F(0) = -\frac{1}{2\pi}(e^r\widehat{F}(r))'|_{r=0} \tag{6.14}$$

(cf (2.5) in this chapter and (2.4) in § 2 of Chapter 1). Assume now that f is an arbitrary smooth compactly supported function on $\mathcal{L}^3$ and

$$\overline{f}(w) = F(z^{-1/2}|w - j|)$$

is the function obtained by averaging f over the non-Euclidean spheres centered at the point j. Since these spheres are orbits of the subgroup of unitary matrices $g = \left(\begin{smallmatrix}\alpha & \beta \\ -\overline{\beta} & \overline{\alpha}\end{smallmatrix}\right)$, $|\alpha|^2 + |\beta|^2 = 1$, it follows that

$$\overline{f}(w) = F(z^{-1/2}|w - j|) = \int f((w\beta + \overline{\alpha})^{-1}(w\alpha - \overline{\beta}))\,d\mu(\alpha,\beta),$$

where $d\mu(\alpha,\beta)$ is the invariant measure on the group of unitary matrices, i.e., on the three-dimensional unit sphere $|\alpha|^2 + |\beta|^2 = 1$, normalized by the condition

$$\int d\mu(\alpha,\beta) = 1.$$

We can readily see that the expression for F can be written in the following convenient form:

$$F(s) = \frac{1}{\pi}\int_{\mathcal{L}^3} z^{1/2}|w - j|^{-1} f(w)\,\delta(z^{-1}|w - j|^2 - s^2)\frac{dx\,dy\,dz}{z^3}. \tag{6.15}$$

Let $\widehat{F}(r) = \mathcal{R}^h \overline{f}(\xi)$, where $r = \rho(j;\xi) = \log(|\xi_i|^2 + |\xi_2|^2)$. Since the transform $\mathcal{R}^h$ commutes with the action of the group $SL(2,\mathbb{C})$, it follows that

$$\begin{aligned}\widehat{F}(r) &= \int_{|\alpha|^2+|\beta|^2=1} \mathcal{R}^h f(e^{r/2}\alpha, e^{r/2}\beta)\, d\mu(\alpha,\beta) \\ &= \pi^2 \left(\frac{i}{2}\right)^2 \int_{\mathbb{C}^2} \mathcal{R}^h f(e^{r/2}\xi_1, e^{r/2}\xi_2)\, \delta(\rho(j;\xi))\, d\xi_1\, d\overline{\xi_1}\, d\xi_2\, d\overline{\xi_2},\end{aligned} \tag{6.16}$$

where $\rho(j;\xi)$ is the distance from j to the horosphere $h(\xi)$. By (6.14),

$$f(j) = \overline{f}(j) = -\frac{1}{2\pi}(e^r \widehat{F}(r))''|_{r=0}.$$

Substituting here the expression for $\widehat{F}(r)$ given in (6.16), we obtain

$$\begin{aligned}f(j) &= -\frac{1}{8\pi} \int_{|\xi_1|^2+|\xi_2|^2=1} L^2 \mathcal{R}^h f(\xi)\, d\mu(\xi) \\ &= -\frac{1}{8\pi^3} \int_{\mathbb{C}^2} L^2 \mathcal{R}^h f(\xi)\, \delta(\rho(j;\xi))\, dv,\end{aligned} \tag{6.17}$$

where

$$L = \xi_1 \frac{\partial}{\partial \xi_1} + \xi_2 \frac{\partial}{\partial \xi_2} + \overline{\xi_1} \frac{\partial}{\partial \overline{\xi_1}} + \overline{\xi_2} \frac{\partial}{\partial \overline{\xi_2}} + 2, \quad dv = \left(\frac{i}{2}\right)^2 d\xi_1\, d\overline{\xi_1}\, d\xi_2\, d\overline{\xi_2}. \tag{6.18}$$

This gives an expression for $f(j)$ in terms of its horospherical transform $\mathcal{R}^h f$. Hence, applying a translation that takes j to an arbitrarily chosen point w, we obtain the following assertion.

THEOREM 6.1. *A function f on $\mathcal{L}^3$ can be expressed in terms of its horospherical transform $\mathcal{R}^h f$ by the following inversion formula:*

$$f(w) = -\frac{1}{2\pi}(e^r \widehat{F}(w,r))''|_{r=0}, \tag{6.19}$$

where

$$\widehat{F}(w,r) = \pi^{-2} \int_{\mathbb{C}^2} \mathcal{R}^h f(e^{r/2}\xi_1, e^{r/2}\xi_2)\, \delta(\rho(w,\xi))\, dv; \tag{6.20}$$

hence,

$$f(w) = -\frac{1}{8\pi^3} \int_{\mathbb{C}^2} L^2 \mathcal{R}^h f(\xi)\, \delta(\rho(w;\xi))\, dv, \tag{6.21}$$

where the operator L is given by (6.18).

REMARK. Since $\mathcal{R}^h f(e^{i\theta}\xi_1, e^{i\theta}\xi_2) = \mathcal{R}^h f(\xi_1,\xi_2)$, we have $L_\xi \mathcal{R}^h f = \overline{L}_\xi \mathcal{R}^h f$, where $L_\xi = \xi_1 \frac{\partial}{\partial \xi_1} + \xi_2 \frac{\partial}{\partial \xi_2} + 1$ and $\overline{L}_\xi = \overline{\xi_1} \frac{\partial}{\partial \overline{\xi_1}} + \overline{\xi_2} \frac{\partial}{\partial \overline{\xi_2}} + 1$. Therefore, $L^2 \mathcal{R}^h f = 4L_\xi^2 \mathcal{R}^h f = 4\overline{L}_\xi^2 \mathcal{R}^h f = 4L_\xi \overline{L}_\xi \mathcal{R}^h f$.

Note that this inversion formula is local (in contrast to the case of hyperbolic plane); namely, to reconstruct a function f at a point w, one must know the integrals of f only over the horospheres infinitely close to w.

6.5. Symmetry relation. We obtain the following result as a by-product, applying a group translation to (6.12).

THEOREM 6.2. *The following* Asgeirsson relation *holds for the hyperbolic space*:

$$e^r \widehat{F}(w;r) = 2\pi \int_{2|\sinh \frac{r}{2}|}^{\infty} F(w;s) s\, ds. \tag{6.22}$$

Here $F(w,s)$ stands for the result of averaging the function f over the spheres centered at the point w, i.e.,

(6.23)

$$\begin{aligned} F(w,s) &= \pi^{-1} \int_{\mathcal{L}^3} z^{1/2} z'^{1/2} |w-w'|^{-1} f(w')\, \delta(z^{-1}z'^{-1}|w-w'|^2 - s^2) \frac{dx'\,dy'\,dz'}{z^3} \\ &= (2\pi)^{-1} \int_{\mathcal{L}^3} \left| \sinh \frac{d(w,w')}{2} \right|^{-1} \delta\left(4\sinh^2 \frac{d(w,w')}{2} - s^2\right) f(w') \frac{dx'\,dy'\,dz'}{z'^3}, \end{aligned}$$

where $d(w,w')$ is the non-Euclidean distance between w and w', and $\widehat{F}(w,r)$ is the result of averaging the function $\mathcal{R}^h f$ over the horospheres equidistant from the point w; see (6.20).

COROLLARY. *The function $e^r \widehat{F}(w,r)$ is even as a function of r,*

$$e^r \widehat{F}(w,r) = e^{-r} \widehat{F}(w,-r).$$

Substituting here the expression for $\widehat{F}$ given by (6.20), we see that $\mathcal{R}^h f$ satisfies the following symmetry relation:

$$\begin{aligned} &e^r \int \mathcal{R}^h f(e^{r/2}\xi_1, e^{r/2}\xi_2)\, \delta(\rho(w;\xi))\, dv \\ &\qquad = e^{-r} \int \mathcal{R}^h f(e^{-r/2}\xi_1, e^{-r/2}\xi_2)\, \delta(\rho(w;\xi))\, dv. \end{aligned} \tag{6.24}$$

6.6. Inversion formula for the horospherical transform in another model of the hyperbolic space. Let us represent the hyperbolic space $\mathcal{L}^3$ as the upper sheet, $x_1 > 0$, of the hyperboloid $[x,x] \equiv x_1^2 - x_2^2 - x_3^2 - x_4^2 = 1$ in $\mathbb{R}^4$.

In this model, the horospheres are the sections of the hyperboloid by the hyperplanes

$$[\xi, x] \equiv \xi_1 x_1 - \xi_2 x_2 - \xi_3 x_3 - \xi_4 x_4 = 1,$$

where ξ is a point of the upper nappe K, $x_1 > 0$, of the cone given by $[\xi,\xi] = 0$. The horospherical transform is given by the following formula:

$$\mathcal{R}^h f(\xi) = \int_{\mathcal{L}^3} f(x)\, \delta([\xi,x]-1)\, dx, \qquad \xi \in K, \tag{6.25}$$

where $dx = x_1^{-1}\, dx_2\, dx_3\, dx_4$ is an invariant measure on $\mathcal{L}^3$.

THEOREM 6.3. *A smooth compactly supported function f on $\mathcal{L}^3$ can be expressed in terms of the horospherical transform $\mathcal{R}^h f$ by the following inversion formula*:

$$f(x) = -\frac{1}{8\pi^2} \int_K \mathcal{R}^h f(\xi)\, \delta''([\xi,x]-1)\, d\xi, \qquad d\xi = \xi_1^{-1}\, d\xi_2\, d\xi_3\, d\xi_4, \tag{6.26}$$

where $\delta''(\cdot)$ is the second derivative of the delta function on $\mathbb{R}$.

Formulas (6.25) and (6.26) are similar to the corresponding formulas of Chapter 1 for the Radon transform and, as in the case of Radon transform, it suffices to prove the inversion formula (6.26) for functions on $\mathcal{L}^3$ depending only on the distance from the point $x^0 = (1, 0, 0, 0)$. One can readily see that, for these functions f, the inversion formula (6.26) coincides with that proven above.

6.7. Integral transform related to completely geodesic surfaces in $\mathcal{L}^3$. Here we consider another integral transform on $\mathcal{L}^3$; it takes functions on $\mathcal{L}^3$ to their integrals over the completely geodesic surfaces in $\mathcal{L}^3$. We denote this transform by $\mathcal{R}^g$.

We represent the hyperbolic space $\mathcal{L}^3$ as the upper sheet of the hyperboloid (6.7) in $\mathbb{R}^4$. In this model, the complete geodesics are the sections of this hyperboloid by the planes

$$\langle \xi, x \rangle \equiv \xi_1 x_1 + \xi_2 x_2 + \xi_3 x_3 + \xi_4 x_4 = 0,$$

where ξ ranges over the hyperboloid of one sheet $[\xi, \xi] = -1$. The integral transform $\mathcal{R}^g$ is given by the following formula:

$$\mathcal{R}^g f(\xi) = \int_{\mathcal{L}^2} f(x)\, \delta(\langle \xi, x \rangle)\, \frac{dx_2\, dx_3\, dx_4}{x_1}, \tag{6.27}$$

where $\delta(\cdot)$ stands for the delta function on $\mathbb{R}$; cf. the definition for the hyperbolic plane in § 5.

As in the case of hyperbolic plane, the inversion formula for the integral transform in question can be obtained as a consequence of the inversion formula for the Radon transform $\mathcal{R}$ on the projective space; see § 8 of Chapter 1. Namely, to a function $f \in \mathcal{L}^3$ we assign the C^∞ function F on $\mathbb{R}^4 \setminus O$ satisfying the homogeneity condition $F(\lambda x) = |\lambda|^{-3} F(x)$ and the conditions $F\,\big|_{[x,x]=1} = f$ and $F = 0$ for $[x, x] < 0$.

It follows from the definition of the operator $\mathcal{R}$ that $\mathcal{R} F(\xi) = \mathcal{R}^g f(\xi)$ for $[\xi, \xi] = -1$ and $\mathcal{R} F(\xi) = 0$ for $[\xi, \xi] > 0$. Therefore, the following assertion is an immediate corollary of the inversion formula (8.6) of Chapter 1 for the integral transform $\mathcal{R}$ (see page 32).

THEOREM 6.4. *A smooth compactly supported function f on $\mathcal{L}^3$ can be expressed in terms of the transform $\mathcal{R}^g f$ related to the completely geodesic surfaces by the following inversion formula*:

$$f(x) = -(2\pi)^{-2} \int_{[\xi,\xi]=-1} \mathcal{R}^g f(\xi)\, \delta''(\langle \xi, x \rangle)\, \omega(\xi), \tag{6.28}$$

where $\delta''(\cdot)$ stands for the second derivative of the delta function on $\mathbb{R}$.

For the meaning of this integral; see § 8 of Chapter 1.

We emphasize that the inversion formula (6.28) is local, in contrast to the case of hyperbolic plane.

Appendix. Horospherical transform for the hyperbolic space of an arbitrary dimension. The *hyperbolic space $\mathcal{L}^n$ of an arbitrary dimension n* is the upper sheet of the hyperboloid in $\mathbb{R}^{n+1}$ given by

$$[x, x] \equiv x_1^2 - x_2^2 - \cdots - x_{n+1}^2 = 1, \qquad x_1 > 0, \tag{6.29}$$

in which the motions are the linear transformations of the space $\mathbb{R}^{n+1}$ that preserve this hypersurface. By the *horospheres* we mean the sections of the hyperboloid by the hyperplanes

$$[\xi, x] \equiv \xi_1 x_1 - \cdots - \xi_{n+1} x_{n+1} = 1, \tag{6.30}$$

where ξ is a point of the upper nappe K ($\xi_1 > 0$) of the cone $[\xi, \xi] = 0$. The *horospherical transform* is given by

$$\mathcal{R}^h f(\xi) = \int_{\mathcal{L}^n} f(x)\, \delta([\xi, x] - 1)\, dx, \qquad \xi \in K, \tag{6.31}$$

where $dx = x_1^{-1}\, dx_2 \cdots dx_{n+1}$ is an invariant measure on $\mathcal{L}^n$.

The structure of the inversion formula depends on the parity of n. We present this formula without proof. If $n = 2m + 1$, then

$$f(x) = \frac{(-1)^m}{2\,(2\pi)^{2m}} \int_K \mathcal{R}^h f(\xi)\, \delta^{(2m)}([\xi, x] - 1)\, d\xi, \tag{6.32}$$

where $d\xi = \xi_1^{-1}\, d\xi_2 \cdots d\xi_{n+1}$ is an invariant measure on the cone K. If $n = 2m$, then

$$f(x) = \frac{(-1)^m (2m-1)!}{(2\pi)^{2m}} \int_K \mathcal{R}^h f(\xi)\, ([\xi, x] - 1)^{-2m}\, d\xi. \tag{6.33}$$

Thus, as in the case of Radon transform, the inversion formula is local for odd n and nonlocal for even n.

7. Analog of the Fourier transform in the hyperbolic space, and its relation to the horospherical transform

7.1. Definition of the Fourier transform. We define the Fourier integral for the hyperbolic space $\mathcal{L}^3$ in the same way as in the case of hyperbolic plane. Consider the Laplace–Beltrami operator Δ on $\mathcal{L}^3$,

$$\Delta = z^2 \left(\frac{\partial^2}{\partial x^2} + \frac{\partial^2}{\partial y^2} + \frac{\partial^2}{\partial y^2} \right) - z \frac{\partial}{\partial z}.$$

It is known that the operator Δ on $L^2(\mathcal{L}^3)$ has continuous Lebesgue spectrum coinciding with the interval $(-\infty, -1)$. Let Z be the unipotent subgroup of complex matrices of the form $\left(\begin{smallmatrix} 1 & 0 \\ t & 1 \end{smallmatrix}\right)$. By the *Fourier integral on* $\mathcal{L}^3$ we mean the expansion of a function $f \in L^2(\mathcal{L}^3)$ in certain special functions related to the operator Δ, namely, in the eigenfunctions that are invariant with respect to the subgroup Z, and their shifts. Every eigenspace of the operator Δ corresponds to an eigenvalue of the form $-\frac{4+s^2}{4}$ ($s \in \mathbb{R}$) and contains two functions invariant with respect to the subgroup Z, namely, the functions $z^{\frac{2 \pm is}{2}}$. Take one of them, say, $z^{\frac{2+is}{2}}$; any translation sends it to a function of the form

$$\Phi_s(w; \xi) = (z|w\xi_2 - \xi_1|^{-2})^{\frac{2+is}{2}} = \exp\left(-\frac{2+is}{2} \rho(w; \xi)\right), \qquad \xi \in \mathbb{C}^2, \tag{7.1}$$

where $\rho(w; \xi)$ is the distance from the point $w = x + yi + zj$ to the horosphere $h(\xi)$. We call the functions $z^{\frac{2 \pm is}{2}}$ the *zonal horospherical functions,* and their shifts (7.1) are simply referred to as *horospherical functions.* If ξ and ξ' are proportional, then the corresponding functions Φ_s differ by factors only, and for $s > 0$, there are no other relations among the functions Φ_s. If s ranges over the positive reals and ξ over a set containing exactly one representative of every complex line in $\mathbb{C}^2 \setminus O$

passing through the origin O, then these functions form a complete set of functions in which every function in $L^2(\mathcal{L}^3)$ can be expanded.

DEFINITION. The *Fourier transform* of a smooth compactly supported function f on the hyperbolic space $\mathcal{L}^3$ is defined by the relation

$$\mathcal{F} f(\xi; s) = \int_{\mathcal{L}^3} f(w)\Phi_{-s}(w;\xi)\frac{dx\,dy\,dz}{z^3} \tag{7.2}$$

(cf. the definition of $\mathcal{F} f$ for $\mathcal{L}^2$ in (3.6)).

Since Φ_s is homogeneous with respect to ξ, it follows that $\mathcal{F} f$ is also a homogeneous function of ξ,

$$\mathcal{F} f(\lambda\xi; s) = |\lambda|^{is-2}\,\mathcal{F} f(\xi; s). \tag{7.3}$$

7.2. Inversion formula. The completeness of the family of horospherical functions Φ_s means that an inversion formula for the Fourier transform $\mathcal{F}$ must exist. We construct this formula by repeating the argument carried out in the case of hyperbolic plane.

First, as in the case of hyperbolic plane, there is a simple relation between the Fourier transform $\mathcal{F} f$ and the horospherical transform $\mathcal{R}^h f$, namely,

$$\mathcal{F} f(\xi; s) = \int_{-\infty}^{+\infty} \mathcal{R}^h f(e^{-\lambda/2}\xi)e^{\frac{is-2}{2}\lambda}\,d\lambda. \tag{7.4}$$

Hence, by the inversion formula for the one-dimensional Fourier transform we obtain

$$\mathcal{R}^h f(\xi) = (2\pi)^{-1}\int_{-\infty}^{+\infty} \mathcal{F} f(\xi; s)\,ds. \tag{7.5}$$

The inversion formula for the Fourier transform $\mathcal{F}$ can be obtained from the inversion formula (6.21) for the horospherical transform $\mathcal{R}^h$ by substituting the expression for $\mathcal{R}^h f$ in terms of $\mathcal{F} f$ (see (7.5)) into (6.21). This gives

$$f(w) = -\frac{1}{16\pi^4}\int_{\mathbb{C}^2}\int_{-\infty}^{+\infty} L^2\,\mathcal{F} f(\xi; s)\,\delta(\rho(w;\xi))\,ds\,dv, \tag{7.6}$$

where

$$L = \xi_1\frac{\partial}{\partial\xi_1} + \xi_2\frac{\partial}{\partial\xi_2} + \overline{\xi_1}\frac{\partial}{\partial\overline{\xi_1}} + \overline{\xi_2}\frac{\partial}{\partial\overline{\xi_2}} + 2.$$

We simplify this expression. It follows from the homogeneity condition (7.3) for $\mathcal{F} f$ that

$$L^2\,\mathcal{F} f(\xi; s) = -s^2\,\mathcal{F} f(\xi; s).$$

Further, we pass in (7.6) from ξ to the new coordinates $\xi = \lambda\xi'$, where $\lambda \in \mathbb{C}\setminus 0$ and ξ' ranges over a surface $\Gamma \subset \mathbb{C}^2\setminus O$ intersecting once almost every complex line passing through the origin O. We obtain

$$f(w) = \frac{1}{16\pi^4}\left(\frac{i}{2}\right)^2\int_{-\infty}^{+\infty} s^2\Big(\int_\Gamma\int_{\mathbb{C}} |\lambda|^{is}\,\mathcal{F} f(\xi; s)\,\delta(2\log|\lambda| + \rho(w;\xi))$$
$$\times\, d\lambda\,d\overline{\lambda}(\xi_1\,d\xi_2 - \xi_2\,d\xi_1)(\overline{\xi_1}\,d\overline{\xi_2} - \overline{\xi_2}\,d\overline{\xi_1})\Big)\,ds.$$

Integrating with respect to λ, we finally obtain the following assertion.

THEOREM 7.1. *A smooth compactly supported function f on $\mathcal{L}^3$ can be expressed in terms of the Fourier transform $\mathcal{F} f$ by the following inversion formula:*

(7.7)

$$f(w) = \frac{1}{16\pi^3}\frac{i}{2}\int_{-\infty}^{+\infty}\int_{\Gamma} s^2\,\mathcal{F} f(\xi;s) \times \exp\left(-\frac{2+is}{2}\rho(w;\xi)\right)(\xi_1\,d\xi_2-\xi_2\,d\xi_1)(\overline{\xi_1}\,d\overline{\xi_2}-\overline{\xi_2}\,d\overline{\xi_1})\,ds$$

$$= \frac{1}{16\pi^3}\frac{i}{2}\int_{-\infty}^{+\infty}\int_{\Gamma} s^2\,\mathcal{F} f(\xi;s)\Phi_s(w;\xi)(\xi_1\,d\xi_2-\xi-2\,d\xi_1)(\overline{\xi_1}\,d\overline{\xi_2}-\overline{\xi_2}\,d\overline{\xi_1})\,ds.$$

7.3. Symmetry relation and the Plancherel formula. Similarly to the case of hyperbolic plane, the shifts of the zonal horospherical functions $z^{\frac{2+is}{2}}$ and $z^{\frac{2-is}{2}}$ can be expressed in terms of each other,

$$z^{\frac{2-is}{2}} = c(s)\frac{i}{2}\int_{\mathbb{C}}(z|w-\xi_1|^{-2})^{\frac{2+is}{2}}\,d\xi_1\,d\overline{\xi_1},$$

where $c(s)=\frac{is}{2\pi}$. This implies the following symmetry relation for $\mathcal{F} f$: *The integral*

$$\int_{\Gamma}\mathcal{F} f(\xi;s)\Phi_s(w;\xi)(\xi_1\,d\xi_2-\xi_2\,d\xi_1)(\overline{\xi_1}\,d\overline{\xi_2}-\overline{\xi_2}\,d\overline{\xi_1})$$

is an even function of s. By this relation, the function f on $\mathcal{L}^3$ can be expressed in terms of the values of the function $\mathcal{F} f(\xi;s)$ for $s>0$ only,

$$f(w) = \frac{1}{8\pi^3}\frac{i}{2}\int_0^{\infty}\int_{\Gamma} s^2\,\mathcal{F} f(\xi;s)\Phi_s(\xi;s)(\xi_1\,d\xi_2-\xi_2\,d\xi_1)(\overline{\xi_1}\,d\overline{\xi_2}-\overline{\xi_2}\,d\overline{\xi_1})\,ds. \tag{7.8}$$

REMARK. As in the case of hyperbolic plane, the symmetry relation for $\mathcal{F} f$ can be obtained from the symmetry relation (6.24) for $\mathcal{R}^h f$.

The next assertion follows from the definition of the Fourier transform $\mathcal{F} f$ and from the inversion formula (7.5).

THEOREM 7.2. *The following* Plancherel formula *holds for any smooth compactly supported function f on $\mathcal{L}^3$:*

$$\int_{\mathcal{L}^3}|f(w)|^2\frac{dxdydz}{z^3} = \frac{1}{8\pi^3}\frac{i}{2}\int_0^{\infty} s^2\left(\int_{\Gamma}|\mathcal{F} f(\xi;s)|^2(\xi_1\,d\xi_2-\xi_2\,d\xi_1)(\overline{\xi_1}\,d\overline{\xi_2}-\overline{\xi_2}\,d\overline{\xi_1})\right)ds \tag{7.9}$$

(cf. the Plancherel formula (3.14) for $\mathcal{L}^2$). Using the Plancherel formula, one can extend the Fourier transform $\mathcal{F} f$ to all functions $f \in L^2(\mathcal{L}^2)$ by the classical scheme.

8. Relation to the representation theory for the group $SL(2,\mathbb{C})$

By analogy with the case of hyperbolic plane, we establish a relation between the results of § 7 and the representation theory for the group $SL(2,\mathbb{C})$.

We have the Hilbert space $L^2(\mathcal{L}^3)$ and a unitary representation T of the group $SL(2,\mathbb{C})$ in this space,

$$T_g f(w) = f(w\circ g), \qquad g\in SL(2,\mathbb{C}),$$

where $w \mapsto w \circ g$ is the motion (of the hyperbolic space) corresponding to g. The problem of representation theory (harmonic analysis for the hyperbolic space) is to decompose $L^2(\mathcal{L}^2)$ into the irreducible invariant subspaces.

One must first describe the irreducible invariant subspaces into which the space $L^2(\mathcal{L}^2)$ can be decomposed. Consider the eigenspace of the Laplace–Beltrami operator Δ on $\mathcal{L}^3$ corresponding to an eigenvalue $-\frac{4+s^2}{4}$. This subspace is spanned by the distributions $\Phi_s(w;\xi)$ (see (4.1)) and the elements of the subspace can be written as

$$f_s(w) = \frac{i}{2}\int_\Gamma u_s(\xi)\Phi_s(z;\xi)(\xi_1\,d\xi_2 - \xi_2\,d\xi_1)(\overline{\xi_1}\,d\overline{\xi_2} - \overline{\xi_2}\,d\overline{\xi_1}), \tag{8.1}$$

where u_s satisfies the homogeneity condition

$$u_s(\lambda\xi) = |\lambda|^{is-2}u_s(\xi), \qquad \lambda \in \mathbb{C}\setminus 0, \tag{8.2}$$

and the integral is taken over an arbitrary surface $\Gamma \subset \mathbb{C}^2 \setminus O$ that intersects once almost every complex line passing through the origin O. We equip the space of functions f_s with the norm

$$\|f_s\|^2 = \int_\Gamma |u_s(\xi)|^2(\xi_1\,d\xi_2 - \xi_2\,d\xi_1)(\overline{\xi_1}\,d\overline{\xi_2} - \overline{\xi_2}\,d\overline{\xi_1}), \tag{8.3}$$

consider the space L^2 with this norm, and denote the latter space by H_s. The motions on $\mathcal{L}^3$ define a unitary representation T_s of the group $SL(2,\mathbb{C})$ in H_s. In terms of functions $u_s(\xi)$, this representation is of the form

$$(T_s(g)u_s)(\xi_1,\xi_2) = u_s(\alpha\xi_1 + \gamma\xi_2, \beta\xi_1 + \delta\xi_2), \qquad g = \begin{pmatrix} \alpha & \beta \\ \gamma & \delta \end{pmatrix}.$$

It is known that this representation is irreducible [**8**]. We introduce the operator $P_s : L^2(\mathcal{L}^3) \to H_s$ by the rule

$$P_s f = f_s,$$

where f_s is given by formula (8.1) in which $u_s(\xi) = \mathcal{F} f(\xi)$. This operator commutes with the action of the group $SL(2,\mathbb{C})$ on the spaces $L^2(\mathcal{L}^3)$ and H_s. Finally, using (7.8) and (7.9), we obtain the expansion of a function $f \in L^2(\mathcal{L}^3)$ with respect to the irreducible subspaces,

$$f = \frac{1}{8\pi^3}\int_0^\infty s^2 f_s\,ds, \qquad f_s = P_s f, \tag{8.4}$$

and the Plancherel formula

$$\int_{\mathcal{L}^3} |f(w)|^2 \frac{dx\,dy\,dz}{z^3} = \frac{1}{8\pi^3}\int_0^{+\infty} s^2\|f_s\|^2\,ds, \tag{8.5}$$

where $\|f_s\|$ is given by (8.3).

9. Wave equation for the hyperbolic plane and hyperbolic space, and the Huygens principle

9.1. Two-dimensional case. Recall that the Laplace–Beltrami operator on the hyperbolic plane $\mathcal{L}^2$ is $\Delta^2 = y^2\left(\frac{\partial^2}{\partial x^2} + \frac{\partial^2}{\partial y^2}\right)$, and the spectrum of this operator in $L^2(\mathcal{L}^2)$ is the interval $(-\infty, -\frac{1}{4})$. Consider the following wave equation on $\mathcal{L}^2$:

$$\frac{\partial^2 u}{\partial t^2} = y^2\left(\frac{\partial^2 u}{\partial x^2} + \frac{\partial^2 u}{\partial y^2}\right) + \frac{1}{4}u. \tag{9.1}$$

The term $\frac{1}{4}u$ is added for the spectrum of the operator on the right-hand side to coincide with the half-line $(-\infty, 0)$.

We show how one can solve the following Cauchy problem for equation (9.1):

$$u(0,z) = 0, \qquad u_t'(0,z) = f(z) \tag{9.2}$$

by using the horocycle transform; for simplicity we assume that $f \in \mathcal{D}(\mathcal{L}^2)$. The argument is parallel to the process of solution of the Cauchy problem for the wave equation on Euclidean plane (see § 3 of Chapter 1). Our solution uses the method of horocycle waves based on the following remarkable fact.

PROPOSITION 9.1. *If $\varphi(s)$ is a smooth function of one variable, then the function*

$$\Phi_\xi(t,z) = e^{\pm t/2}\varphi(\rho(z; e^{\pm t/2}\xi)) = e^{\pm t/2}\varphi(\rho(z;\xi) \pm t), \tag{9.3}$$

where $\rho(z;\xi) = \log(y^{-1}|\xi_2 z - \xi_1|^2)$ is the distance from z to the horocycle $h(\xi)$, is a solution of the wave equation (9.1) *for each $\xi \in \mathbb{R}^2$.*

The proof can be obtained by a direct verification.

The functions (9.3) are called *horocycle waves.* The *method of horocycle waves* consists in representing an arbitrary solution of (9.1) as a superposition of horocycle waves.

We now proceed with the construction of a solution of the Cauchy problem (9.2). Let $\mathcal{R}^h f$ be the horocycle transform of a function f and let $\widehat{F}(z;r)$ be the result of averaging the function $\mathcal{R}^h f$ over the horocycles equidistant from a point $z \in \mathcal{L}^2$; see (2.11). Consider the function

$$u(t,z) = -\frac{1}{4\pi}\int_{-\infty}^{+\infty} \frac{e^{\frac{r+t}{2}}\widehat{F}(z,r+t)}{\sinh\frac{r}{2}}\,dr. \tag{9.4}$$

It follows from (2.11) that

$$e^{\frac{r+t}{2}}\widehat{F}(z,r+t) = \frac{1}{\pi}e^{-\frac{r+t}{2}}\int_{\mathbb{R}^2} \mathcal{R}^h f(\xi)\,\delta(\rho(z;\xi) - r - t)\,d\xi_1\,d\xi_2,$$

and thus $u(t,z)$ is a solution of (9.1) as a superposition of horocycle waves. We show that u is the desired solution of the Cauchy problem (9.2). By the symmetry relation (2.16), $r \mapsto e^{r/2}\widehat{F}(z,r)$ is an even function of r; hence,

$$u(0,z) = -\frac{1}{4\pi}\int_{-\infty}^{+\infty} \frac{e^{r/2}\widehat{F}(z;r)}{\sinh\frac{r}{2}}\,dr = 0.$$

Further, by the inversion formula (2.10) for the horocycle transform we have

$$u_t'(0,z) = -\frac{1}{4\pi}\int_{-\infty}^{+\infty} \frac{(e^{r/2}\widehat{F}(z,r))_r'}{\sinh\frac{r}{2}}\,dr = f(z).$$

We want to return in (9.4) to the original function f. To do this, we use the Asgeirsson relation (2.15) for the hyperbolic plane. From (2.15), for $t > 0$ we obtain

$$\begin{aligned} u(t,z) &= -\frac{1}{2\pi}\int_{-\infty}^{+\infty}\int_{2|\sinh\frac{r+t}{2}|}^{\infty} \frac{F(z,s)s\,ds\,dr}{\sinh\frac{r}{2}\sqrt{s^2 - 4\sinh^2\frac{r+t}{2}}} \\ &= -\frac{1}{2\pi}\int_0^{\infty}\int_{2|\sinh\frac{r}{2}|<s} \frac{F(z,s)s\,dr\,ds}{\sinh\frac{r-t}{2}\sqrt{s^2 - 4\sinh^2\frac{r}{2}}}, \end{aligned} \tag{9.5}$$

where $F(z,s)$ is the result of averaging the function f over the non-Euclidean circles in $\mathcal{L}^2$ centered at the point z (see (2.14)); the integral with respect to z must be understood in the principal value sense.

We carry out the integration with respect to r in (9.5) by using the following relation:[5]

$$J \equiv \int_{-b}^{b} \frac{dx}{\sinh(x-a)\sqrt{\sinh^2 b - \sinh^2 x}} = \begin{cases} 0 & \text{for} \quad a < b, \\ \frac{-\pi}{\sqrt{\sinh^2 a - \sinh^2 b}} & \text{for} \quad a > b. \end{cases}$$

Here $a, b > 0$. We obtain

$$u(t,z) = \frac{1}{2} \int_0^{2\sinh\frac{t}{2}} \frac{F(z,s)\,ds}{\sqrt{\sinh^2\frac{t}{2} - s^2/4}}. \tag{9.6}$$

Substituting here the expression for the function $F(z,s)$ in terms of f (see (2.14)), we arrive at the following result.

THEOREM 9.1. *The solution of the Cauchy problem* (9.2) *for the wave equation* (9.1) *is of the form*

$$u(t,z) = \frac{1}{4\pi} \int_{d(z,z')<t} \frac{f(z')}{\sqrt{\sinh^2\frac{t}{2} - \sinh^2\frac{d(z,z')}{2}}} \frac{dx'dy'}{y'^2}, \tag{9.7}$$

where $d(z,z')$ stands for the non-Euclidean distance between the points z and z' (see (1.2)*).*

The formula thus obtained is similar to the Poisson formula for the wave equation on the Euclidean plane; see § 3 of Chapter 1. As in the Euclidean case, the solution of the equation at a point z at a moment t depends only on the initial data in the non-Euclidean disk of radius t, centered at the point z (due to the existence of the forward front of the wave).

9.2. Three-dimensional case. We now pass to the hyperbolic space $\mathcal{L}^3$. Recall that the Laplace–Beltrami operator on $\mathcal{L}^3$ is

$$\Delta = z^2\left(\frac{\partial^2}{\partial x^2} + \frac{\partial^2}{\partial y^2} + \frac{\partial^2}{\partial z^2}\right) - z\frac{\partial}{\partial z},$$

---

[5]**Computation of the integral** J. Obviously, J is an odd function of a. Hence,

$$J = \frac{1}{2}\int_{-b}^{b} \frac{(\sinh^{-1}(x-a) - \sinh^{-1}(x+a))\,dx}{\sqrt{\sinh^2 b - \sinh^2 x}} = \int_{-b}^{b} \frac{\sinh a \cosh x\,dx}{(\sinh^2 x - \sinh^2 a)\sqrt{\sinh^2 b - \sinh^2 x}}.$$

Therefore, by elementary manipulations we get

$$J = \int_{-\sinh b}^{\sinh b} \frac{\sinh a\,dt}{(t^2 - \sinh^2 a)\sqrt{\sinh^2 b - t^2}} = \int_{-1}^{1} \frac{\sinh a\,dt}{(t^2\sinh^2 b - \sinh^2 a)\sqrt{1-t^2}}$$
$$= \sinh a \int_{-\pi/2}^{\pi/2} \frac{dt}{\sinh^2 b \sin^2 t - \sinh^2 a} = -\int_{-\pi/2}^{\pi/2} \frac{dt}{\sinh b \sin t + \sinh a}.$$

The last integral can be obtained directly: P.V. $J = 0$ for $a < b$, and $J = -\frac{\pi}{\sqrt{\sinh^2 a - \sinh^2 b}}$ for $a > b$ (see formula 3.613.1 in [**25**]).

and the spectrum of this operator in $L^2(\mathcal{L}^3)$ is the interval $(-\infty, -1)$. Consider the following wave equation on $\mathcal{L}^3$:

$$\frac{\partial^2 u}{\partial t^2} = z^2\left(\frac{\partial^2 u}{\partial x^2} + \frac{\partial^2 u}{\partial y^2} + \frac{\partial^2 u}{\partial z^2}\right) - z\frac{\partial u}{\partial z} + u. \tag{9.8}$$

As in the case of $\mathcal{L}^2$, the term u is added for the spectrum of the operator on the right-hand side to coincide with the half-line $(-\infty, 0)$. We solve the following Cauchy problem for (9.8) by the method of horospherical waves:

$$u(0,w) = 0, \qquad u'_t(0,w) = f(w) \qquad (w = x + ui + zj). \tag{9.9}$$

The method uses the following property.

PROPOSITION 9.2. *If $\varphi(s)$ is a smooth function of one variable, then the function*

$$\Phi_\xi(t,w) = e^{\pm t}\varphi(\rho(w;\xi) \pm t), \tag{9.10}$$

where $\rho(w;\xi) = \log(z^{-1}|w\xi_2 - \xi_1|^2)$ is the distance from a point w to a horosphere $h(\xi)$, is a solution of the wave equation (9.8) for each $\xi \in \mathbb{C}^2$.

The proof reduces to a direct verification.

We refer to the solutions of the form (9.10) as *horospherical waves.*

Let us proceed with solving the Cauchy problem (9.9). Let $\mathcal{R}^h f$ be the horospherical transform of the function f and let $\widehat{F}(w;r)$ be the result of averaging the function $\mathcal{R}^h f$ over the horospheres equidistant from the point $w \in \mathcal{L}^3$; see (6.20). Consider the function

$$u(t,w) = -\frac{1}{2\pi}\frac{d}{dt}(e^t\widehat{F}(w,t)). \tag{9.11}$$

It follows from (6.20) that

$$e^t\widehat{F}(w,t) = \pi^{-2}e^{-t}\int_{\mathbb{C}^2}\mathcal{R}^h f(\xi)\,\delta(\rho(w;\xi))\,dv.$$

Since the function $u(t,w)$ is a superposition of horospherical waves, it is a solution of the wave equation (9.8). We prove that u is the desired solution of the Cauchy problem (9.9). It follows from the symmetry relation (6.23) that $e^t\widehat{F}(w,t)$ is an even function of t; hence,

$$u(0,w) = \frac{d}{dt}(e^t\widehat{F}(w,t))|_{t=0} = 0.$$

Further, using the inversion formula (6.19), we see for the horospherical transform that

$$u'_t(0,w) = -\frac{1}{2\pi}\frac{d^2}{dt^2}(e^t\widehat{F}(w,t))|_{t=0} = f(w).$$

We wish to return in (9.11) to the original function f. To do this, we use the Asgeirsson relation (6.22) for the hyperbolic space. From (6.22), for $t > 0$ we obtain

$$u(t,w) = \sinh t\, F(w, 2\sinh\frac{t}{2}),$$

where $F(w,s)$ is the result of averaging the function f over the spheres in $\mathcal{L}^3$ centered at the point w; see (6.23). Using (6.23), we obtain the following result.

THEOREM 9.2. *The solution of the Cauchy problem* (9.9) *for the wave equation* (9.8) *is of the form*

$$\begin{aligned}u(t,w) &= \frac{1}{8\pi}\sinh t\int_{\mathcal{L}^3}\sinh^{-1}\frac{d(w,w')}{2}\\&\qquad\times\delta\left(4\sinh^2\frac{d(w,w')}{2}-4\sinh^2\frac{t}{2}\right)f(w')\frac{dx'dy'dz'}{z'^3}\\&=\frac{1}{16\pi}\tanh^{-1}\frac{t}{2}\int_{\mathcal{L}^3}f(w')\delta\left(\sinh\frac{d(w,w')}{2}-\sinh\frac{t}{2}\right)\frac{dx'dy'dz'}{z'^3}.\end{aligned}$$

The formula thus obtained is similar to the Kirchhoff formula in § 3 of Chapter 1 for the wave equation in Euclidean space. As in the Euclidean case, the value of the solution of the wave equation at a point w at a moment t depends only on the initial data at the points w' that are at the non-Euclidean distance t from the point w (the Huygens principle).

CHAPTER 4

Integral Geometry and Harmonic Analysis on the Group $G = SL(2,\mathbb{C})$

In Chapter 3 we constructed integral geometry and harmonic analysis on the hyperbolic space equipped with the natural action of the group $G = SL(2,\mathbb{C})$, i.e., of the group of complex matrices $g = \begin{pmatrix} \alpha & \beta \\ \gamma & \delta \end{pmatrix}$ with unit determinant, $\alpha\delta - \beta\gamma = 1$. In this chapter we present similar constructions on the group G itself. The theory thus obtained naturally includes the theory constructed in Chapter 3 for the hyperbolic space. We regard the group G as a homogeneous space of the group $G \times G$, and from this point of view, we can simply treat the group G as the hyperboloid $\alpha\delta - \beta\gamma = 1$ in $\mathbb{C}^4$ equipped with the action of the subgroup of linear transformations of $\mathbb{C}^4$ that preserve this hyperboloid.

1. Geometry on the group G

1.1. Group G as a homogeneous space. We regard G as a homogeneous space of the group $G \times G$ acting on G by left and right translations,

$$g \mapsto g_1^{-1} g g_2, \quad (g_1, g_2) \in G \times G.$$

Since the isotropy subgroup of the identity element $e \in G$ is the diagonal subgroup, $\operatorname{diag}(G \times G) \cong G$, it follows that this homogeneous space is isomorphic to $G \times G/G$, where $G \subset G \times G$ is represented as the subgroup of elements (g, g), $g \in G$.

REMARK. The homogeneous space G is symmetric because the isotropy subgroup $\operatorname{diag}(G \times G)$ is the subgroup of elements invariant under the involution $\theta\,(g_1, g_2) \mapsto (g_2, g_1)$. The space G is not Riemannian, but it is pseudo-Riemannian, i.e., the metric on this space is nondegenerate but not positive definite. This fact is of importance because integral geometry and harmonic analysis for symmetric spaces can be constructed in a more complete form than those for arbitrary homogeneous spaces.

Note that the action of $G \times G$ on G is not effective, and the ineffective kernel, i.e., the subgroup preserving all elements of G, consists of two elements (E, E) and $-(E, E)$. Therefore, the quotient group $G \times G/\{\pm(E, E)\}$ effectively acts on G.

As was already said, in what follows, when considering G, we ignore the group structure and view G only as the hyperboloid in $\mathbb{C}^4$ with coordinates $(\alpha, \beta, \gamma, \delta)$ such that

$$\alpha\delta - \beta\gamma = 1.$$

The group of linear transformations of $\mathbb{C}^3$ that preserve the quadratic form $\alpha\delta - \beta\gamma$, i.e., the group $SO(4,\mathbb{C})$, acts on G effectively and transitively. The transformations corresponding to the elements of $G \times G$ obviously preserve the form $\alpha\delta - \beta\gamma$, and we obtain an embedding of $G \times G/\{\pm(E, E)\}$ in $SO(4, \mathbb{C})$. It follows from the dimension considerations (the dimensions of these groups coincide) that we have in fact an isomorphism

$$SL(2, \mathbb{C}) \times SL(2, \mathbb{C})/\{\pm(E, E)\} \cong SO(4, \mathbb{C}).$$

Under the action of $SO(4, \mathbb{C})$ on G, the isotropy subgroup coincides with $SO(3, \mathbb{C})$, and another classical isomorphism arises,

$$SO(3, \mathbb{C}) \cong SL(2, \mathbb{C})/\{\pm E\}.$$

1.2. Plane sections of the hyperboloid G. We intend to relate the action of the group $G \times G$ on the hyperboloid G to the geometry of this hyperboloid. Namely, we study the sections of G by hyperplanes in $\mathbb{C}^4$ as orbits of some subgroups of $G \times G$.

It is convenient to present hyperplanes in $\mathbb{C}^4$ by equations of the form

$$\langle g_0, g \rangle = c, \qquad g_0 = (\alpha_0, \beta_0, \gamma_0, \delta_0), \tag{1.1}$$

where $\langle\ ,\ \rangle$ is the polarization of the quadratic form $\alpha\delta - \beta\gamma$.

In the detailed notation,

$$\frac{1}{2}(\delta_0\alpha - \gamma_0\beta - \beta_0\gamma + \alpha_0\delta) = c.$$

Thus, each hyperplane is given by a pair g_0, c, where $g_0 \in \mathbb{C}^4 \setminus O$ and $c \in \mathbb{C}$, and this pair is defined up to a common factor. The group $G \times G$ naturally acts on the set of hyperplanes; namely, an element $(g_1, g_2) \in G \times G$ sends a hyperplane $\langle g_0, g \rangle = c$ to the hyperplane $\langle g_1^{-1} g_0 g_2, g \rangle = c$.

We first describe the orbits of $G \times G$ in the set of hyperplanes. In general position ($\langle g_0, g_0 \rangle \neq 0$) one can normalize g_0 by the condition $\langle g_0, g_0 \rangle = 1$. Obviously, for any chosen number c, the hyperplanes (1.1) with $\langle g_0, g_0 \rangle = 1$ form an orbit of $G \times G$ isomorphic to the original homogeneous space G. The remaining subset of hyperplanes (for which $\langle g_0, g_0 \rangle = 0$) is partitioned into two orbits of $G \times G$, namely, the family of hyperplanes

$$\langle g_0, g \rangle = 0, \qquad \langle g_0, g_0 \rangle = 0, \tag{1.2}$$

tangent to the asymptotic cone K given by $\alpha\delta - \beta\gamma = 0$, and the family of hyperplanes

$$\langle g_0, g \rangle = 1, \qquad \langle g_0, g_0 \rangle = 0, \tag{1.3}$$

parallel to the previous hyperplanes. Obviously, the second orbit is isomorphic (as a homogeneous space of the group $G \times G$) to the asymptotic cone K with deleted vertex; the isotropy subgroup of this orbit consists of the pairs $(\delta z_1, \delta^{-1} z_2)$, where z_1 and z_2 are matrices of the form $\left(\begin{smallmatrix} 1 & t \\ 0 & 1 \end{smallmatrix}\right)$ and δ is a diagonal matrix. The first orbit is isomorphic to the set of generatrices of the asymptotic cone K and has a stabilizer which is the subgroup of all pairs $(\delta_1 z_1, \delta_2 z_2)$ with z_1 and z_2 as above and diagonal δ_1 and δ_2.

Thus, the set of plane sections of the hyperboloid G is decomposed into the following orbits of the group $G \times G$: the one-parameter family of orbits isomorphic to the homogeneous space G, the orbit isomorphic to the asymptotic cone K with deleted vertex, and the orbit isomorphic to the set of generatrices of the cone K.

We now study the structure of plane sections of the hyperboloid G. It follows from the invariance considerations that it suffices to consider a single representative on every orbit of $G \times G$ in the set of plane sections. For the representatives on the orbits in general position we take the sections of G by hyperplanes of the form

$$\alpha + \delta = c. \tag{1.4}$$

If $c \neq \pm 2$, then the hyperplane (1.4) intersects G along a two-dimensional hyperboloid. The hyperplane (1.4) is tangent to G in the special case $c = \pm 2$, and the intersection of this hyperplane with G is a cone with the vertex at the tangency point.

The sections of G by the hyperplanes (1.4) have a simple group-theoretic sense. Namely, since $\alpha + \delta = \operatorname{Tr} g$ for $g = \left(\begin{smallmatrix} \alpha & \beta \\ \gamma & \delta \end{smallmatrix}\right)$, the section of G by a hyperplane $\alpha + \delta = c$ is a conjugacy class; if $c \neq \pm 2$, then the conjugacy class is generic, and for $c = \pm 2$, the conjugacy class consists of elements of Jordan type (the vertex of the cone is assumed to be deleted).

For representatives of the two special orbits of $G \times G$ one can take the sections of G by the hyperplanes $\gamma = 0$ and $\gamma = 1$, respectively. Obviously, the first section is a cylinder and the other is a paraboloid. Thus, there are four types of sections of G by hyperplanes in $\mathbb{C}^4$, namely, hyperboloids, cones, paraboloids, and cylinders. Naturally, the sections of parabolic and cylindrical types are obtained by passing to the limit from sections of hyperbolic and conic types.

Every plane section of the hyperboloid G can be equipped with the structure of a homogeneous space. Let us first consider the sections by planes of the form $\alpha + \delta = c$. The subgroup $\operatorname{diag}(G \times G) \cong G$ transitively acts on each of these sections (in the case of conic section it is assumed that the vertex of the cone is deleted); thus, all these sections, and hence all their shifts as well, are homogeneous spaces of G. One can readily see that the group D of diagonal matrices is a stabilizer of any section of hyperbolic type, and the group $\pm Z$, where Z is the subgroup of unipotent matrices $\left(\begin{smallmatrix} 1 & t \\ 0 & 1 \end{smallmatrix}\right)$, is a stabilizer of every section of conic type.

We now consider the sections of G by the hyperplanes $\gamma = 1$ and $\gamma = 0$. The subgroup $Z \times Z$ acts freely and transitively on the first section; thus, this section is equipped with the structure of the homogeneous space $Z \times Z/\{E\}$. The group $Z \times Z$ acts on the other section as well, but this action is no longer transitive. One can readily see that the subgroup $U \subset Z \times Z$ of all pairs $(\delta z_1, \delta^{-1} z_2)$, where $\delta \in D$ and $z_1, z_2 \in Z$, acts on the section transitively, and the subgroup $\operatorname{diag}(\pm Z, \pm Z)$ of pairs $(\pm z, \pm z)$, $z \in Z$, which is isomorphic to $\pm Z$, is a stabilizer of this section.

Thus, the sections of G of hyperbolic, conic, parabolic, and cylindrical type are equipped with the structures of the homogeneous spaces G/D, $G/\pm Z$, $Z \times Z/\{E\}$, and $U/\operatorname{diag}(\pm Z, \pm Z)$, respectively.

1.3. Manifold of horospheres. Consider the set of generatrices of the hyperboloid G. In our theory these lines play the role similar to that of horospheres in the hyperbolic space, and we refer to these objects as horospheres by analogy with the hyperbolic space. Note that the horospheres are the generatrices of all plane sections of the hyperboloid G.

The group $G \times G$ naturally acts on the set of horospheres H, and this action is transitive (to prove this assertion, it suffices to consider the set of horospheres passing through an arbitrarily chosen point $g \in G$ and to show that the stabilizer of this point transitively acts on this set of horospheres).

The horospheres on G have a simple group-theoretic interpretation. Namely, the subgroup $Z \subset G$ of matrices of the form $\left(\begin{smallmatrix} 1 & t \\ 0 & 1 \end{smallmatrix}\right)$ is one of the horospheres. Since all horospheres are obtained from this horosphere by translations, it follows that the set of horospheres H coincides with the set of double cosets $g_1^{-1}Zg_2$. Therefore,

$$H \cong (G \times G)/V,$$

where V is the subgroup preserving the horosphere Z. One can readily see that the elements of V are all possible pairs $(\delta z_1, \delta z_2)$, where $z_1, z_2 \in Z$, $\delta \in D$. Since we obviously have $\dim V = 3$, it follows that $\dim H = 3$, i.e., the dimension of the space of horospheres is equal to that of G.

We introduce a coordinate system on H. To this end, we first consider the set H_0 of the generatrices of the asymptotic cone K given by $\alpha\delta - \beta\gamma = 0$. The cone K admits two one-parameter families of two-dimensional planes, namely, the family

$$\begin{aligned} \xi_1\alpha + \xi_2\gamma &= 0, \\ \xi_1\beta + \xi_2\delta &= 0 \end{aligned} \tag{1.5}$$

and the family

$$\begin{aligned} \eta_1\delta - \eta_2\gamma &= 0, \\ -\eta_1\beta + \eta_2\alpha &= 0. \end{aligned} \tag{1.6}$$

In the matrix notation, this gives the family of planes $\xi g = 0$ and the family of planes $g\eta^\perp = 0$, where $\eta^\perp = \binom{\eta_2}{-\eta_1}$.

Every generatrix of the cone K can be obtained as the intersection of a pair of planes (1.5), (1.6) and is thus given by a pair of vectors $\xi, \eta \in \mathbb{C}^2 \setminus O$, each defined up to a factor; we denote this generatrix by $l(\xi, \eta)$.

The manifold of horospheres H forms a natural bundle over H_0, with the fiber over $l(\xi, \eta)$ being the set of all horospheres parallel to $l(\xi, \eta)$. We claim that every horosphere parallel to $l(\xi, \eta)$ is given in $\mathbb{C}^4$ by the system of equations

$$\begin{aligned} \xi_1'\alpha + \xi_2'\gamma &= \eta_1', \\ \xi_1'\beta + \xi_2'\delta &= \eta_2', \\ \eta_1'\delta - \eta_2'\gamma &= \xi_1', \\ -\eta_1'\beta + \eta_2'\alpha &= \xi_2', \end{aligned} \tag{1.7}$$

where $\xi', \eta' \in \mathbb{C}^2 \setminus O$ are vectors proportional to ξ and η, respectively.

Indeed, consider the first two and the last two equations of (1.7). On G we can represent them in the form $\xi' g = \eta'$, $\eta' g^{-1} = \xi'$; hence, these two systems of equations are equivalent on G. Hence, system (1.7) defines a line in $\mathbb{C}^4$, and this line belongs to G. Further, it is clear that this line is parallel to $l(\xi, \eta)$.

REMARK. The horospheres parallel to the line $l(\xi, \eta)$ are generatrices of the cylinder obtained as the section of G by the hyperplane which is tangent to the asymptotic cone K along the line $l(\xi, \eta)$. The equation of this hyperplane is

$$\eta_2(\xi_1\alpha + \xi_2\gamma) - \eta_1(\xi_1\beta + \xi_2\delta) = 0, \tag{1.8}$$

or $\xi g \eta^\perp = 0$ in the matrix form.

Thus, every horosphere on G can be given by a system of equations $\xi g = \eta$, i.e.,

$$\begin{aligned} \xi_1\alpha + \xi_2\gamma &= \eta_1, \\ \xi_1\beta + \xi_2\delta &= \eta_2, \end{aligned}$$

where $\xi, \eta \in \mathbb{C}^2 \setminus O$, or by the equivalent system $\eta g^{-1} = \xi$, i.e.,

$$\eta_1\delta - \eta_2\gamma = \xi_1,$$
$$-\eta_1\beta + \eta_2\alpha = \xi_2.$$

Denote this horosphere by $h(\xi, \eta)$ and regard ξ, η as parameters on the manifold of horospheres H. Since we obviously have $h(\lambda\xi, \lambda\eta) = h(\xi, \eta)$ for each $\lambda \neq 0$, it is natural to refer to ξ, η as *homogeneous coordinates* on H.

The action of an element $(g_1, g_2) \in G \times G$ on H is given in the homogeneous coordinates ξ, η as follows:

$$h(\xi, \eta) \mapsto h(\xi g_1, \eta g_2).$$

Note that the horosphere Z has the homogeneous coordinates $\xi_0 = \eta_0 = (0, 1)$; hence, every horosphere $g_1^{-1} Z g_2$ has the homogeneous coordinates $\xi = \xi_0 g_1$, $\eta = \xi_0 g_2$.

It is of prime importance in what follows that, along with the action of the group $G \times G$, the manifold of horospheres H is equipped with an action of $\mathbb{C}^\times$; namely, corresponding to each number $\lambda \in \mathbb{C}^\times$ is the transformation

$$h(\xi, \eta) \mapsto h(\xi, \lambda\eta)$$

which takes every horosphere to a parallel horosphere. These transformations are called *left translations on* H. Obviously, they commute with the action of the group $G \times G$.

We now describe the families of generatrices (horospheres) on the plane sections Γ of the hyperboloid G in the homogeneous coordinates ξ, η.

By virtue of the aforesaid, each cylindrical section Γ is the intersection of G with a hyperplane of the form (1.8), and the horospheres on Γ are $h(\xi, \lambda\eta)$, where λ ranges over $\mathbb{C}^\times$.

Consider now the plane sections Γ of the other types. For each of these sections, the equation of the secant hyperplane can be written as

$$2\langle g_0, g\rangle = 1 + \langle g_0, g_0\rangle. \tag{1.9}$$

Hence, since $\langle g, g\rangle = 1$ on G, it follows that the section Γ itself is defined on G by the equation

$$\langle g - g_0, g - g_0\rangle = 0,$$

i.e.,

$$(\alpha - \alpha_0)(\delta - \delta_0) - (\beta - \beta_0)(\gamma - \gamma_0) = 0. \tag{1.10}$$

This section is a cone, a paraboloid, or a hyperboloid if $\det g_0 = 1$, $\det g_0 = 0$, or $\det g_0 \neq 0, 1$, respectively.

The equations of generatrices (horospheres) on Γ immediately follow from (1.10): the horospheres of one family are given by the equations

$$\xi_1(\alpha - \alpha_0) + \xi_2(\gamma - \gamma_0) = 0,$$
$$\xi_1(\beta - \beta_0) + \xi_2(\delta - \delta_0) = 0,$$

i.e., $\xi g = \xi g_0$, and the horospheres of the other family by the equations

$$\eta_1(\delta - \delta_0) - \eta_2(\gamma - \gamma_0) = 0,$$
$$-\eta_1(\beta - \beta_0) + \eta_2(\alpha - \alpha_0) = 0,$$

i.e., $\eta g^{-1} = \eta \widehat{g}_0$, where $\widehat{g}_0 = \begin{pmatrix} \delta_0 & -\beta_0 \\ -\gamma_0 & \alpha_0 \end{pmatrix}$. Thus, the following assertion is proved.

PROPOSITION 1.1. *The families of horospheres belonging to the section of the hyperboloid G by a hyperplane* (1.10) *are the family of horospheres $h(\xi, \xi g_0)$ and the family of horospheres $h(\eta \widehat{g_0}, \eta)$, where $\widehat{g_0} = \begin{pmatrix} \delta_0 & -\beta_0 \\ -\gamma_0 & \alpha_0 \end{pmatrix}$ and ξ, η range over the set $\mathbb{C}^2 \setminus O$.*

COROLLARY 1. *For any chosen $g_0 \in G$ and $\lambda \neq 0$, the families of horospheres belonging to the section of G by the hyperplane*

$$2\lambda\langle g_0, g\rangle = 1 + \lambda^2$$

are formed by the horospheres $h(\xi, \lambda\xi g_0)$ and $h(\xi, \lambda^{-1}\xi g_0)$, where ξ ranges over $\mathbb{C}^2 \setminus O$.

COROLLARY 2. *The horospheres passing through an arbitrarily chosen point $g_0 \in G$ are the horospheres $h(\xi, \xi g_0)$, where ξ ranges over $\mathbb{C}^2 \setminus O$.*

1.4. Embedding the manifold of horospheres H in the projective space. Assigning to each horosphere $h(\xi, \eta)$ the point of the projective space $\mathbb{CP}^3$ with the homogeneous coordinates $(\xi_1, \xi_2, \eta_1, \eta_2)$, we obtain an embedding

$$H \hookrightarrow \mathbb{CP}^3.$$

The image of H under this embedding is the entire space $\mathbb{CP}^3$ from which a pair of disjoint lines is deleted, namely, the line $\xi = 0$ and the line $\eta = 0$. Thus, $\mathbb{CP}^3$ is a natural compactification of the manifold of horospheres H.

The points at infinity on H have the following simple meaning. The set of points at infinity on G is a two-dimensional hyperboloid having two families of generatrices. Corresponding to these generatrices are the points of the lines $\xi = 0$ and $\eta = 0$ in $\mathbb{CP}^3$, i.e., the points at infinity on H.

REMARK. We explain how this construction must be modified when passing from the group $G = SL(2, \mathbb{C})$ to its quotient group by the center $PSL(2, \mathbb{C}) = G/\{\pm E\}$, i.e., when antipodal points of the hyperboloid $\alpha\delta - \beta\gamma = 1$ are identified. The manifold $\widetilde{G} = G/\{\pm E\}$ can be embedded in the projective space $\mathbb{CP}^3$ by assigning to each matrix $\begin{pmatrix} \alpha & \beta \\ \gamma & \delta \end{pmatrix}$ the point of $\mathbb{CP}^3$ with the homogeneous coordinates $(\alpha, \beta, \gamma, \delta)$. The image of $\widetilde{G}$ under this embedding is the entire projective space $\mathbb{CP}^3$ except for the quadric Γ given by the equation $\alpha\delta - \beta\gamma = 0$; the points of this quadric can be treated as the points at infinity on $\widetilde{G}$. One can readily see that the horospheres on $\widetilde{G}$ are taken by this embedding to all possible lines in $\mathbb{CP}^3$ tangent to the quadric Γ. The equation of each of these lines can be written in the following form:

$$\frac{\xi_1\alpha + \xi_2\gamma}{\eta_1} = \frac{\xi_1\beta + \xi_2\delta}{\eta_2} = \frac{\eta_1\delta - \eta_2\gamma}{\xi_1} = \frac{-\eta_1\beta + \eta_2\alpha}{\xi_2}, \tag{1.11}$$

where $\xi \neq 0$ and $\eta \neq 0$ (the rank of this system is equal to 2).

A compactification of this manifold of lines can be obtained by adding all generatrices of the quadric Γ; these generatrices are defined by (1.11) with either $\xi = 0$ or $\eta = 0$ (two families of generatrices). The resulting variety is isomorphic to the projective space $\mathbb{CP}^3$; namely, corresponding to each line (1.11) is a point of $\mathbb{CP}^3$ with the homogeneous coordinates $(\xi_1, \xi_2, \eta_1, \eta_2)$.

1.5. Line complex in $\mathbb{C}^3$ associated with the manifold of horospheres. It is convenient to pass from the manifold of lines (horospheres) on the hyperboloid G to the manifold of lines in $\mathbb{C}^3$. To do this, we use the projection of the hyperboloid $\alpha\delta - \beta\gamma = 1$ to the coordinate subspace $\mathbb{C}^3$ with the coordinates (α, β, δ). The image of the hyperboloid is the entire space $\mathbb{C}^3$ except for the points $(\alpha, 0, \delta)$ such that $\alpha\delta \neq 1$.

When projecting the hyperboloid G to $\mathbb{C}^3$, the generatrices of G that are not parallel to the hyperplane $\beta = 0$ are taken to the lines intersecting the hyperbola $\beta = 0$, $\alpha\delta = 1$ and not belonging to the plane $\beta = 0$. Thus, with the manifold of horospheres on the hyperboloid G we associated a three-dimensional manifold K of lines in $\mathbb{C}^3$. In differential geometry, three-dimensional manifolds of lines in $\mathbb{C}^3$ are usually called *line complexes.*

REMARK. Only one point of the hyperboloid lies over any point $(\alpha, \beta, \delta) \in \mathbb{C}^3$ with $\beta \neq 0$. On the other hand, over any point such that $\beta = 0$ and $\alpha\delta = 1$, lies the entire line. Thus, the hyperboloid is obtained by pasting lines, by means of a blow-up process, in place of points of the hyperbola given by the equations $\beta = 0$, $\alpha\delta = 1$. The lines of our complex are just the lines that can be lifted to the hyperboloid as a result of this blow-up process.

1.6. Manifold of paraboloids. Now we describe another version of horospheres in G. Consider paraboloids in G, i.e., two-dimensional surfaces cut out in G by isotropic hyperplanes in $\mathbb{C}^4$ not passing through O,

$$\langle g_0, g\rangle = p, \qquad g \in G,$$

where g_0 is a nonzero point of the asymptotic cone

$$\langle g_0, g_0\rangle \equiv \det(g_0) = 0,$$

and $p \neq 0$ is an arbitrary nonzero number.[1]

We introduce coordinates in the space H^p of all paraboloids; these coordinates are similar to the homogeneous coordinates in the space of horospheres H. Set $\zeta^\perp = \binom{\zeta_2}{-\zeta_1}$ for each 2-vector $\zeta = (\zeta_1, \zeta_2)$. Thus, $\xi\zeta^\perp = -\zeta\xi^\perp = \xi_1\zeta_2 - \xi_2\zeta_1$ is an antisymmetric bilinear form on $\mathbb{C}^2$. Obviously, this form is invariant with respect to the group G, i.e.,

$$(\xi g)(\eta g)^\perp = \xi\eta^\perp \quad \text{for each} \quad g \in G.$$

To each pair of vectors $\xi, \zeta \in \mathbb{C}^2 \setminus O$ and every number $p \neq 0$ we assign an isotropic hyperplane in the four-dimensional space of matrices g. This hyperplane is given by the equation

$$\xi g \zeta^\perp = p, \qquad p \neq 0, \tag{1.12}$$

or, in the detailed notation, by the equation

$$(\xi_1\alpha + \xi_2\gamma)\zeta_2 - (\xi_1\beta + \xi_2\delta)\zeta_1 = p, \tag{1.13}$$

where $\xi = (\xi_1, \xi_2)$, $\zeta = (\zeta_1, \zeta_2)$. One can readily see that the equations of isotropic hyperplanes not passing through the origin O and only these equations can be represented in the form (1.12).

---

[1]Sometimes horospheres in our sense are called horocycles, and paraboloids are called horospheres.

Denote by $\widetilde{h}(\xi,\zeta,p)$ the paraboloid in G given by (1.12). Obviously,

$$\widetilde{h}(\lambda\xi,\ \mu\zeta,\ \lambda\mu p) = \widetilde{h}(\xi,\zeta,p) \quad \text{for any numbers} \quad \lambda,\mu \neq 0. \tag{1.14}$$

REMARK. In (1.13) one might assume that $p=1$ and write $\widetilde{h}(\xi,\zeta)$ instead of $\widetilde{h}(\xi,\zeta,1)$. Then $\widetilde{h}(\lambda\xi,\lambda^{-1}\zeta) = \widetilde{h}(\xi,\zeta)$ for each $\lambda \neq 0$. However, it is more convenient for us not to do this.

In the coordinates ξ,ζ,p, an element $(g_1,g_2) \in G\times G$ acts on H^p as follows:

$$\widetilde{h}(\xi,\ \zeta,\ p) \mapsto \widetilde{h}(\xi g_1,\ \zeta g_2,\ p).$$

For instance, the paraboloid ZsZ, where $s = \left(\begin{smallmatrix} 0 & -1 \\ 1 & 0 \end{smallmatrix}\right)$, has the coordinates $\xi_0 = \zeta_0 = (0,1)$, $p_0 = 1$. Hence, the paraboloid $g_1^{-1} Z\, s\, Z\, g_2$ has the coordinates $(\xi_0 g_1, \zeta_0 g_2, 1)$.

It is of prime importance that the space of paraboloids H^p (as well as the space of horospheres H) is equipped with an action of $\mathbb{C}^\times$ that commutes with the action of $G\times G$; namely, corresponding to any $\lambda \in \mathbb{C}^\times$ is the transform

$$\widetilde{h}(\xi,\zeta,p) \mapsto \widetilde{h}(\xi,\zeta,\lambda p),$$

which sends every plane parabolic section to a parallel section.

Every paraboloid $\widetilde{h}(\xi,\zeta,p)$ admits two one-parameter families of horospheres, namely, the horospheres parallel to the planes $\xi g = 0$ and $g\zeta^\perp = 0$, respectively, which belong to the asymptotic cone in $\mathbb{C}^4$.

We describe these families of planes in the homogeneous coordinates on the manifold of horospheres H. The first family consists of the horospheres $h(\xi,\eta)$ where η satisfies the relation $\eta\zeta^\perp = p$. The other one consists of the horospheres $h(\eta,\zeta)$ where η satisfies the relation $\xi\eta^\perp = p$.

Indeed, a horosphere $h(\xi,\eta)$ is the set of matrices $g\in G$ satisfying the relation $\xi g = \eta$, and the condition that the horosphere is contained in a paraboloid $\widetilde{h}(\xi,\zeta,p)$ is equivalent to the relation $\eta\zeta^\perp = p$. A horosphere $h(\eta,\zeta)$ is the set of matrices $g \in G$ satisfying the relation $\eta g = \zeta$, or, equivalently, the relation $\eta^\perp = g\,\zeta^\perp$. Therefore, the condition that this horosphere is contained in a paraboloid $\widetilde{h}(\xi,\zeta,p)$ is equivalent to the relation $\xi\eta^\perp = p$.

Note that exactly one horosphere of each family passes through any given point of the paraboloid.

Conversely, two one-parameter families of paraboloids pass through any given horosphere $h(\xi,\eta)$, namely, the family of paraboloids $\widetilde{h}(\xi,\zeta,1)$ where $\eta\zeta^\perp = 1$, and the family of paraboloids $\widetilde{h}(\zeta,\eta,1)$ where $\zeta\xi^\perp = 1$.

We now discuss the analogy between paraboloids and horospheres. Similarly to the case of horospheres, every paraboloid is an orbit of the subgroup $Z\times Z$ of $G\times G$ or of a subgroup conjugate to $Z\times Z$. Conversely, every orbit of the subgroup $Z\times Z$ or its conjugate subgroup in G is either a horosphere or a paraboloid. Indeed, let $L\subset G$ be an arbitrary orbit of $Z\times Z$, i.e., the set of matrices of the form $z_1\, g^0\, z_2$, where $g^0 = \begin{pmatrix} \alpha_0 & \beta_0 \\ \gamma_0 & \delta_0 \end{pmatrix}$ is an arbitrarily chosen element of G, and z_1 and z_2 range over the subgroup Z. Obviously, if $\gamma_0 = 0$, then L is a horosphere. If $\gamma_0 \neq 0$, then L is the section of G by the hyperplane $\gamma = \gamma_0$, i.e., a paraboloid.

Thus, the manifold of orbits in $G\times G$ of the subgroup $Z\times Z$ and of the subgroups conjugate to $Z\times Z$ is the union of two homogeneous spaces of the group $G\times G$, namely, the space of horospheres H and the space of paraboloids H^p.

Let us compare the manifold of horospheres H with the manifold of paraboloids H^p. Both manifolds are equipped with the structures of a bundle,

$$H \to H^c, \qquad H^p \to H^c,$$

with the fiber $\mathbb{C}^\times$ over the manifold H^c of cylindrical sections. The mapping $H \to H^c$ takes every horosphere to the cylindrical surface passing through this horosphere, and every fiber of the bundle consists of horospheres parallel to one another. The mapping $H^p \to H^c$ takes every parabolic section to the parallel cylindrical section, and every fiber of the bundle consists of parabolic sections parallel to one another. In the coordinate form, to every horosphere $h(\xi,\eta)$ and every paraboloid $\widetilde{h}(\xi,\eta,p)$ we assign the cylinder given by the equation $\xi g \eta^\perp = 0$.

The group $G \times G$ permutes the fibers of each of these bundles, and the group $\mathbb{C}^\times$ acts on the fibers.

Note that H and H^p are not isomorphic when regarded as homogeneous spaces of the group $G \times G$. Indeed, their stabilizers in $G \times G$ are the group V of pairs $(\delta z_1, \delta z_2)$ and the group U of pairs $(\delta^{-1} z_1, \delta z_2)$, respectively, where $\delta \in D$ and z_1, $z_2 \in Z$. These subgroups are not conjugate in $G \times G$; they are not even isomorphic.

REMARK. The space H^c of cylindrical sections is naturally isomorphic to the product $\mathbb{CP}^1 \times \mathbb{CP}^1$ of two complex projective lines, and the bundles H and H^p over H^c with the fiber $\mathbb{C} \setminus 0$ coincide with the line bundles $\mathcal{O}(-1,+1)$ and $\mathcal{O}(-1,-1)$, well known in algebraic geometry.

2. Integral geometry on the group $G = SL(2,\mathbb{C})$

2.1. Integral transforms related to the space H of horospheres and the complex of lines K. To the manifold of horospheres H we associate the integral transform sending a compactly supported C^∞ function on G to the integrals of this function over the horospheres.

Each horosphere $h(\xi,\eta)$ is a two-sided coset $g_1^{-1} Z g_2$, where Z is the subgroup of matrices $\left(\begin{smallmatrix} 1 & t \\ 0 & 1 \end{smallmatrix}\right)$ and $g_1, g_2 \in G$ are arbitrary matrices satisfying the condition

$$\xi_0 g_1 = \xi, \quad \xi_0 g_2 = \eta, \quad \text{where} \quad \xi_0 = (0,1). \tag{2.1}$$

DEFINITION. Let us define the integral of a function f over the horosphere $h(\xi,\eta)$ by the formula

$$\mathcal{R} f(\xi,\eta) = \frac{i}{2} \int f(g_1^{-1} z g_2)\, dt\, d\bar{t}, \qquad z = \begin{pmatrix} 1 & t \\ 0 & 1 \end{pmatrix}, \tag{2.2}$$

where $g_1, g_2 \in G$ are arbitrary matrices satisfying (2.1). The function $\mathcal{R} f(\xi,\eta)$ is called the *horospherical transform* of f.

Formula (2.2) can be written as

$$\mathcal{R} f(\xi,\eta) = \int_G f(g)\, \delta(\xi g - \eta)\, d\mu(g), \tag{2.3}$$

where $\delta(\,\cdot\,)$ is the delta function on $\mathbb{C}^2$ and $d\mu(g)$ is an invariant measure on G; in the coordinates α, β, δ on G, this measure is

$$d\mu(g) = \left(\frac{i}{2}\right)^3 |\beta|^{-2}\, d\alpha\, d\bar{\alpha}\, d\beta\, d\bar{\beta}\, d\delta\, d\bar{\delta}.$$

The equality of integrals (2.2) and (2.3) is verified directly.

It follows from the definition that the function $\mathcal{R} f(\xi,\eta)$ satisfies the homogeneity condition

$$\mathcal{R} f(\lambda\xi,\lambda\eta) = |\lambda|^{-4}\mathcal{R} f(\xi,\eta) \quad \text{for each} \quad \lambda \neq 0. \tag{2.4}$$

Thus, the function $\mathcal{R} f(\xi,\eta)$ depends not only on the horosphere $h(\xi,\eta)$ but also on the homogeneous coordinates ξ,η in which this horosphere is given; therefore, we can regard $\mathcal{R} f$ as a section of a one-dimensional vector bundle over the manifold of horospheres H.

It also follows from (2.2) that the operator $\mathcal{R}$ commutes with the action of the group $G \times G$, i.e., if $f_1(g) = f(g_1^{-1} g g_2)$, then $\mathcal{R} f_1(\xi,\eta) = \mathcal{R} f(\xi g_1, \eta g_2)$.

Introducing affine coordinates on H, one can define the horospherical transform as a function on H with numerical values rather than as a section of a vector bundle over H. It is convenient to introduce the following affine coordinates on H:

$$x_1 = \frac{\xi_1}{\xi_2}, \qquad x_2 = \frac{\eta_1}{\eta_2}, \qquad \lambda = \frac{\eta_2}{\xi_2}. \tag{2.5}$$

In these coordinates every horosphere $h(\xi,\eta)$, where $\xi_2 \neq 0$ and $\eta_2 \neq 0$, is given by the equations

$$\alpha = x_2\beta + \lambda^{-1}, \qquad \delta = -x_1\beta + \lambda.$$

REMARK. The condition $\xi_2 \neq 0$, $\eta_2 \neq 0$ is equivalent to the condition that the horosphere $h(\xi,\eta)$ is not parallel to the hyperplane $\beta = 0$ in $\mathbb{C}^4$.

Taking α,β,δ as the local coordinates on G, one can define the *horospherical transform* of any function $f(g) = f(\alpha,\beta,\delta)$ by the following relation:

$$\mathcal{R}_0 f(x_1,x_2,\lambda) = \frac{i}{2}\int f(x_2\beta+\lambda^{-1},\beta,-x_1\beta+\lambda)\,d\beta\,d\overline{\beta}. \tag{2.6}$$

Geometrically, the passage from $\mathcal{R}$ to $\mathcal{R}_0$ means that instead of the integral transform related to the manifold of horospheres H, we consider the integral transform related to the line complex K in $\mathbb{C}^3$ associated with H, i.e., the transform taking a function on $\mathbb{C}^3$ to the integrals of this function over the lines in $\mathbb{C}^3$ intersecting the hyperbola $\beta = 0$, $\alpha\delta = 1$.

Let us establish the relation between the functions $\mathcal{R} f$ and $\mathcal{R}_0 f$. In (2.2) one can set $g_1 = \begin{pmatrix} \xi_2^{-1} & 0 \\ \xi_1 & \xi_2 \end{pmatrix}$ and $g_2 = \begin{pmatrix} \eta_2^{-1} & 0 \\ \eta_1 & \eta_2 \end{pmatrix}$; then the relation $g = g_1^{-1} z g_2$, where $g = \begin{pmatrix} \alpha & \beta \\ \gamma & \delta \end{pmatrix}$ and $z = \begin{pmatrix} 1 & t \\ 0 & 1 \end{pmatrix}$, implies that

$$\alpha = \eta_1\xi_2 t + \frac{\xi_2}{\eta_2}, \qquad \beta = \xi_2\eta_2 t, \qquad \delta = -\xi_1\eta_2 t + \frac{\eta_2}{\xi_2}.$$

Finally, the expression for $\mathcal{R} f$ in (2.2) becomes

$$\mathcal{R} f(\xi,\eta) = |\xi_2\eta_2|^{-2}\frac{i}{2}\int f(x_2\beta+\lambda^{-1},\beta,-x_1\beta+\lambda)\,d\beta\,d\overline{\beta}, \tag{2.7}$$

where x_1, x_2, λ are connected with ξ,η by (2.5). Comparing this expression with that given by (2.6) for $\mathcal{R}_0 f$, we obtain the desired relations between $\mathcal{R} f$ and $\mathcal{R}_0 f$:

$$\mathcal{R} f(\xi,\eta) = |\xi_2\eta_2|^{-2}\mathcal{R}_0 f\left(\frac{\xi_1}{\xi_2},\frac{\eta_1}{\eta_2},\frac{\eta_2}{\xi_2}\right), \tag{2.8}$$

$$\mathcal{R}_0 f(x_1,x_2,\lambda) = |\lambda|^2\mathcal{R} f(x_1,1,\lambda x_2,\lambda). \tag{2.9}$$

2.2. Symmetry relations for the horospherical transform. Every function $\varphi = \mathcal{R}f$ satisfies some additional symmetry relations. Let us describe the structure of these relations. Assume that Γ is an arbitrary section of the hyperboloid G by a hyperplane in $\mathbb{C}^4$. Since Γ is a quadratic surface, it follows that in general position there are two families of generatrices (horospheres) on Γ. A symmetry relation is that the integrals of the function $\varphi = \mathcal{R}f$ over these families of horospheres (with respect to an appropriate measure) coincide because both integrals are equal to the integral of the original function f over the surface Γ. We write these relations in an explicit form.

Choose an element $g_0 \in G$ and a number $\lambda \in \mathbb{C}\setminus 0$ arbitrarily and consider two families of horospheres, namely, $h(\xi, \lambda\xi g_0)$ and $h(\lambda\eta g_0^{-1}, \eta)$. It was already shown in 1.3 that these horospheres are generatrices of the surface $\Gamma \subset G$ obtained as the section of G by the hyperplane

$$2\lambda\langle g_0, g\rangle = 1 + \lambda^2.$$

The surface Γ is a hyperboloid for $\lambda \neq \pm 1$ and a cone for $\lambda = \pm 1$.

PROPOSITION 2.1. *The function $\varphi = \mathcal{R}f$ satisfies the* symmetry relation

$$\int_{\gamma_\xi} \varphi(\xi, \lambda\xi g_0)\omega(\xi)\overline{\omega(\xi)} = \int_{\gamma_\eta} \varphi(\lambda\eta g_0^{-1}, \eta)\omega(\eta)\overline{\omega(\eta)} \tag{2.10}$$

for any $g_0 \in G$ and $\lambda \in \mathbb{C}\setminus 0$.

Here $\omega(\xi)$ stands for $\xi_1\, d\xi_2 - \xi_2\, d\xi_1$, and γ_ξ and γ_η are arbitrary surfaces in $\mathbb{C}^2 \setminus O$ that intersect once almost every line passing through the origin O; it follows from the homogeneity condition (2.4) for φ that the integrals do not depend on the choice of γ_ξ and γ_η.

For the proof, it suffices to substitute the expression for $\mathcal{R}f$ in terms of f (see (2.3)) into each integral (2.10). Simple manipulations transform each of these integrals to the same integral

$$\int_G f(g)\, \delta(-2\lambda\langle g_0, g\rangle + 1 + \lambda^2)\, d\mu(g) = \int_G f(g)\, \delta(\langle g - \lambda g_0, g - \lambda g_0\rangle)\, d\mu(g).$$

Note that relation (2.10) is meaningful only for $\lambda \neq \pm 1$, in which case the section Γ is not a cone.

REMARK. The symmetry relation (2.10) can also be represented in the following form:

$$\int_{\gamma_\xi} \varphi(\xi, \xi g_0)\omega(\xi)\overline{\omega(\xi)} = \int_{\gamma_\eta} \varphi(\eta\widehat{g_0}, \eta)\omega(\eta)\overline{\omega(\eta)},$$

where $g_0 = \begin{pmatrix} \alpha_0 & \beta_0 \\ \gamma_0 & \delta_0 \end{pmatrix}$ is an arbitrary nonsingular matrix and $\widehat{g_0} = \begin{pmatrix} \delta_0 & -\beta_0 \\ -\gamma_0 & \alpha_0 \end{pmatrix}$. In this form the relation remains valid for singular matrices $g_0 \neq 0$ as well. Note that, for any singular matrix g_0, the horospheres $h(\xi, \xi g_0)$ and $h(\eta\widetilde{g_0}, \eta)$ are generatrices of the paraboloid obtained as the section of G by the hyperplane $2\langle g_0, g\rangle = 1$.

2.3. Inversion formula for the integral transform $\mathcal{R}_0$ related to the line complex K in $\mathbb{C}^3$. Consider the integral transform $\mathcal{R}_0$ given by (2.6); this transform takes every compactly supported function $f(\alpha, \beta, \delta)$ on $\mathbb{C}^3$ to the integrals of this function over the lines in $\mathbb{C}^3$ intersecting the hyperbola $\beta = 0$, $\alpha\delta = 1$. We intend to reconstruct f from the function $\varphi = \mathcal{R}_0 f$. Recall that in Chapter 2 we solved a more general problem of reconstructing a function on $\mathbb{C}^3$ from the integrals of this function over the lines intersecting an arbitrarily chosen algebraic curve in $\Lambda \subset \mathbb{C}^3$. Therefore, it is possible to apply the results of Chapter 2. However, we give here another solution, which is independent of Chapter 2. We first present the solution of the problem under the additional assumption that $f = 0$ if β is sufficiently small. Under this assumption, the forthcoming manipulations are correct. In what follows we get rid of this unnecessary assumption.

We first pass from the function $\varphi(x_1, x_2, \lambda)$ to its Fourier transform with respect to x_1 and x_2,

$$\widetilde{\varphi}(\xi_1, \xi_2, \lambda) = (2\pi)^{-2}\left(\frac{i}{2}\right)^2 \int \varphi(x_1, x_2, \lambda) e^{i\operatorname{Re}(\xi_1 x_1 + \xi_2 x_2)}\, dx_1\, d\overline{x_1}\, dx_2\, d\overline{x_2}.$$

Let us find an expression for $\widetilde{\varphi}$ in terms of the original function f. We have

(2.11)

$$\begin{aligned}\widetilde{\varphi}(\xi_1, \xi_2, \lambda) &= (2\pi)^{-2}\left(\frac{i}{2}\right)^3 \int f(x_2 t + \lambda^{-1}, t, -x_1 t + \lambda) \\ &\qquad\times e^{i\operatorname{Re}(\xi_1 x_1 + \xi_2 x_2)}\, dt\, d\bar{t}\, dx_1\, d\overline{x_1}\, dx_2\, d\overline{x_2} \\ &= (2\pi)^{-2}\left(\frac{i}{2}\right)^3 \int f(\alpha, \beta, \delta) e^{i\operatorname{Re}\left(\frac{\lambda-\delta}{\beta}\xi_1 + \frac{\alpha - \lambda^{-1}}{\beta}\xi_2\right)} |\beta|^{-4}\, d\alpha\, d\bar{\alpha}\, d\beta\, d\bar{\beta}\, d\delta\, d\bar{\delta}.\end{aligned}$$

We now pass from coordinates α, β, δ on $\mathbb{C}^3$ to the new coordinates

$$p_1 = -\frac{\delta}{\beta}, \qquad p_2 = \frac{\alpha}{\beta}, \qquad p_3 = \frac{1}{\beta}$$

and set

$$f(\alpha, \beta, \delta) = |p_3|^4 f_1(p_1, p_2, p_3).$$

In this case, formula (2.11) becomes

$$\begin{aligned}\widetilde{\varphi}(\xi_1, \xi_2, \lambda) &= (2\pi)^{-2}\left(\frac{i}{2}\right)^3 \int f_1(p_1, p_2, p_3) \\ &\qquad\times e^{i\operatorname{Re}(p_1\xi_1 + p_2\xi_2 + p_3(\lambda\xi_1 - \lambda^{-1}\xi_2))}\, dp_1\, d\overline{p_1}\, dp_2\, d\overline{p_2}\, dp_3\, d\overline{p_3}.\end{aligned}$$

We thus obtain a simple relation between the function $\widetilde{\varphi}$ and the Fourier transform F of f_1,

$$\begin{aligned}F(\xi_1, \xi_2, \xi_3) &= (2\pi)^{-3}\left(\frac{i}{2}\right)^3 \int f(p_1, p_2, p_3) \\ &\qquad\times e^{i\operatorname{Re}(p_1\xi_1 + p_2\xi_2 + p_3\xi_3)}\, dp_1\, d\overline{p_1}\, dp_2\, d\overline{p_2}\, dp_3\, d\overline{p_3},\end{aligned}$$

namely,

$$\widetilde{\varphi}(\xi_1, \xi_2, \lambda) = 2\pi F(\xi_1, \xi_2, \lambda\xi_1 - \lambda^{-1}\xi_2). \tag{2.12}$$

The desired inversion formula readily follows from (2.12). For that, one must apply the inversion formula for the Fourier transform of f_1,

$$\begin{aligned}|\beta|^4 f(\alpha, \beta, \delta) &= f_1(p_1, p_2, p_3) \\ &= (2\pi)^{-3}\left(\frac{i}{2}\right)^3 \int F(\xi_1, \xi_2, \xi_3) \\ &\qquad\times e^{-i\operatorname{Re}(\xi_1 p_1 + \xi_2 p_2 + \xi_3 p_3)}\, d\xi_1\, d\overline{\xi_1}\, d\xi_2\, d\overline{\xi_2}\, d\xi_3\, d\overline{\xi_3}.\end{aligned}$$

Making here the change of variables $\xi_3 = \lambda\xi_1 - \lambda^{-1}\xi_2$, we see by (2.12) that

$$
\begin{aligned}
f(\alpha,\beta,\delta) &= \frac{1}{2}(2\pi)^{-4}\left(\frac{i}{2}\right)^3 |\beta|^{-4}\int \widetilde{\varphi}(\xi_1,\xi_2,\lambda)|\xi_1+\lambda^{-2}\xi_2|^2 \\
&\qquad\times e^{-i\operatorname{Re}\left(-\xi_1\frac{\delta}{\beta}+\xi_2\frac{\alpha}{\beta}+(\lambda\xi_1-\lambda^{-1}\xi_2)\frac{1}{\beta}\right)}\,d\xi_1\,d\overline{\xi_1}\,d\xi_2\,d\overline{\xi_2}d\lambda\,d\overline{\lambda} \\
&= \frac{1}{2}(2\pi)^{-2}\frac{i}{2}|\beta|^{-4}\int L\overline{L}\varphi\left(\frac{\lambda-\delta}{\beta},\frac{\alpha-\lambda^{-1}}{\beta},\lambda\right)\,d\lambda\,d\overline{\lambda},
\end{aligned}
\tag{2.13}
$$

where $L = \frac{\partial}{\partial x_1} + \lambda^{-2}\frac{\partial}{\partial x_2}$ and $\overline{L} = \frac{\partial}{\partial \overline{x_1}} + \overline{\lambda}^{-2}\frac{\partial}{\partial \overline{x_2}}$.

It remains to simplify the expression obtained. Since $\beta^{-1}L + \frac{\partial}{\partial\lambda}(\overline{\beta}^{-1}\overline{L} + \frac{\partial}{\partial\overline{\lambda}})$ is the full derivative of the function $\varphi(\frac{\lambda-\delta}{\beta},\frac{\alpha-\lambda^{-1}}{\beta},\lambda)$ with respect to λ (to $\overline{\lambda}$), one can replace $\beta^{-1}L$ and $\overline{\beta}^{-1}\overline{L}$ in (2.13) with $-\frac{\partial}{\partial\lambda}$ and $-\frac{\partial}{\partial\overline{\lambda}}$, respectively. We finally obtain

$$
f(\alpha,\beta,\delta) = \frac{1}{8\pi^2}\frac{i}{2}|\beta|^{-2}\int \varphi''_{\lambda,\overline{\lambda}}\left(\frac{\lambda-\delta}{\beta},\frac{\alpha-\lambda^{-1}}{\beta},\lambda\right)\,d\lambda\,d\overline{\lambda}.
$$

The desired inversion formula follows by changing the variable, $x = \frac{\lambda-\delta}{\beta}$. We now formulate the final result.

THEOREM 2.1. *Let $f(\alpha,\beta,\gamma)$ be a compactly supported C^∞ function on $\mathbb{C}^3$ that vanishes for the sufficiently small values of β, and let $\varphi = \mathcal{R}_0 f$ be the integral transform of f given by* (2.6). *Then the following inversion formula holds*:

$$
f(\alpha,\beta,\delta) = \frac{1}{8\pi^2}\frac{i}{2}\int \varphi''_{\lambda,\overline{\lambda}}\left(x,\frac{\alpha x+\gamma}{\beta x+\delta},\beta x+\delta\right)\,dx\,d\overline{x}.
\tag{2.14}
$$

Note that the differentiation with respect to λ in this formula is the differentiation with respect to the set of parallel lines intersecting the hyperbola $\beta = 0$, $\alpha\delta = 1$, and the integral is taken over the set of lines intersecting this hyperbola and passing through the point (α,β,δ).

As in the case of the hyperbolic space, the inversion formula thus obtained is local, namely, to reconstruct the value of a function f at a point (α,β,δ) one must know only the integrals of this function over the lines intersecting the hyperbola $\beta = 0$, $\alpha\delta = 1$ that are infinitely close to this point.

REMARK. The line complex K is admissible in the sense of Chapter 2, §5. Therefore, the inversion formula (2.13) can be obtained using the operator κ.

2.4. Inversion formula for the horospherical transform. Assume that $f(g)$, $g = \left(\begin{smallmatrix}\alpha & \beta\\ \gamma & \delta\end{smallmatrix}\right)$, is a compactly supported C^∞ function on G. The problem is to express f in terms of the horospherical transform $\mathcal{R}f(\xi,\eta)$ defined by (2.3). We temporarily assume that $f = 0$ for all sufficiently small values of β. In this case, the function f, viewed as a function of α, β, δ, can be extended to a compactly supported C^∞ function on the entire space $\mathbb{C}^3$, and this extension vanishes for all sufficiently small values of β. Hence, the extended function can be expressed in terms of $\varphi = \mathcal{R}_0 f$ by the inversion formula (2.14). Hence, to obtain the desired result, it suffices to use the expression for $\mathcal{R}_0 f$ in terms of $\mathcal{R}f$; see (2.9). We thus obtain

$$
f(g) = \frac{1}{8\pi^2}\frac{i}{2}\int L\overline{L}\mathcal{R}f(x,1,\alpha x+\gamma,\beta x+\delta)\,dx\,d\overline{x},
\tag{2.15}
$$

where $L = (\alpha x + \gamma)\frac{\partial}{\partial \eta_1} + (\beta x + \delta)\frac{\partial}{\partial \eta_2} + 1$. Using the homogeneity condition (2.4) for $\mathcal{R} f$, one can write the inversion formula in a more symmetric form,

$$f(g) = \frac{1}{8\pi^3}\frac{i}{2}\int_\gamma L_\eta \overline{L_\eta}\, \mathcal{R} f(\xi, \eta)\,|_{\eta=\xi g}\, \omega(\xi)\overline{\omega(\xi)}, \tag{2.16}$$

where $L_\eta = \eta_1\frac{\partial}{\partial \eta_1} + \eta_2\frac{\partial}{\partial \eta_2} + 1$, $\omega(\xi) = \xi_1\, d\xi_2 - \xi_2\, d\xi_1$, and the integral is taken over an arbitrary surface $\gamma \subset \mathbb{C}^2 \setminus O$ which intersects once almost every line passing through the origin O.

Since the horospherical transform $\mathcal{R}$ is invariant with respect to the action of $G \times G$, it follows that the initial assumption on f is unessential; namely, if the inversion formula (2.16) holds for the functions vanishing for all sufficiently small values of β, then this formula remains valid for any compactly supported C^∞ function on G. This proves the following assertion.

THEOREM 2.2. *Let f be an arbitrary compactly supported C^∞ function on G. If $\mathcal{R} f$ is the horospherical transform of f defined by* (2.3), *then the inversion formula* (2.16), *which expresses f in terms of $\mathcal{R} f$, holds.*

In formula (2.16), the integral is taken over the set of horospheres $h(\xi, \xi g)$, i.e., the horospheres passing through a point $g \in G$. The operator $L_\eta = \eta_1\frac{\partial}{\partial \eta_1} + \eta_2\frac{\partial}{\partial \eta_2} + 1$ occurring in the formula can be given as follows:

$$L_\eta \varphi(\xi, \eta) = \frac{\partial}{\partial \lambda}(\lambda\, \varphi(\xi, \lambda\eta))\,\Big|_{\lambda=1}.$$

Recall that $h(\xi, \eta) \mapsto h(\xi_0, \lambda\eta)$ is the left translation taking every horosphere to a parallel horosphere; see 1.3. Thus, L_η is the operator of the infinitesimal left translation.

REMARK. It follows from the homogeneity condition (2.4) for $\mathcal{R} f$ that

$$(L_\xi + L_\eta)\,\mathcal{R} f = (\overline{L_\xi} + \overline{L_\eta})\,\mathcal{R} f = 0,$$

where $L_\xi = \xi_1\frac{\partial}{\partial \xi_1} + \xi_2\frac{\partial}{\partial \xi_2} + 1$. Therefore, the operators L_η and $\overline{L_\eta}$ in the inversion formula (2.16) can be replaced with $L = \frac{1}{2}\left(\xi_1\frac{\partial}{\partial \xi_1} + \xi_2\frac{\partial}{\partial \xi_2} - \eta_1\frac{\partial}{\partial \eta_1} - \eta_2\frac{\partial}{\partial \eta_2}\right)$ and $\overline{L}$, respectively.

2.5. Inversion formula for the horospherical transform on the hyperbolic space $\mathcal{L}^3$. The inversion formula for the horospherical transform $\mathcal{R}$ on G implies an inversion formula for the horospherical transform $\mathcal{R}^h$ on the hyperbolic space $\mathcal{L}^3$ (for the definition of $\mathcal{R}^h$, see Chapter 3). Namely, since $\mathcal{L}^3 = U\backslash G$, where $U = SU(2)$, it follows that each function f on $\mathcal{L}^3$ can be lifted to G, i.e., f can be viewed as a function $\widetilde{f}$ on G satisfying the condition

$$\widetilde{f}(ug) = \widetilde{f}(g) \text{ for each } u \in U.$$

Since U is compact, it follows that the image $\widetilde{f}$ of a compactly supported function f on $\mathcal{L}^3$ is a compactly supported function on G; therefore, the space of compactly supported functions on $\mathcal{L}^3$ is embedded in the space of compactly supported functions on G. Further, one can readily see that, under the mapping $G \to U \setminus G = \mathcal{L}^3$, the horospheres on G are taken to horospheres on $\mathcal{L}^3$ and the integrals $\mathcal{R}\widetilde{f}$ of a function $\widetilde{f}$ over the horospheres on G are equal (under an appropriate agreement between the measures on the horospheres) to the integrals

$\mathcal{R}^h f$ of the function f over the images of these horospheres. Thus, to obtain an inversion formula for $\mathcal{R}^h$, one must only find an explicit expression for $\mathcal{R}\widetilde{f}$ in terms of $\mathcal{R}^h f$ and substitute this expression into the inversion formula (2.16) for $\mathcal{R}$.

Let us state several assertions readily following from the definitions of the horospheres and the horospherical transforms on G and $\mathcal{L}^3$.

1. Under the mapping $G \to U \setminus G = \mathcal{L}^3$, every horosphere $h(\xi, \eta)$ on G goes to the horosphere $h(\eta)$ on $\mathcal{L}^3$.

2. The transform $\mathcal{R}\widetilde{f}(\xi, \eta)$, as a function of ξ, depends only on the variable $|\xi| = (|\xi_1|^2 + |\xi_2|^2)^{1/2}$.

3. The functions $\mathcal{R}\widetilde{f}$ and $\mathcal{R}^h f$ are related as follows:

$$\mathcal{R}\widetilde{f}(-1, 0; \eta_1, \eta_2) = \mathcal{R}^h f(\eta_1, \eta_2).$$

The desired expression for $\mathcal{R}\widetilde{f}$ in terms of $\mathcal{R}^h f$ follows from 2) and 3) and from the homogeneity condition for $\mathcal{R} f$,

$$\mathcal{R}\widetilde{f}(\xi, \eta) = |\xi|^{-4}\mathcal{R}^h\left(\frac{\eta}{|\xi|}\right). \tag{2.17}$$

Substituting this expression into the inversion formula (2.16) for the horospherical transform $\mathcal{R}$ on G for $g = e$, after elementary manipulations we obtain the following inversion formula for the horospherical transform $\mathcal{R}^h$ on $\mathcal{L}^3$:

$$f(w_0) = \frac{1}{8\pi^2}\int_{\gamma_\eta} L_\eta \overline{L_\eta}\,\mathcal{R}^h f(\eta)\,\omega(\eta)\,\overline{\omega(\eta)}, \tag{2.18}$$

where $w_0 \in \mathcal{L}^3$ is the point corresponding to the identity coset in $K \setminus G$. The expression for f at an arbitrary point of $w \in \mathcal{L}^3$ is obtained from (2.18) by the group translation. The resulting inversion formula coincides with the inversion formula (6.17) of Chapter 3.

2.6. Integral transform related to paraboloids on G. Recall (see 1.6) that each paraboloid on G can be obtained as the section of G by a hyperplane in $\mathbb{R}^4$ of the following form:

$$(\xi_1\alpha + \xi_2\gamma)\zeta_2 - (\xi_1\beta + \xi_2\delta)\zeta_1 = p, \qquad p \neq 0,$$

or, in the matrix form,

$$\xi g \zeta^\perp = p, \qquad p \neq 0, \tag{2.19}$$

where $\xi = (\xi_1, \xi_2)$ and $\zeta = (\zeta_1, \zeta_2)$ are nonzero vectors in $\mathbb{C}^2$, and $\zeta^\perp$ stands for the column vector $\left(\begin{smallmatrix}\zeta_2\\ -\zeta_1\end{smallmatrix}\right)$. Thus, each paraboloid on G is given by a pair of vectors $\xi, \zeta \in \mathbb{C}^2 \setminus O$ and a number $p \neq 0$ and is defined as the set of matrices $g \in G$ satisfying (2.19). We denote this paraboloid by $\widetilde{h}(\xi, \zeta, p)$. Obviously, $\widetilde{h}(\lambda_1\xi, \lambda_2\zeta, \lambda_1\lambda_2 p) = \widetilde{h}(\xi, \zeta, p)$ for any $\lambda_1, \lambda_2 \neq 0$.

DEFINITION. We define the integral of a C^∞ smooth function f on G over the paraboloid $\widetilde{h}(\xi, \zeta, p)$ by the formula

$$\mathcal{R}^p f(\xi, \zeta, p) = \int_G f(g)\,\delta(\xi g \zeta^\perp - p)\,d\mu(g), \tag{2.20}$$

where $\delta(\cdot)$ is the delta function on $\mathbb{C}$.

Obviously, the function $\mathcal{R}^p f$ satisfies the following homogeneity condition:

$$\mathcal{R}^p f(\lambda_1\xi, \lambda_2\zeta, \lambda_1\lambda_2 p) = |\lambda_1\lambda_2|^{-2}\mathcal{R}^p f(\xi,\zeta,p). \tag{2.21}$$

Thus, $\mathcal{R}^p f(\xi,\zeta,p)$ depends not only on the paraboloid $\widetilde{h}(\xi,\zeta,p)$ itself but also on the parameters ξ, ζ, p defining this paraboloid.

There is a simple relation between the integral transform $\mathcal{R}^p$ and the horospherical transform $\mathcal{R}$. Namely, the passage from the function $f(g)$ to $\mathcal{R}^p f(\xi,\zeta,p)$ can be carried out in two steps. We first integrate f over the horospheres $h(\xi,\eta)$ for fixed ξ; we thus obtain the function $\mathcal{R} f(\xi,\eta)$. Then we integrate $\mathcal{R} f(\xi,\eta)$ over the set of horospheres into which the paraboloid $\widetilde{h}(\xi,\zeta,p)$ is fibered, i.e., over the set of horospheres $h(\xi,\eta)$, where ξ is fixed and η satisfies the relation $\eta\zeta^\perp = p$. We express the result by an explicit formula.

THEOREM 2.3. *The transform $\mathcal{R}^p f$ can be expressed in terms of the horospherical transform $\mathcal{R} f$ as follows:*

$$\mathcal{R}^p f(\xi,\zeta,p) = \left(\frac{i}{2}\right)^2 \int_{\mathbb{C}^2} \mathcal{R} f(\xi,\eta)\,\delta(\eta\zeta^\perp - p)\,d\eta\,d\overline{\eta}. \tag{2.22}$$

It follows from (2.22) that, for every ξ, the function $\mathcal{R}^p f(\xi,\eta,p)$ is the two-dimensional Radon transform of the function $\mathcal{R} f(\xi,\eta)$ viewed as a function of η. Thus, the operator $\mathcal{R}^p$ is the composition of the operator $\mathcal{R}$ of horospherical transform and of the two-dimensional Radon transform on $\mathbb{C}^2$.

We clarify the relation between the transform $\mathcal{R}^p$ and the Radon transform. Consider the family H_ξ of the horospheres $h(\xi,\eta)$ with a chosen value of ξ. This family consists of all horospheres parallel to some two-dimensional space of the form (1.5) on the surface of the asymptotic cone K. The mapping $h(\xi,\eta) \mapsto \eta$ defines an isomorphism $H_\xi \to \mathbb{C}^2 \setminus O$. On every paraboloid $\widetilde{h}(\xi,\zeta,p)$ one of the families of generatrices is contained in H_ξ, and the isomorphism $H_\xi \to \mathbb{C}^2\setminus O$ takes this family to a line on $\mathbb{C}^2$. Thus, due to this isomorphism, the integration of the function $\mathcal{R} f(\xi,\eta)$ over the set of generatrices of the paraboloid $\widetilde{h}(\xi,\zeta,p)$ can be interpreted as the integration of a function on $\mathbb{C}^2$ over a line in $\mathbb{C}^2$.

Using (2.22) and the inversion formula for the two-dimensional Radon transform in $\mathbb{C}^2$ (see § 9 of Chapter 1), we obtain the following expression for $\mathcal{R} f$ in terms of $\mathcal{R}^p f$:

$$\begin{aligned}\mathcal{R} f(\xi,\eta) &= -\pi^{-2}\frac{i}{2}\int_{\gamma_\zeta} \frac{\partial^2}{\partial p\,\partial\overline{p}}\mathcal{R}^p f(\xi,\zeta,p)\Big|_{p=\eta\zeta^\perp}\omega(\zeta)\,\overline{\omega(\zeta)} \\ &= -\pi^2\left(\frac{i}{2}\right)^2\int_{\gamma_\zeta}\int_{\mathbb{C}}\mathcal{R}^p f(\xi,\zeta,p)\,\delta^{(1,1)}(\eta\zeta^\perp - p)\,dp\,d\overline{p}\,\omega(\zeta)\,\overline{\omega(\zeta)},\end{aligned} \tag{2.23}$$

where $\delta^{(1,1)}(s) = \frac{\partial^2}{\partial s\partial\overline{s}}\delta(s)$.

REMARK. In (2.22) the function $\mathcal{R} f$ is averaged over one of the families of horospheres on the paraboloid $\widetilde{h}(\xi,\zeta,p)$. Averaging the function over the other family of horospheres, we obtain a relation similar to (2.22):

$$\mathcal{R}^p f(\xi,\zeta,p) = \left(\frac{i}{2}\right)^2 \int_{\mathbb{C}^2} \mathcal{R} f(\eta,\zeta)\,\delta(\xi\eta^\perp - p)\,d\eta\,d\overline{\eta}. \tag{2.24}$$

By the inversion formula for the Radon transform in $\mathbb{C}^2$, this implies that

$$\mathcal{R} f(\xi,\eta) = -\pi^{-2}\frac{i}{2}\int_{\gamma_\zeta} \frac{\partial^2}{\partial p \partial \overline{p}} \mathcal{R}^p f(\zeta,\eta,p)\Big|_{p=\xi\zeta^\perp} \omega(\zeta)\,\overline{\omega(\zeta)}.$$

We now find an inversion formula for the integral transform $\mathcal{R}^p$. Since $\mathcal{R}^p$ is the composition of the horospherical transform $\mathcal{R}$ and the two-dimensional Radon transform, it follows that the inversion formula for $\mathcal{R}^p$ can be obtained as the composition of the inversion formulas for these two transforms. Substituting (2.23) into the inversion formula (2.16) for $\mathcal{R}$, we obtain

$$f(g) = -\frac{1}{8\pi^5}\left(\frac{i}{2}\right)^3 \int_{\gamma_\xi}\int_{\gamma_\xi}\int_{\mathbb{C}} \mathcal{R}^p f(\xi,\zeta,p) \times \frac{\partial^2}{\partial p\partial\overline{p}} L_\eta \overline{L_\eta}\delta(\eta\zeta^\perp - p)\Big|_{\eta=\xi g} dp\, d\overline{p}\, \omega(\zeta)\,\overline{\omega(\zeta)}\,\omega(\xi)\,\overline{\omega(\xi)}.$$

After elementary manipulations we obtain the following assertion.

THEOREM 2.4. *A function f on G can be expressed in terms of the integral transform $\mathcal{R}^p f$ by the following inversion formula*:

$$f(g) = -\frac{1}{8\pi^5}\left(\frac{i}{2}\right)^2 \int_{\gamma_\zeta}\int_{\gamma_\xi} \frac{\partial^2}{\partial p\partial\overline{p}}\Big(|p|^2 \frac{\partial^2}{\partial p\partial\overline{p}} \mathcal{R}^p f(\xi,\zeta,p)\Big)\Big|_{p=\xi g\zeta^\perp} \times \omega(\xi)\,\overline{\omega(\xi)}\,\omega(\zeta)\overline{\omega(\zeta)}. \tag{2.25}$$

REMARK. The integral transform $\mathcal{R}^p$ of functions on G produces an integral transform of functions on the hyperbolic space $\mathcal{L}^3$. Namely, when projecting G to the hyperbolic space $\mathcal{L}^3 = SU(2)\backslash G$, every paraboloid $\widetilde{h}(\xi,\zeta,1)$ (which is a manifold of real dimension four) is mapped onto the domain $\mathcal{L}_\zeta \subset \mathcal{L}^3$ which is the exterior of the horosphere (in the hyperbolic space) given by the equation $\frac{|w\zeta_2-\zeta_1|^2}{z} = 1$. The integral transform $\mathcal{R}^p$ on G becomes the integral transform of functions on $\mathcal{L}^3$ that is given by the following relation:

$$\widetilde{\mathcal{R}}^h f(\zeta_1,\zeta_2) = \int_{\mathcal{L}_\zeta}\left(\frac{|w\zeta_2-\zeta_1|^2}{z}\right)^{-1} f(w)\frac{dx\,dy\,dz}{z^3},$$

where $w = x+yi+zj$ (see Chapter 3). The inversion formula for $\widetilde{\mathcal{R}}^h$ readily follows from the inversion formula for $\mathcal{R}^p$.

Appendix. Remark on the horospherical transform on the group $SL(2,\mathbb{R})$. The integral transforms $\mathcal{R}$ and $\mathcal{R}_0$ introduced in this section can be extended to the real case without any modifications. However, in contrast to the complex case, in the real case the mappings $f \mapsto \mathcal{R} f$ and $f \mapsto \mathcal{R}_0 f$ have nonzero kernel, and therefore a function f cannot be reconstructed from its image. It suffices to show this fact in the case of the transform $\mathcal{R}_0$.

In the real case, the operator $\mathcal{R}_0$ takes a function $f(\alpha,\beta,\delta)$ on $\mathbb{R}^3$ to the integrals of f over the lines in $\mathbb{R}^3$ intersecting the hyperbola $\beta = 0$, $\alpha\delta = 1$, i.e.,

$$\mathcal{R}_0 f(x_1,x_2,\lambda) = \int_{-\infty}^{+\infty} f(x_2 t + \lambda^{-1}, t, -x_1 t + \lambda)\,dt.$$

The arguments used in 2.3 to obtain the inversion formula can be repeated for the real case. We pass to new coordinates $p_1 = -\frac{\delta}{\beta}$, $p_2 = \frac{\alpha}{\beta}$, $p_3 = \frac{1}{\beta}$ on $\mathbb{R}^3$

and set $f(\alpha,\beta,\delta) = |p_3|^2 f_1(p_1,p_2,p_3)$. Denote by $F(\xi_1,\xi_2,\xi_3)$ the Fourier transform of f_1 and by $\widetilde{\varphi}(\xi_1,\xi_2,\lambda)$ the Fourier transform of the function $\varphi(x_1,x_2,\lambda) = \mathcal{R}_0 f(x_1,x_2,\lambda)$ with respect to the variables x_1, x_2. The following relation can be established as in the complex case:

$$\widetilde{\varphi}(\xi_1,\xi_2,\lambda) = (2\pi)^{1/2} F(\xi_1,\xi_2,\lambda\xi_1 - \lambda^{-1}\xi_2). \tag{2.26}$$

We face the question of whether one can reconstruct the function F from the function $\widetilde{\varphi}$. It is clear that this is impossible in general. The condition $\xi_3 = \lambda\xi_1 - \lambda^{-1}\xi_2$, where λ is real, implies the relation $\xi_3^2 + 4\xi_1\xi_2 > 0$, and hence the function F is defined by (2.26) in the domain $\xi_3^2 + 4\xi_1\xi_2 > 0$ only.

Note that if f is a compactly supported function vanishing for the small values of β, then it can be reconstructed from the function $\mathcal{R}_0 f$. Indeed, in this case the above function $f_1(p_1,p_2,p_3)$ is also compactly supported, and hence its Fourier transform F is an entire analytic function. Therefore, the function F, which is defined on the domain $\xi_3^2 + 4\xi_1\xi_2 > 0$ on $\mathbb{R}^3$, can be thus defined on the entire space $\mathbb{R}^3$.

It is also possible to construct a complex version of the parabolic transform on $SL(2,\mathbb{R})$. This transform has no kernel and admits an explicit inversion formula (see [**17**]).

3. Harmonic analysis on the group $G = SL(2,\mathbb{C})$

In this section, following the same scheme as the one used in the case of hyperbolic space (see Chapter 3), we construct an analog of the Fourier transform on the group G. The construction uses the eigenfunctions (of the Laplace operators on G) that are constant on the orbits of the subgroup $Z \times Z \subset G \times G$, where Z is the subgroup of matrices $\left(\begin{smallmatrix} 1 & t \\ 0 & 1 \end{smallmatrix}\right)$. There are two types of such orbits (see § 1), namely, the one-dimensional orbits (horospheres) and the two-dimensional orbits (paraboloids). Correspondingly, one may consider eigenfunctions (of the Laplace operators) constant on the orbits of the first or second type. This leads to two versions of the Fourier transform on G.

In this section we present in detail the first version of the Fourier transform, i.e., the one related to horospheres. The main result is the expansion of a function on G in an analog of the Fourier integral; it is obtained below as a corollary of the inversion formula for the horospherical transform; see § 2. The other version of the Fourier transform on G is considered briefly in the next section.

3.1. Laplace–Beltrami operator on the group G. We begin with presenting two second-order differential operators Δ and $\overline{\Delta}$ on G that commute with each other and with the action of the group $G \times G$.

Operators of this type in the space of all matrices $g = \left(\begin{smallmatrix} \alpha & \beta \\ \gamma & \delta \end{smallmatrix}\right)$, i.e., in $\mathbb{C}^4$, indeed exist, namely,

$$\Delta = \frac{\partial^2}{\partial\alpha\partial\delta} - \frac{\partial^2}{\partial\beta\partial\gamma}, \qquad \overline{\Delta} = \frac{\partial^2}{\partial\overline{\alpha}\partial\overline{\delta}} - \frac{\partial^2}{\partial\overline{\beta}\partial\overline{\gamma}}. \tag{3.1}$$

To define such operators on the hyperboloid G given by $\alpha\delta - \beta\gamma = 1$, we introduce generalized spherical coordinates $x = \rho y$ on $\mathbb{C}^4$, where $\rho \in \mathbb{C}$ and $y \in G$, and consider our operators only on the functions that do not depend on ρ. The operators Δ and $\overline{\Delta}$ thus defined are called the *Laplace–Beltrami operators* on G.

We present the expressions (omitting the calculations) for the operators Δ and $\overline{\Delta}$ in the coordinates s, t, λ on G that are given by the Gauss decomposition

$$\begin{pmatrix} \alpha & \beta \\ \gamma & \delta \end{pmatrix} = \begin{pmatrix} 1 & 0 \\ s & 1 \end{pmatrix} \begin{pmatrix} \lambda & 0 \\ 0 & \lambda^{-1} \end{pmatrix} \begin{pmatrix} 1 & t \\ 0 & 1 \end{pmatrix}.$$

It follows from this decomposition that $\lambda = \alpha$, $s = \frac{\gamma}{\alpha}$, and $t = \frac{\beta}{\alpha}$. We obtain

$$\Delta = -\frac{1}{4}\left(\lambda\frac{\partial}{\partial\lambda} + 1\right)^2 - \lambda^2 \frac{\partial^2}{\partial s \partial t} + \frac{1}{4}, \tag{3.2}$$

and a similar expression for $\overline{\Delta}$.

3.2. Horospherical functions on G. Following the scheme presented in Chapter 3 for the hyperbolic space, we must first introduce an analog of the subgroup $Z \subset G$ of matrices of the form $\left(\begin{smallmatrix} 1 & t \\ 0 & 1 \end{smallmatrix}\right)$ for the group $G \times G$. For such a subgroup it is natural to take the group $Z \times Z$.

Let us find on G the eigenfunctions of the operators Δ and $\overline{\Delta}$ that are invariant with respect to the subgroup $Z \times Z \subset G \times G$. To do this, note that the orbits of the group $Z \times Z$ in G are the two-dimensional subsets of matrices $\left(\begin{smallmatrix} \alpha & \beta \\ \gamma & \delta \end{smallmatrix}\right)$ with a chosen $\gamma \neq 0$ (paraboloids) and the one-dimensional subsets of matrices $\left(\begin{smallmatrix} \alpha & \beta \\ 0 & \alpha^{-1} \end{smallmatrix}\right)$ with a chosen α (horospheres). Hence, the distributions Φ on G invariant with respect to the subgroup $Z \times Z$ are

$$\Phi(g) = u(\gamma) + \delta(\gamma) v(\alpha), \qquad g = \begin{pmatrix} \alpha & \beta \\ \gamma & \delta \end{pmatrix},$$

where u and v are arbitrary functions and $\delta(\cdot)$ is the delta function. We must find out what functions $u(\gamma)$ and $\delta(\gamma)v(\alpha)$ are eigenfunctions of the Laplace–Beltrami operators. After simple manipulations we obtain the following assertion.

PROPOSITION 3.1. *The following distributions on* G *are* $Z \times Z$*-invariant eigenfunctions of the Laplace–Beltrami operators* Δ *and* $\overline{\Delta}$:

$$\Phi_\chi(g) = \delta(\gamma)\,\chi(\alpha), \tag{3.3}$$

$$\Psi_\chi(g) = \chi^{-1}(\gamma)\,|\gamma|^{-2}, \qquad g = \begin{pmatrix} \alpha & \beta \\ \gamma & \delta \end{pmatrix}, \tag{3.4}$$

where $\delta(\cdot)$ *is the delta function on* $\mathbb{C}$ *and* χ *is an arbitrary multiplicative character on* $\mathbb{C}$, *i.e., a function of the form*

$$\chi(t) = t^{n_1}\bar{t}^{n_2} = |t|^{2n_2} t^{n_1 - n_2}, \quad n_1 - n_2 \in \mathbb{Z}. \tag{3.5}$$

Moreover, the functions Φ_χ, $\Phi_{\chi^{-1}}$, Ψ_χ, *and* $\Psi_{\chi^{-1}}$ *belong to (the same) eigenspace of* Δ *and* $\overline{\Delta}$ *corresponding to the eigenvalues* $-\frac{1}{4}(n_1^2 - 1)$ *and* $-\frac{1}{4}(n_2^2 - 1)$, *respectively.*

Thus, we have two types of functions, namely, the functions Φ_χ concentrated on a two-dimensional submanifolds of G and the functions Ψ_χ. Correspondingly, there are two versions of harmonic analysis on G; the first version uses the functions Φ_χ and the second the functions Ψ_χ. Here we present the first version in detail. The second version is considered in the next section.

DEFINITION. The distributions $\Phi_\chi(g)$, where χ is an arbitrary character of the form (3.5), are called the *zonal horospherical functions* on G and their shifts $\Phi_\chi(g_1 g g_2^{-1})$, $g_1, g_2 \in G$, are simply referred to as *horospherical functions* on G.

As we shall see later, the horospherical functions $\Phi_\chi(g_1 g g_2^{-1})$ where χ ranges over the unitary characters $(\chi(t)\,\overline{\chi(t)} \equiv 1)$, form a complete family in $L^2(G)$, i.e., each function in $L^2(G)$ can be expanded in these functions.

By (3.3), the zonal horospherical functions are concentrated on the cylindrical surface $\gamma = 0$ in G formed by the horospheres of the form δZ, $\delta \in D$, and the functions are constant on each of these horospheres. Hence, every horospherical function $g \mapsto \Phi_\chi(g_1 g g_2^{-1})$ is concentrated on one of the cylindrical surfaces in G and is constant on the generatrices of this surface.

DEFINITION. We define the *horospherical functions* $\Phi_\chi(g, \xi, \eta)$ on the manifold of triples formed by $g \in G$ and $\xi, \eta \in \mathbb{C}^2 \setminus O$ by the following relation:

$$\Phi_\chi(g, \xi, \eta) = \Phi_\chi(g_1 g g_2^{-1}), \tag{3.6}$$

where $g_1, g_2 \in G$ are arbitrary matrices satisfying the conditions

$$\xi = \xi_0 g_1, \qquad \eta = \xi_0 g_2, \quad \text{where} \quad \xi_0 = (0, 1). \tag{3.7}$$

The right-hand side does not depend on the choice of g_1 and g_2. In particular, $\Phi_\chi(g, \xi_0, \xi_0) = \Phi_\chi(g)$.

For any chosen vectors ξ and η, the distribution $(g, \xi, \eta) \mapsto \Phi_\chi(g, \xi, \eta)$ is concentrated on a unique cylindrical surface in G passing through the horosphere $h(\xi, \eta)$, and it is constant on all horospheres $h(\xi, \lambda\eta)$, $\lambda \in \mathbb{C} \setminus 0$, forming this surface. For a chosen $g \in G$ and for $|\xi_1|^2 + |\xi_2|^2 = 1$, this distribution reduces to a function on the manifold of horospheres H, as can be seen from the following formula (3.9).

PROPOSITION 3.2. *The horospherical functions $\Phi_\chi(g, \xi, \eta)$ satisfy the relations*

$$\Phi_\chi(g_1 g g_2^{-1}, \xi, \eta) = \Phi_\chi(g; \xi g_1, \eta g_2) \tag{3.8}$$

for any $g_1, g_2 \in G$ and

$$\Phi_\chi(g; \lambda_1 \xi, \lambda_2 \eta) = \chi(\lambda_1^{-1} \lambda_2)\, |\lambda_1 \lambda_2|^{-2}\, \Phi_\chi(g, \xi, \eta) \tag{3.9}$$

for any $\lambda_1, \lambda_2 \in \mathbb{C} \setminus 0$.

We give an explicit expression for the horospherical functions.

PROPOSITION 3.3. *The horospherical functions are of the form*

$$\Phi_\chi(g, \xi, \eta) = \delta\big(\eta_2(\xi_1\alpha + \xi_2\gamma) - \eta_1(\xi_1\beta + \xi_2\delta)\big)\, \chi\left(\frac{-\eta_1\beta + \eta_2\alpha}{\xi_2}\right), \tag{3.10}$$

where $\delta(\cdot)$ is the delta function on $\mathbb{C}$.

Indeed, one can set $g_1 = \begin{pmatrix} \xi_2^{-1} & 0 \\ \xi_1 & \xi_2 \end{pmatrix}$ and $g_2 = \begin{pmatrix} \eta_2^{-1} & 0 \\ \eta_1 & \eta_2 \end{pmatrix}$ in (3.6) and (3.7). This gives $g_1 g g_2^{-1} = \begin{pmatrix} \alpha' & \beta' \\ \gamma' & \delta' \end{pmatrix}$, where $\alpha' = \xi_2^{-1}(-\eta_1\beta + \eta_2\alpha)$ and $\gamma' = \eta_2(\xi_1\alpha + \xi_2\gamma) - \eta_1(\xi_1\beta + \xi_2\delta)$. Hence, using (3.6) and (3.3), we obtain

$$\Phi_\chi(g, \xi, \eta) = \delta(\gamma')\, \chi(\alpha'),$$

which gives (3.10).

REMARK. On the cylindrical surface $\gamma' = \eta_2(\xi_1\alpha + \xi_2\gamma) - \eta_1(\xi_1\beta + \xi_2\delta) = 0$ we have

$$\frac{\xi_1\alpha + \xi_2\gamma}{\eta_1} = \frac{\xi_1\beta + \xi_2\delta}{\eta_2} = \frac{\xi_1}{\eta_1\delta - \eta_2\gamma} = \frac{\xi_2}{-\eta_1\beta + \eta_2\alpha}.$$

Therefore, the ratio $\frac{\xi_2}{-\eta_1\beta+\eta_2\alpha}$ (the argument of the character χ^{-1} in (3.10)) can be replaced by each of these ratios.

PROPOSITION 3.4. *The distribution $\Phi_\chi(g,\xi,\eta)$ can be expressed by the integral*

$$\Phi_\chi(g,\xi,\eta) = \frac{i}{2}\int_{\mathbb{C}} \delta(\xi g - \lambda\eta)\,\chi^{-1}(\lambda)\,d\lambda\,d\overline{\lambda}, \tag{3.11}$$

where $\delta(\cdot)$ is the delta function on $\mathbb{C}^2$.

Indeed, it suffices to prove this relation for a zonal horospherical function, i.e., for $\xi = \eta = (0,1)$. However, for these values of ξ and η, the right-hand side is equal to

$$\frac{i}{2}\int_{\mathbb{C}} \delta(\gamma,\, \delta-\lambda)\,\chi^{-1}(\lambda)\,d\lambda\,d\overline{\lambda} \;=\; \delta(\gamma)\,\chi^{-1}(\delta) \;=\; \delta(\gamma)\,\chi(\alpha) \;=\; \Phi_\chi(g,\xi,\eta).$$

3.3. Fourier transform on G.

DEFINITION. Let us define the *Fourier transform* $\mathcal{F} f$ of a compactly supported C^∞ function f on G as follows:

$$\mathcal{F} f(\xi,\eta,\chi) = \langle \Phi_\chi(\cdot,\xi,\eta),\, f\rangle = \int_G \Phi_\chi(g,\xi,\eta)\, f(g)\, d\mu(g), \tag{3.12}$$

where $\Phi_\chi(g,\xi,\eta)$ is the horospherical function on G.

Using the explicit expression (3.10) and (3.11) for the horospherical functions, we can represent $\mathcal{F} f$ as follows:
(3.13)

$$\mathcal{F} f(\xi,\eta,\chi) = \int_G \delta(\eta_2(\xi_1\alpha+\xi_2\gamma) - \eta_1(\xi_1\beta+\xi_2\delta))\,\chi\left(\frac{-\eta_1\beta+\eta_2\alpha}{\xi_2}\right) f(g)\,d\mu(g),$$

where $g = \begin{pmatrix} \alpha & \beta \\ \gamma & \delta \end{pmatrix}$, or

$$\mathcal{F} f(\xi,\eta,\chi) = \frac{i}{2}\int_G\int_{\mathbb{C}} \delta(\xi g - \lambda\eta)\,\chi^{-1}(\lambda)\, f(g)\, d\lambda\, d\overline{\lambda}\, d\mu(g). \tag{3.14}$$

We emphasize that the integral (3.12) is in fact two-dimensional; namely, it is taken over the cylindrical surface in G on which the function $\Phi_\chi(\,\cdot\,;\xi,\eta)$ is concentrated.

Some facts immediately follow from the definitions.

1. The function $\mathcal{F} f$ satisfies the homogeneity condition

$$\mathcal{F} f(\lambda_1\xi, \lambda_2\eta, \chi) = \chi^{-1}(\lambda_1)\,|\lambda_1|^{-2}\,\chi(\lambda_2)\,|\lambda_2|^{-2}\,\mathcal{F} f(\xi,\eta;\chi). \tag{3.15}$$

2. The operator $\mathcal{F}$ commutes with the action of the group $G\times G$ on G and on the space of pairs (ξ,η); namely,

$$\text{if}\quad f_1(g) = f(g_1^{-1} g g_2), \quad\text{then}\quad \mathcal{F} f_1(\xi,\eta;\chi) = \mathcal{F} f(\xi g_1, \eta g_2;\chi). \tag{3.16}$$

We now establish another property of the operator $\mathcal{F}$. Choose an arbitrary unitary character χ (thus, $\chi\overline{\chi} \equiv 1$) and consider the space of functions $\varphi(\xi,\eta;\chi)$ satisfying the homogeneity condition (3.15). Equip this space with the structure of a C^*-algebra by setting

$$\varphi^*(\xi,\eta;\chi) = \overline{\varphi(\eta,\xi;\chi)},$$

$$(\varphi_1 * \varphi_2)(\xi,\eta;\chi) = \frac{i}{2}\int_{\gamma_\zeta} \varphi(\xi,\zeta;\chi)\,\varphi(\zeta,\eta;\chi)\,\omega(\zeta)\,\overline{\omega(\zeta)},$$

where $\omega(\zeta) = \zeta_1\, d\zeta_2 - \zeta_2\, d\zeta_1$ and the integral is taken over an arbitrary surface γ_ζ in $\mathbb{C}^2 \setminus O$ that intersects once almost every line passing through the origin O.

PROPOSITION 3.5. *The mapping $f \mapsto \mathcal{F} f(\,\cdot\,;\chi)$ is a C^*-algebra homomorphism, i.e., $\mathcal{F}(f_1 * f_2) = \mathcal{F} f_1 * \mathcal{F} f_2$ and $(\mathcal{F} f)^* = \mathcal{F} f^*$, where*

$$(f_1 * f_2)(g) = \int f_1(\widetilde{g})\, f_2(\widetilde{g}^{-1} g)\, d\mu(\widetilde{g})$$

and $f^*(g) = \overline{f(g^{-1})}$.

PROOF. The relation $(\mathcal{F} f)^* = \mathcal{F} f^*$ immediately follows from the definition of the Fourier transform. We now prove the first formula. By (3.16) we have

(3.17)

$$\begin{aligned}\mathcal{F}(f_1 * f_2)(\xi, \eta; \chi) &= \int_G f_1(\widetilde{g})\, \mathcal{F} f_2(\xi\widetilde{g}, \eta; \chi)\, d\mu(\widetilde{g}) \\ &= \left(\frac{i}{2}\right)^2 \int_G \int_{\mathbb{C}^2} f_1(\widetilde{g})\, \mathcal{F} f_2(\zeta, \eta; \chi) \delta(\xi\widetilde{g} - \zeta)\, d\zeta d\overline{\zeta} d\mu(\widetilde{g}) \\ &= \left(\frac{i}{2}\right)^2 \int_G \int_{\gamma_\zeta} \int_{\mathbb{C}} f_1(\widetilde{g})\, \mathcal{F} f_2(\zeta, \eta; \chi) \delta(\xi\widetilde{g} - \lambda\zeta) \chi^{-1}(\lambda)\, d\lambda d\overline{\lambda} \omega(\zeta) \overline{\omega(\zeta)} d\mu(\widetilde{g}),\end{aligned}$$

where $d\zeta = d\zeta_1\, d\zeta_2$; the last formula follows by passing to the "polar" coordinates with respect to ζ, namely, $\zeta = \lambda\zeta'$, where $\lambda \in \mathbb{C}$ and $\zeta' \in \gamma_\zeta$. Since

$$\frac{i}{2} \int_G \int_{\mathbb{C}} f_1(g) \delta(\xi g - \lambda\zeta)\, \chi^{-1}(\lambda)\, d\lambda\, d\overline{\lambda}\, d\mu(g) = \mathcal{F} f_1(\xi, \zeta; \chi),$$

it follows from (3.17) that

$$\begin{aligned}\mathcal{F}(f_1 * f_2)(\xi, \eta; \chi) &= \frac{i}{2} \int_{\gamma_\zeta} \mathcal{F} f_1(\xi, \zeta; \chi)\, \mathcal{F} f_2(\zeta, \eta; \chi)\, \omega(\zeta)\, \overline{\omega(\zeta)} \\ &= (\mathcal{F} f_1 * \mathcal{F} f_2)(\xi, \eta; \chi).\end{aligned}$$

COROLLARY. *We have*

$$\mathcal{F}(f * f^*)(\xi, \eta; \chi) = \frac{i}{2} \int_{\gamma_\zeta} \mathcal{F} f(\xi, \zeta; \chi)\, \overline{\mathcal{F} f(\eta, \zeta, \chi)}\, \omega(\zeta)\, \overline{\omega(\zeta)}. \tag{3.18}$$

3.4. Relation between the Fourier transform on G and the horospherical transform.

THEOREM 3.1. *The Fourier transform $\mathcal{F} f$ of a function f is related to the horospherical transform $\mathcal{R} f$ of f as follows:*

$$\mathcal{F} f(\xi, \eta; \chi) = \frac{i}{2} \int_{\mathbb{C}} \mathcal{R} f(\xi, \lambda\eta)\, \chi^{-1}(\lambda)\, d\lambda\, d\overline{\lambda}. \tag{3.19}$$

Indeed, using the explicit expression for $\mathcal{R} f$ given by (2.3) (see page 119), we can represent (3.14) in the form (3.19).

Thus, to pass from a function f on G to the Fourier transform $\mathcal{F} f$, we first integrate f over the horospheres on G, i.e., we pass from f to the horospherical transform $\mathcal{R} f$, and then integrate the function $\mathcal{R} f(\xi, \lambda\eta)\, \chi^{-1}(\lambda)$ with respect to λ.

REMARK. Using the homogeneity condition

$$\mathcal{R} f(\lambda\xi, \lambda\eta) = |\lambda|^{-4}\, \mathcal{R} f(\xi, \eta),$$

where λ is an arbitrary nonzero number, one can represent (3.19) as follows:

$$\mathcal{F}\, f(\xi,\eta;\chi) = \frac{i}{2}\int_{\mathbb{C}} \mathcal{R}\, f(\lambda\xi,\eta)\,\chi(\lambda)\, d\lambda\, d\bar{\lambda}. \tag{3.20}$$

We see that $\mathcal{F}\, f$ is obtained from $\mathcal{R}\, f$ by using the complex Mellin transform. Since $\chi(\lambda) = e^{i(\rho s + m\varphi)}$, where $s = \log|\lambda|$ and $\varphi = \arg\lambda$, this transform is reduced to the composition of the Fourier transforms on the real line and on the circle, and therefore, conversely, applying the classical inversion formula for the Fourier transform on $\mathbb{R}$ and on the circle S, we can express $\mathcal{R}\, f$ in terms of $\mathcal{F}\, f$. Namely, it follows from (3.19) that the following assertion holds.

COROLLARY. *The horospherical transform $\mathcal{R}\, f$ of a function f can be expressed in terms of the Fourier transform $\mathcal{F}\, f$ by the formula*

$$\mathcal{R}\, f(\xi,\eta) = (2\pi)^{-2}\int_X \mathcal{F}\, f(\xi,\eta;\chi)\, d\chi, \tag{3.21}$$

where the integral is taken over the set X of unitary characters $\chi(t) = t^{\frac{i\rho+m}{2}}\bar{t}^{\frac{i\rho-m}{2}}$, $\rho\in\mathbb{R}$, $m\in\mathbb{Z}$; the integration over X is understood as the usual integration over ρ and summation over m.

3.5. Symmetry relation for the Fourier transform. In § 2 (see (2.10)) we proved the symmetry relation for the horospherical transform $\mathcal{R}\, f$. Substituting the expression for $\mathcal{R}\, f$ in terms of $\mathcal{F}\, f$ (see (3.21)) into this relation, we obtain the following formula for the function $\mathcal{F}\, f$:

$$\int_X\int_{\gamma_\xi} \mathcal{F}\, f(\xi,\lambda\xi g;\chi)\,\omega(\xi)\,\overline{\omega(\xi)}\, d\chi = \int_X\int_{\gamma_\eta} \mathcal{F}\, f(\lambda\eta g^{-1},\eta;\chi)\,\omega(\eta)\,\overline{\omega(\eta)}\, d\chi$$

for any $g\in G$ and $\lambda\in\mathbb{C}\setminus 0$.

Let us write this relation in a simpler form. Replacing η by ξg and χ by χ^{-1} on the right-hand side and applying the homogeneity condition (3.15) for $\mathcal{F}\, f$, we see that

$$\int_X Z(g,\chi)\,\chi(\lambda)\,|\lambda|^{-2}\, d\chi = \int_X Z(g,\chi^{-1})\,\chi(\lambda)\,|\lambda|^{-2}\, d\chi,$$

where

$$Z(g,\chi) = \int_{\gamma_\xi} \mathcal{F}\, f(\xi,\xi g;\chi)\,\omega(\xi)\,\overline{\omega(\xi)}. \tag{3.22}$$

Since this relation holds for each $\lambda \neq 0$, the following assertion is true.

PROPOSITION 3.6. *The function $Z(g,\chi)$ given by (3.22) satisfies the functional equation*

$$Z(g,\chi) = Z(g,\chi^{-1}) \tag{3.23}$$

for any $g\in G$ and $\chi\in X$.

3.6. Inversion formula for the Fourier transform. The completeness of the family of horospherical functions $\Phi_\chi(g;\xi,\eta)$, where χ ranges over the unitary characters, means that there must be an inversion formula for the Fourier transform. Such a formula exists indeed, and it immediately follows from the inversion formula for the horospherical transform $\mathcal{R} f$; see (2.16). Namely, substituting the expression for $\mathcal{R} f$ in terms of $\mathcal{F} f$ into this formula, we obtain

$$
\begin{aligned}
f(g) &= (2\pi)^{-5}\frac{i}{2}\int_X\int_{\gamma_\xi} L_\eta\overline{L_\eta}\,\mathcal{F} f(\xi,\eta;\chi)|_{\eta=\xi g}\omega(\xi)\overline{\omega(\xi)}\,d\chi \\
&= -2^{-7}\pi^{-5}\frac{i}{2}\int_X\int_{\gamma_\xi} c(\chi)\,\mathcal{F} f(\xi,\xi g;\chi)\,\omega(\xi)\,\overline{\omega(\xi)}\,d\chi,
\end{aligned}
\tag{3.24}
$$

where $c(\chi) = \rho^2 + m^2$ for $\chi(t) = t^{\frac{i\rho+m}{2}}\bar{t}^{\frac{i\rho-m}{2}}$; the last relation follows from the homogeneity condition (3.15) for $\mathcal{F} f$ because condition (3.15) implies that $L_\eta\overline{L_\eta}\,\mathcal{F} f(\xi,\eta;\chi) = -\frac{1}{4}(\rho^2+m^2)\,\mathcal{F} f(\xi,\eta;\chi)$.

By the symmetry relation (3.23) for $\mathcal{F} f$, the integral over X can be replaced by the double integral over an arbitrarily chosen subset $X_0 \subset X$ containing exactly one representative of almost every pair of characters χ and χ^{-1}. This gives the following assertion.

THEOREM 3.2. *Each compactly supported C^∞ function f on G can be expressed in terms of the Fourier transform $\mathcal{F} f$ by the following inversion formula:*

$$
f(g) = -2^{-6}\pi^{-5}\frac{i}{2}\int_{X_0}\int_{\gamma_\xi} c(\chi)\,\mathcal{F} f(\xi,\xi g;\chi)\,\omega(\xi)\,\overline{\omega(\xi)}\,d\chi,
\tag{3.25}
$$

where $c(\chi) = \rho^2 + m^2$ for $\chi(t) = t^{\frac{i\rho+m}{2}}\bar{t}^{\frac{i\rho-m}{2}}$ and $X_0 \subset X$ is an arbitrarily chosen subset that contains exactly one representative of almost every pair of characters χ and χ^{-1}.

Recall that (as was assumed everywhere in §§ 2 and 3) the surface γ_ξ in $\mathbb{C}\setminus 0$ intersects once almost every line passing through the point 0; by the homogeneity condition for $\mathcal{F} f$, the integral does not depend on the choice of this surface.

One can write the inversion formula (3.25) similarly to formula (3.12) expressing $\mathcal{F} f$ in terms of f. Namely,

$$
\begin{aligned}
f(g) &= -2^{-6}\pi^{-5}\left(\frac{i}{2}\right)^2\int_{X_0}\int_{\gamma_\eta}\int_{\gamma_\xi} c(\chi)\Phi_{\chi^{-1}}(g;\xi,\eta) \\
&\qquad\qquad \times \mathcal{F} f(\xi,\eta;\chi)\,\omega(\xi)\overline{\omega(\xi)}\omega(\eta)\overline{\omega(\eta)}\,d\chi \\
&= -2^{-6}\pi^{-5}\left(\frac{i}{2}\right)^2\int_{X_0}\int_{\gamma_\eta}\int_{\gamma_\xi} \delta(\eta_2(\xi_1\alpha+\xi_2\gamma)-\eta_1(\xi_1\beta+\xi_2\delta)) \\
&\qquad\qquad \times\chi\left(\frac{\xi_1\alpha+\xi_2\gamma}{\eta_1}\right)\mathcal{F} f(\xi,\eta;\chi)\,\omega(\xi)\,\overline{\omega(\xi)}\omega(\eta)\overline{\omega(\eta)}\,d\chi,
\end{aligned}
\tag{3.26}
$$

where $g = \left(\begin{smallmatrix}\alpha & \beta\\ \gamma & \delta\end{smallmatrix}\right)$. To prove (3.26), it suffices to show that

$$
\begin{aligned}
&\frac{i}{2}\int_{\gamma_\xi}\mathcal{F} f(\xi,\xi g;\chi)\,\omega(\xi)\,\overline{\omega(\xi)} \\
&\quad = \left(\frac{i}{2}\right)^2\int_{\gamma_\eta}\int_{\gamma_\xi}\Phi_{\chi^{-1}}(g;\xi,\eta)\,\mathcal{F} f(\xi,\eta;\chi)\,\omega(\xi)\,\overline{\omega(\xi)}\omega(\eta)\overline{\omega(\eta)}.
\end{aligned}
\tag{3.27}
$$

We have

$$\begin{aligned}\frac{i}{2}\int_{\gamma_\xi} &\mathcal{F} f(\xi, \xi g; \chi)\, \omega(\xi)\, \overline{\omega(\xi)} \\ &= \left(\frac{i}{2}\right)^3 \int_{\gamma_\xi}\int_{\mathbb{C}^2} \mathcal{F} f(\xi, \eta; \chi)\, \delta(\xi g - \eta)\, d\eta\, d\overline{\eta}\, \omega(\xi)\overline{\omega(\xi)} \\ &= \left(\frac{i}{2}\right)^3 \int_{\gamma_\xi}\int_{\gamma_\eta}\int_{\mathbb{C}} \mathcal{F} f(\xi, \eta; \chi)\, \delta(\xi g - \lambda\eta)\, \chi(\lambda)\, d\lambda\, d\overline{\lambda}\, \omega(\eta)\overline{\omega(\eta)}\omega(\xi)\overline{\omega(\xi)},\end{aligned}$$

where the last equality is obtained by passing to the "polar" coordinates $\eta = \lambda\eta'$, where $\lambda \in \mathbb{C}$ and $\eta' \in \gamma_\eta$. Integrating over λ, we obtain (3.27) by using (3.11).

3.7. Analog of the Plancherel formula. The following assertion is obtained as a corollary to the inversion formula for the Fourier transform.

THEOREM 3.3. *The following relation holds for each compactly supported C^∞ function f on G:*

$$\begin{aligned}(3.28)\quad &\int_G |f(g)|^2\, d\mu(g) \\ &= -2^{-6}\pi^{-5}\left(\frac{i}{2}\right)^2 \int_{X_0}\int_{\gamma_\eta}\int_{\gamma_\xi} c(\chi)\, |\mathcal{F} f(\xi, \eta; \chi)|^2 \omega(\xi)\overline{\omega(\xi)}\omega(\eta)\overline{\omega(\eta)}\, d\chi\end{aligned}$$

(*the notation is the same as in* 3.6).

For the proof, consider the integral $\int_G f(g)\overline{f(g)}\, d\mu(g)$ and replace the factor $\overline{f(g)}$ in the integrand by the expression for $\overline{f(g)}$ in terms of $\mathcal{F} f$ given by the inversion formula (3.26). Since $\overline{\Phi_{\chi^{-1}}(g; \xi, \eta)} = \Phi_\chi(g; \xi, \eta)$, we obtain

$$\begin{aligned}\int_G |f(g)|^2\, d\mu(g) &= -2^{-6}\pi^{-5}\left(\frac{i}{2}\right)^2 \int_G\int_{X_0}\int_{\gamma_\eta}\int_{\gamma_\xi} c(\chi)\, f(g)\Phi_\chi(g; \xi, \eta) \\ &\qquad\qquad \times \overline{\mathcal{F} f(\xi, \eta; \chi)}\omega(\xi)\overline{\omega(\xi)}\omega(\eta)\overline{\omega(\eta)}\, d\chi\, d\mu(g) \\ &= -2^{-6}\pi^{-5}\left(\frac{i}{2}\right)^2 \int_{X_0}\int_{\gamma_\eta}\int_{\gamma_\xi} c(\chi)\, \mathcal{F} f(\xi, \eta; \chi)\overline{\mathcal{F} f(\xi, \eta; \chi)} \\ &\qquad\qquad \times \omega(\xi)\overline{\omega(\xi)}\omega(\eta)\overline{\omega(\eta)}\, d\chi.\end{aligned}$$

Denote by H_{X_0} the Hilbert space of functions $\varphi(\xi, \eta; \chi)$, where $\xi, \eta \in \mathbb{C}^2 \setminus O$ and $\chi \in X_0$, satisfying the same homogeneity conditions (3.15) as those for $\mathcal{F} f$, where H_{X_0} is equipped with the norm

$$\|\varphi\|^2 = -2^{-6}\pi^{-5}\left(\frac{i}{2}\right)^2 \int_{X_0}\int_{\gamma_\eta}\int_{\gamma_\xi} c(\chi)\, |\varphi(\xi, \eta; \chi)|^2 \omega(\xi)\overline{\omega(\xi)}\omega(\eta)\overline{\omega(\eta)}\, d\chi.$$

By the Plancherel formula, the operator $\mathcal{F}$, which was originally defined only on the compactly supported C^∞ functions, extends to the entire space $L^2(G)$ and defines an isometric mapping

$$\mathcal{F}\colon\ L^2(G) \to H_{X_0}.$$

One can prove that this mapping is in fact a Hilbert space isomorphism.

REMARK. One can define the space $H_{X_0} = H$ without choosing $X_0 \subset X$, namely, as the function space on the entire set X of unitary characters with the norm

$$\|\varphi\|^2 = -2^{-7}\pi^{-5}\left(\frac{i}{2}\right)^2 \int_X \int_{\gamma_\eta} \int_{\gamma_\xi} c(\chi)\, |\varphi(\xi,\eta;\chi)|^2\, \omega(\xi)\, \overline{\omega(\xi)}\, \omega(\eta)\, \overline{\omega(\eta)}\, d\chi,$$

assuming that the functions φ satisfy the additional symmetry condition (3.23).

3.8. Relation between the Fourier transform on G and the representations of the group $G \times G$. Consider the Hilbert space $L^2(G)$. It is equipped with the natural unitary representation of the group $G \times G$ by the translation operators:

$$(T(g_1, g_2)f)(g) = f(g_1^{-1} g g_2).$$

The problem of harmonic analysis on G is to decompose this representation into irreducible ones. We first construct the irreducible representations of the group $G \times G$. Take an eigenspace of the Laplace–Beltrami operators Δ and $\overline{\Delta}$ generated by the horospherical functions $\Phi_{\chi^{-1}}(\,\cdot\,; \xi, \eta)$, where $\chi(t) = t^{\frac{i\rho+m}{2}} \bar{t}^{\frac{i\rho-m}{2}}$. The elements of this space can be represented as

$$f_\chi(g) = \left(\frac{i}{2}\right)^2 \int_{\gamma_\eta} \int_{\gamma_\xi} u_\chi(\xi,\eta) \Phi_{\chi^{-1}}(g;\xi,\eta)\, \omega(\xi)\, \overline{\omega(\xi)}\, \omega(\eta)\, \overline{\omega(\eta)}, \tag{3.29}$$

where $u_\chi(\xi,\eta)$ satisfies the homogeneity condition

$$u_\chi(\lambda_1\xi, \lambda_2\eta) = \chi^{-1}(\lambda_1)\, |\lambda_1|^{-2} \chi(\lambda_2)\, |\lambda_2|^{-2} u_\chi(\xi,\eta)$$

for any nonzero λ_1 and λ_2. The integration is carried out over arbitrary surfaces γ_η and γ_ξ in $\mathbb{C} \setminus 0$ that intersect once almost every line passing through the point 0; the integral does not depend on the choice of these surfaces.

In the space formed by the functions f_χ we introduce the norm

$$\|f_\chi\|^2 = \left(\frac{i}{2}\right)^2 \int_{\gamma_\eta} \int_{\gamma_\xi} |u_\chi(\xi,\eta)|^2 \omega(\xi)\, \overline{\omega(\xi)}\, \omega(\eta)\, \overline{\omega(\eta)}, \tag{3.30}$$

consider the L^2 space with this norm, and denote the latter space by $\mathcal{H}_\chi$. The translations by elements of G define a unitary representation T_χ of $G \times G$ in $\mathcal{H}_\chi$, which acts on the functions $u_\chi(\xi,\eta)$ as follows:

$$(T_\chi(g_1, g_2) u_\chi)(\xi,\eta) = u_\chi(\xi g_1, \eta g_2).$$

Obviously, T_χ is the tensor product of the unitary representations of the factors of $G \times G$. Namely,

$$\mathcal{H}_\chi = H_{\chi^{-1}} \otimes H_\chi,$$
$$T_\chi(g_1, g_2) = \tau_{\chi^{-1}}(g_1) \otimes \tau_\chi(g_2),$$

where H_χ is the Hilbert space of functions $\varphi(\xi)$ on $\mathbb{C}^2 \setminus O$ satisfying the homogeneity condition

$$\varphi(\lambda\xi) = \chi(\lambda)\, |\lambda|^{-2} \varphi(\xi), \qquad \lambda \neq 0,$$

with the norm

$$\|\varphi\|^2 = \frac{i}{2} \int_{\gamma_\xi} |\varphi(\xi)|^2 \omega(\xi) \overline{\omega(\xi)},$$

and $\tau_{\chi^{-1}}$ and τ_χ are the natural representations of G in the spaces $H_{\chi^{-1}}$ and H_χ, respectively,

$$(\tau_\chi(g)\varphi)(\xi) = \varphi(\xi g). \tag{3.31}$$

It is known from the representation theory of the group $G = SL(2,\mathbb{C})$ (see [**8**]) that the representations τ_χ are irreducible, and two representations τ_{χ_1} and τ_{χ_2}, $\chi_1 \neq \chi_2$, are equivalent if and only if $\chi_2 = \chi_1^{-1}$. Hence, the representations T_χ of $G \times G$ are irreducible, and the representations T_{χ_1} and T_{χ_2}, $\chi_1 \neq \chi_2$, are equivalent if and only if $\chi_2 = \chi_1^{-1}$.

We want to decompose a function $f \in L^2(G)$ in the functions $f_\chi \in \mathcal{H}_\chi$. To this end, we define an operator $P_\chi\colon L^2(G) \to \mathcal{H}_\chi$ by the rule

$$P_\chi f = f_\chi,$$

where f_χ is given by formula (3.29) in which

$$u_\chi(\xi,\eta) = \mathcal{F} f(\xi,\eta;\chi).$$

Obviously, this operator commutes with the action of $G \times G$ on the spaces $L^2(G)$ and $\mathcal{H}_X$. Moreover, one can readily see that $P_\chi \neq 0$ for any χ. The next assertion immediately follows from the inversion formula (3.26) and the Plancherel formula (3.28) for the Fourier transform $\mathcal{F}$.

THEOREM 3.4. *The functions $f \in L^2(G)$ can be decomposed in the functions $f_\chi = P_\chi f$; namely, the following relations hold*:

$$f = -2^{-6}\pi^{-5} \int_{X_0} c(\chi)\, f_\chi \, d\chi, \tag{3.32}$$

$$\int_G |f(g)|^2 \, d\mu(g) = -2^{-6}\pi^{-5} \int_{X_0} c(\chi) \|f_\chi\|^2 \, d\chi, \tag{3.33}$$

where $c(\chi) = \rho^2 + m^2$ for $\chi(t) = t^{\frac{i\rho+m}{2}} \bar{t}^{\frac{i\rho-m}{2}}$; the integration is carried out over an arbitrarily chosen character subset $X_0 \subset X$ containing exactly one representative of almost every pair of characters χ, χ^{-1}.

COROLLARY. *The representation T of $G \times G$ in the space $L^2(G)$ has simple spectrum, i.e., it is decomposed into pairwise nonequivalent irreducible representations.*

Note that the simplicity of the spectrum of T (i.e., the fact that the integration in (3.32) and (3.33) is only over X_0 rather than over the entire character set X) follows from the symmetry relation for the Fourier transform $\mathcal{F}$.

REMARK. Simplicity of the spectrum is a usual fact observed when one decomposes representations related to symmetric spaces (see Chapter 3, where the representations related to the hyperbolic plane and hyperbolic space were considered).

3.9. Relation to the representations of the group G. We now discuss the previous results from the viewpoint of the representation theory for the group G. Consider the unitary representation of G in the space $L^2(G)$ by the operators of right translation,

$$(\tau(g_0)\varphi)(g) = \varphi(gg_0)$$

(the so-called *right regular representation*). The projection $\widetilde{\tau}_\chi$ of this representation to the space $\mathcal{H}_\chi = H_{\chi^{-1}} \otimes H_\chi$ is the tensor product

$$\widetilde{\tau}_\chi = I \otimes \tau_\chi,$$

where I is the identity representation in $H_{\chi^{-1}}$, and τ_χ is the irreducible unitary representation in H_χ given by (3.31). Hence, $\widetilde{\tau}_\chi$ is a multiple of the irreducible unitary representation τ_χ of G. Thus, relations (3.32) and (3.33) define the decomposition of the (right) regular representation of the group G into pairwise nonequivalent representations which are multiples of irreducible representations.

In terms of representations of the group G the Fourier transform $\mathcal{F} f$ has the following meaning. For any $\chi \in X$, consider the irreducible representation τ_χ of G in the space H_χ. To each compactly supported C^∞ function f on G we assign the following operator on the space H_χ:

$$\widehat{\tau}_\chi(f) = \int_G f(g)\tau_\chi(g)\, d\mu(g). \tag{3.34}$$

This operator acts on the functions $\varphi \in H_\chi$ by the rule

$$(\widehat{\tau}_\chi(f)\varphi)(\xi) = \int_G f(g)\, \varphi(\xi g)\, d\mu(g). \tag{3.35}$$

PROPOSITION 3.7. *The integral kernel of the operator $\widehat{\tau}_\chi(f)$ (in the homogeneous coordinates ξ, η) is the function $\mathcal{F} f(\xi, \eta; \chi^{-1})$, i.e.,*

$$(\widehat{\tau}_\chi(f)\varphi)(\xi) = \frac{i}{2} \int_{\gamma_\eta} \mathcal{F} f(\xi, \eta; \chi^{-1})\, \varphi(\eta)\, \omega(\eta)\, \overline{\omega(\eta)}. \tag{3.36}$$

Indeed, we integrate over G in (3.35) in two steps, namely, we first integrate over the horospheres $\xi g = \eta$, which gives the horospherical transform $\mathcal{R} f(\xi, \eta)\, \varphi(\eta)$, and then over η. We obtain

$$(\widehat{\tau}_\chi(f)\varphi)(\xi) = \left(\frac{i}{2}\right)^2 \int_{\mathbb{C}^2} \mathcal{R} f(\xi, \eta)\, \varphi(\eta)\, d\eta\, d\overline{\eta}, \qquad d\eta = d\eta_1 d\eta_2.$$

Here we pass to the "polar" coordinates $\eta = \lambda\eta'$, where $\lambda \in \mathbb{C}$ and $\eta' \in \gamma_\eta$. Since $\varphi(\lambda\eta) = \chi(\lambda)|\lambda|^{-2}\varphi(\eta)$ and $d\eta = \lambda\, d\lambda \omega(\eta)$, it follows that we obtain (3.36) after integrating with respect to λ.

COROLLARY. *The operator $\widehat{\tau}_\chi(f)$ has a trace, and this trace is equal to*

$$\operatorname{Tr} \widehat{\tau}_\chi(f) = \frac{i}{2} \int_{\gamma_\xi} \mathcal{F} f(\xi, \xi; \chi^{-1})\, \omega(\xi)\, \overline{\omega(\xi)}. \tag{3.37}$$

Using this result, we can evaluate the trace of a unitary operator $\tau_\chi(g)$, $g \in G$. Certainly, this operator has no trace in the usual sense. However, it is natural to define $\operatorname{Tr} \tau_\chi(g)$ as a distribution on the space of test functions on G,

$$\langle \operatorname{Tr} \tau_\chi(\cdot),\, f \rangle = \operatorname{Tr} \widehat{\tau}_\chi(f).$$

PROPOSITION 3.8.

$$\operatorname{Tr} \tau_\chi(g) = \frac{\chi(\lambda_g^{(1)}) + \chi(\lambda_g^{(2)})}{|\lambda_g^{(1)} - \lambda_g^{(2)}|}, \tag{3.38}$$

where $\lambda_g^{(1)}$ and $\lambda_g^{(2)}$ are the eigenvalues of the matrix g.

PROOF. According to the definition of $\mathcal{F} f$, we have

$$\frac{i}{2}\int_{\gamma_\xi} \mathcal{F} f(\xi,\xi,\chi^{-1})\,\omega(\xi)\,\overline{\omega(\xi)} = \frac{i}{2}\int_{\gamma_\xi}\int_G \delta(\xi_2(\xi_1\alpha+\xi_2\gamma)-\xi_1(\xi_1\beta+\xi_2\delta)) \times \chi\left(\frac{\xi_1\beta+\xi_2\delta}{\xi_2}\right) f(g)\,d\mu(g)\,\omega(\xi)\,\overline{\omega(\xi)}.$$

Hence, if for γ_ξ we take the line $\xi_1 = t$, $\xi_2 = 1$, then

$$\text{(3.39)}\qquad \begin{aligned}\operatorname{Tr}\,\tau_\chi(g) &= \frac{i}{2}\int \delta(t\alpha+\gamma-t(t\beta+\delta))\,\chi(t\beta+\delta)\,dt\,d\bar{t}\\ &= \frac{i}{2}\int \delta((s-\lambda_g^{(1)})(s-\lambda_g^{(2)}))\,\chi(s)\,ds\,d\bar{s}\end{aligned}$$

(the last equality is obtained by changing the variable $t\beta+\delta = s$). Formula (3.38) immediately follows from (3.39).

Since $\mathcal{F} f(\xi,\eta;\chi^{-1})$ is the integral kernel of the operator $\widehat{\tau}_\chi(f)$, we can modify the Plancherel formula. Namely, consider the operator $\widehat{\tau}_\chi(f)\widehat{\tau}^*_\chi(f) = \widehat{\tau}_\chi(f * f^*)$, where $\widehat{\tau}^*_\chi(f)$ is the operator conjugate to $\widehat{\tau}_\chi(f)$. By (3.18), the integral kernel of this operator is

$$\varphi(\xi,\eta;\chi) = \frac{i}{2}\int_{\gamma_\xi} \mathcal{F} f(\xi,\zeta;\chi)\ \mathcal{F} f(\eta,\zeta,\chi)\,\omega(\zeta)\,\overline{\omega(\zeta)}.$$

Hence,

$$\operatorname{Tr}\,(\widehat{\tau}_\chi(f)\widehat{\tau}^*_\chi(f)) = \left(\frac{i}{2}\right)^2 \int_{\gamma_\eta}\int_{\gamma_\xi} |\mathcal{F} f(\xi,\zeta;\chi)|^2\,\omega(\xi)\,\overline{\omega(\xi)}\,\omega(\zeta)\,\overline{\omega(\zeta)}.$$

Thus, the Plancherel formula (3.28) can be written as

$$\text{(3.40)}\qquad \int |f(g)|^2\,d\mu(g) = 2^{-6}\pi^{-5}\int_{X_0} c(\chi)\operatorname{Tr}\,(\widehat{\tau}_\chi(f)\widehat{\tau}^*_\chi(f))\,d\chi.$$

REMARK. The expansion (3.25) makes sense not only for the test functions but also for the distributions on G. Note that it is natural to view the delta function $\delta(g)$ on G as the character of the regular representation. Therefore, the inversion formula (3.25) for $\delta(g)$ can be written as

$$\text{(3.41)}\qquad \delta(g) = 2^{-6}\pi^{-5}\int_{X_0} c(\chi)\operatorname{Tr}\,\tau_\chi(g)\,d\chi.$$

4. Another version of the Fourier transform on $G = SL(2,\mathbb{C})$

In § 3 we have constructed the projections of functions on G to the eigenspaces of the Laplace–Beltrami operators Δ and $\overline{\Delta}$. The projection was carried out by using the zonal horospherical functions $\Phi_\chi(g) = \delta(\gamma)\,\chi(\alpha)$. However, one can do it in a different way, by using the zonal functions Ψ_χ given by

$$\text{(4.1)}\qquad \Psi_\chi(g) = \chi^{-1}(\gamma)\,|\gamma|^{-2}, \quad \text{where} \quad g = \begin{pmatrix}\alpha & \beta\\ \gamma & \delta\end{pmatrix}$$

(see 3.2). This leads to another version of the Fourier transform on G, which we now briefly describe.

4.1. Functions $\Psi_\chi(g;\xi,\zeta)$. Define analogs of the functions $\Phi_\chi(g;\xi,\eta)$ (see § 3) by the formula

$$\Psi_\chi(g_1 g g_2^{-1}) = \Psi_\chi(g;\xi,\zeta), \tag{4.2}$$

where $\xi,\zeta \in \mathbb{C}^2 \setminus O$ are given by the relations

$$\xi = \xi_0 g_1, \qquad \zeta = \zeta_0 g_2, \quad \text{where} \quad \xi_0 = \zeta_0 = (0,1). \tag{4.3}$$

One can readily see that the left-hand side of (4.2) does not depend on the choice of the matrices $g_1, g_2 \in G$ satisfying (4.3). In particular,

$$\Psi_\chi(g) = \Psi_\chi(g;\xi_0,\zeta_0).$$

Formulas (4.1) and (4.2) lead to the following explicit expression for these functions:

$$\Psi_\chi(g;\xi,\zeta) = \chi^{-1}(\xi g \zeta^\perp)\,|\xi g \zeta^\perp|^{-2}. \tag{4.4}$$

We note the following two properties of Ψ_χ (which are immediate consequences of the definitions):

$$\Psi_\chi(g_1 g g_2^{-1};\xi,\zeta) = \Psi_\chi(g;\xi g_1,\zeta g_2) \tag{4.5}$$

for any $g_1, g_2 \in G$;

$$\Psi_\chi(g\,;\lambda_1\xi,\lambda_2\zeta) = \chi^{-1}(\lambda_1\lambda_2)\,|\lambda_1\lambda_2|^{-2}\Psi_\chi(g\,;\xi,\zeta) \tag{4.6}$$

for any $\lambda_1, \lambda_2 \in \mathbb{C} \setminus 0$.

THEOREM 4.1. *The functions Ψ_χ are related to the horospherical functions Φ_χ (see § 3) as follows:*

$$\Psi_\chi(g;\xi,\zeta) = \frac{i}{2}\int_{\gamma_\eta} \Phi_\chi(g;\xi,\eta)\,\chi^{-1}(\eta\zeta^\perp)\,|\eta\zeta^\perp|^{-2}\omega(\eta)\,\overline{\omega(\eta)}; \tag{4.7}$$

$$\Phi_\chi(g;\xi,\eta) = (2\pi)^{-4}\,c(\chi)\,\frac{i}{2}\int_{\gamma_\zeta} \Psi_\chi(g;\xi,\zeta)\,\chi(\eta\zeta^\perp)\,|\eta\zeta^\perp|^{-2}\omega(\zeta)\,\overline{\omega(\zeta)}, \tag{4.8}$$

where $c(\chi) = \rho^2 + m^2$ for $\chi(t) = t^{\frac{i\rho+m}{2}}\bar{t}^{\frac{i\rho-m}{2}}$. The integration is carried out over arbitrary two-dimensional real surfaces γ_η and γ_ζ in $\mathbb{C}^2 \setminus O$ intersecting once almost every complex line passing through the origin O.

Both integrals must be understood as integrals of distributions. We do not present rigorous definitions of these integrals and restrict ourselves to a formal proof of (4.7) and (4.8).

We first note that it suffices to prove these relations for a chosen element g and for a chosen $\xi = \xi_0$. By setting $g = e$ and $\xi_0 = (1,0)$, we obtain

$$\Phi_\chi(e,\xi_0,\eta) = \delta(\eta_2)\,\chi(\eta_1); \qquad \Psi_\chi(e,\xi_0,\zeta) = \chi^{-1}(\zeta_2)\,|\zeta_2|^{-2}.$$

If in (4.7) with $g = e$ and $\xi = \xi_0$ we take for γ_η the line $\eta_1 = 1$, then the integral becomes

$$\frac{i}{2}\int_{\mathbb{C}} \delta(\eta_2)\,\chi^{-1}(\zeta_2 - \eta_2\zeta_1)\,|\zeta_2 - \eta_2\zeta_1|^{-2}\,d\eta_2\,d\overline{\eta}_2.$$

Obviously, this integral is equal to $\chi^{-1}(\zeta_2)\,|\zeta_2|^{-2}$, i.e., to $\Psi_\chi(e,\xi_0,\zeta)$.

Similarly, if in (4.8) with $g = e$ and $\xi = \xi_0$ we take for γ_ζ the line $\zeta_2 = 1$, then the integral becomes

$$\frac{i}{2}\int_{\mathbb{C}} \chi(\eta_1 - \eta_2\zeta_1)\,|\eta_1 - \eta_2\zeta_1|^{-2}\,d\zeta_1\,d\overline{\zeta}_1$$
$$= \chi(\eta_1)\,\frac{i}{2}\int_{\mathbb{C}} \chi(1 - \eta_2\zeta_1)\,|1 - \eta_2\zeta_1|^{-2}\,d\zeta_1\,d\overline{\zeta}_1.$$

Obviously, the distribution

$$\varphi(\eta_2) = \frac{i}{2}\int_{\mathbb{C}} \chi(1 - \eta_2\zeta_1)\,|1 - \eta_2\zeta_1|^{-2}\,d\zeta_1\,d\overline{\zeta}_1$$

satisfies the homogeneity condition $\varphi(\lambda\eta_2) = |\lambda|^{-2}\,\varphi(\eta_2)$. This implies that $\varphi(\eta_2) = c\,\delta(\eta_2)$.

We leave the evaluation of the factor c to the reader as a useful exercise on integration of distributions.

4.2. Fourier transform on G.

DEFINITION. We define the *Fourier transform* $\widehat{\mathcal{F}}f$ of a compactly supported C^∞ function f on G by the rule

$$\widehat{\mathcal{F}}f(\xi,\zeta;\chi) = \langle \Psi_\chi(\cdot,\xi,\zeta),\, f\rangle = \int_G f(g)\,\chi^{-1}(\xi g\zeta^\perp)\,|\xi g\zeta^\perp|^{-2}\,d\mu(g). \tag{4.9}$$

There is a simple relation between the Fourier transform $\widehat{\mathcal{F}}f$ and the integral transform $\mathcal{R}^p f$ in § 2 related to paraboloids on G. Namely, comparing the definition of $\widehat{\mathcal{F}}f$ with the definition of $\mathcal{R}^p f$ (see (2.20)), we obtain the following assertion.

THEOREM 4.2. *The operators $\widehat{\mathcal{F}}$ and $\mathcal{R}^p$ are related as follows:*

$$\widehat{\mathcal{F}}f(\xi,\zeta;\chi) = \frac{i}{2}\int \mathcal{R}^p f(\xi,\zeta;p)\,\chi^{-1}(p)|p|^{-2}\,dp\,d\overline{p}. \tag{4.10}$$

Thus, $\widehat{\mathcal{F}}f$ is obtained from $\mathcal{R}^p f$ by using the Mellin transform with respect to p; cf. the relation between $\mathcal{F}f$ and $\mathcal{R}f$ in 3.4. Hence, applying the inversion formula for the Mellin transform, we obtain the following assertion.

COROLLARY. *The function $\mathcal{R}^p f$ can be expressed in terms of $\widehat{\mathcal{F}}f$ as follows:*

$$\mathcal{R}^p f(\xi,\zeta,p) = (2\pi)^{-2}\int_X \widehat{\mathcal{F}}f(\xi,\zeta,\chi)\,\chi(p)\,d\chi, \tag{4.11}$$

where the integral is taken over the set X of unitary characters.

4.3. Relation between the above two versions of the Fourier transform. It follows from the relations between the horospherical functions Ψ_χ and Φ_χ (see 4.1) that the functions $\widehat{\mathcal{F}}f$ and $\mathcal{F}f$ satisfy the same relations, where $\mathcal{F}$ is the Fourier operator introduced in § 3.

THEOREM 4.3. *The Fourier operators $\widehat{\mathcal{F}}$ and $\mathcal{F}$ are related as follows:*

$$\widehat{\mathcal{F}}f(\xi,\zeta;\chi) = \frac{i}{2}\int_{\gamma_\eta} \mathcal{F}f(\xi,\eta;\chi)\,\chi^{-1}(\eta\zeta^\perp)\,|\eta\zeta^\perp|^{-2}\,\omega(\eta)\,\overline{\omega(\eta)}; \tag{4.12}$$

$$\mathcal{F}f(\xi,\eta;\chi) = (2\pi)^{-4}c(\chi)\frac{i}{2}\int_{\gamma_\zeta} \widehat{\mathcal{F}}f(\xi,\zeta;\chi)\,\chi(\eta\zeta^\perp)\,|\eta\zeta^\perp|^{-2}\,\omega(\zeta)\,\overline{\omega(\zeta)}, \tag{4.13}$$

where $c(\chi) = \rho^2 + m^2$ for $\chi(t) = t^{\frac{i\rho+m}{2}}\overline{t}^{\frac{i\rho-m}{2}}$.

REMARK. These relations can also be established immediately by using the known relations between the integral transforms $\mathcal{R}$ and $\mathcal{F}$, between $\mathcal{R}$ and $\mathcal{R}^p$, and between $\mathcal{R}^p$ and $\widehat{\mathcal{F}}$.

4.4. Symmetry relation. The next assertion follows from (4.12) and the symmetry relation for $\mathcal{F}f$.

THEOREM 4.4. *The function $\mathcal{F}f$ satisfies the symmetry relation*

$$\widehat{Z}(g,\chi) = \widehat{Z}(g,\chi^{-1}) \quad \textit{for any} \quad g \in G, \quad \chi \in X, \tag{4.14}$$

where

$$\widehat{Z}(g,\chi) = \left(\frac{i}{2}\right)^2 \int_{\gamma_\zeta}\int_{\gamma_\xi} \widehat{\mathcal{F}}f(\xi,\zeta;\chi)\,\chi(\xi g\zeta^\perp)\,|\xi g\zeta^\perp|^{-2}\omega(\xi)\,\overline{\omega(\xi)}.$$

4.5. Inversion formula and Plancherel formula for the Fourier transform $\widehat{\mathcal{F}}$.

THEOREM 4.5. *Every compactly supported C^∞ function f on G can be expressed in terms of the Fourier transform $\widehat{\mathcal{F}}f$ by the following inversion formula*:

$$\begin{aligned} f(g) = -2^{-8}\pi^{-7}\left(\frac{i}{2}\right)^2 \int_{X_0}\int_{\gamma_\zeta}\int_{\gamma_\xi} & c^2(\chi)\widehat{\mathcal{F}}f(\xi,\zeta;\chi)\Psi_{\chi^{-1}}(g;\xi,\zeta) \\ & \times \omega(\xi)\,\overline{\omega(\xi)}\,\omega(\zeta)\,\overline{\omega(\zeta)}\,d\chi, \end{aligned} \tag{4.15}$$

where X_0 is an arbitrary subset of the set X of the unitary characters $\chi(t) = t^{\frac{i\rho+m}{2}}\bar{t}^{\frac{i\rho-m}{2}}$, $\rho \in \mathbb{R}$, $m \in \mathbb{Z}$, such that X_0 contains precisely one representative of almost every pair of characters χ, χ^{-1}.

Indeed, it immediately follows from the inversion formula for the integral transform $\mathcal{R}^p$ related to paraboloids on G (see (2.25)) and from the above relation between $\mathcal{R}^p f$ and $\widehat{\mathcal{F}}f$ that

$$\begin{aligned} f(g) &= -2^{-9}\pi^{-7}\left(\frac{i}{2}\right)^2 \int_{X}\int_{\gamma_\zeta}\int_{\gamma_\xi} c^2(\chi)\widehat{\mathcal{F}}f(\xi,\zeta;\chi)\chi(\xi g\zeta^\perp)\,|\xi g\zeta^\perp|^{-2} \\ &\qquad \times \omega(\xi)\,\overline{\omega(\xi)}\,\omega(\zeta)\,\overline{\omega(\zeta)}\,d\chi \\ &= -2^{-9}\pi^{-7}\left(\frac{i}{2}\right)^2 \int_{X}\int_{\gamma_\zeta}\int_{\gamma_\xi} c^2(\chi)\widehat{\mathcal{F}}f(\xi,\zeta;\chi)\Psi_{\chi^{-1}}(g;\xi,\zeta) \\ &\qquad \times \omega(\xi)\,\overline{\omega(\xi)}\,\omega(\zeta)\,\overline{\omega(\zeta)}\,d\chi. \end{aligned} \tag{4.16}$$

By the symmetry relation (4.14) for $\widehat{\mathcal{F}}f$, the integral over X in (4.16) can be replaced by the doubled integral over the subset $X_0 \subset X$.

Note that there is an analogy between the inversion formula (4.15) and the original formula (4.9) expressing $\widehat{\mathcal{F}}f$ in terms of f.

As in the case of the Fourier transform $\mathcal{F}$, the inversion formula (4.15) implies the following assertion.

COROLLARY. *The following Plancherel formula holds for functions on G:*

$$(4.17)\quad \int_G |f(g)|^2\,d\mu(g) = -2^{-8}\pi^{-7}\left(\frac{i}{2}\right)^2 \int_{X_0}\int_{\gamma_\zeta}\int_{\gamma_\xi} c^2(\chi)\,|\widehat{\mathcal{F}}f(\xi,\zeta,\chi)|^2\,\omega(\xi)\,\overline{\omega(\xi)}\,\omega(\zeta)\,\overline{\omega(\zeta)}\,d\chi.$$

According to the corollary, the Fourier transform $\widehat{\mathcal{F}}$, which was originally defined for compactly supported C^∞ functions on G only, can be extended to all functions in $L^2(G)$.

4.6. Relations with representation theory. The arguments in § 3 about the relation between the Fourier transform $\mathcal{F}$ and representation theory can be applied to the Fourier transform $\widehat{\mathcal{F}}$ without substantial modifications. Namely, to decompose the representation T of $G\times G$ in the space $L^2(G)$ into irreducible representations, we consider the eigenspace (of the Laplace–Beltrami operators Δ and $\overline{\Delta}$) generated by the horospherical functions $\Psi_{\chi^{-1}}(\,\cdot\,;\xi,\zeta)$, where $\chi(t) = t^{\frac{i\rho+m}{2}}\bar{t}^{\frac{i\rho-m}{2}}$. The elements of this subspace can be written as

$$(4.18)\qquad \widehat{f}_\chi(g) = \left(\frac{i}{2}\right)^2 \int_{\gamma_\zeta}\int_{\gamma_\xi} \widehat{u}_\chi(\xi,\zeta)\,\Psi_{\chi^{-1}}(g;\xi,\zeta)\,\omega(\xi)\,\overline{\omega(\xi)}\,\omega(\zeta)\,\overline{\omega(\zeta)},$$

where $\widehat{u}_\chi(\xi,\zeta)$ satisfies the homogeneity condition

$$\widehat{u}_\chi(\lambda_1\xi,\lambda_2\zeta) = \chi^{-1}(\lambda_1\lambda_2)|\lambda_1\lambda_2|^{-2}\widehat{u}_\chi(\xi,\zeta)$$

for any $\lambda_1,\lambda_2\in\mathbb{C}\setminus 0$. Equip the space of functions $\widehat{f}_\chi$ with the norm

$$\|\widehat{f}_\chi\|^2 = \left(\frac{i}{2}\right)^2 \int_{\gamma_\zeta}\int_{\gamma_\xi} |\widehat{u}_\chi(\xi,\zeta)|^2\omega(\xi)\,\overline{\omega(\xi)}\,\omega(\zeta)\,\overline{\omega(\zeta)},$$

consider the L^2 space with this norm, and denote this space by $\widehat{H}_\chi$. The translations on G define an irreducible unitary representation T_χ of $G\times G$ in $\widehat{H}_\chi$, and this representation acts on the functions $\widehat{u}_\chi(\xi,\zeta)$ as follows:

$$(T_\chi(g_1,g_2)\widehat{u}_\chi)(\xi,\zeta) = \widehat{u}_\chi(\xi g_1,\zeta g_2).$$

We further construct the operator $\widehat{P}_\chi\colon L^2(G)\to\widehat{H}_\chi$ by the rule $\widehat{P}_\chi f = \widehat{f}_\chi$, where $\widehat{f}_\chi$ is given by relation (4.18) in which $\widehat{u}_\chi(\xi,\zeta) = \widehat{\mathcal{F}}f(\xi,\zeta;\chi)$.

The desired decomposition of the representation T is given by the formulas

$$(4.19)\qquad f = -2^{-8}\pi^{-7}\int_{X_0} c^2(\chi)\widehat{f}_\chi\,d\chi, \qquad \widehat{f}_\chi = \widehat{P}_\chi f,$$

$$(4.20)\qquad \int_G |f(g)|^2\,d\mu(g) = 2^{-8}\pi^{-7}\int_{X_0} c^2(\chi)\,\|\widehat{f}_\chi\|^2\,d\chi.$$

These relations immediately follow from (4.15) and from the Plancherel formula (4.17) for the Fourier transform $\widehat{\mathcal{F}}$.

Since the decomposition of the representation T of $G\times G$ into irreducible representations is unique (because the spectrum is simple), it follows that the decomposition thus obtained coincides with the one found in 3.8. Namely,

$$\widehat{f}_\chi = c^{-1}(\chi)\,f_\chi, \qquad \|\widehat{f}_\chi\| = c^{-1/2}(\chi)\|f_\chi\|.$$

Note that the relation $\widehat{f_\chi} = c^{-1}(\chi)\, f_\chi$ is equivalent to the above relations between $\mathcal{F} f$ and $\widehat{\mathcal{F}} f$.

CHAPTER 5

Integral Geometry on Quadrics

This chapter is devoted to the problem of reconstructing a function f on a quadric $X \subset \mathbb{R}^{n+1}$ from the integrals $\widehat{f}$ of f over the hyperplane sections of the quadric. This is an example of an overdetermined problem of integral geometry; namely, we reconstruct a function of n variables (n is the dimension of X) from a function of $(n+1)$ variables ($n+1$ is the dimension of the space of sections). Another example of an overdetermined problem (the John transform in $\mathbb{R}^3$ and $\mathbb{C}^3$) was already treated in Chapter 2.

As in any overdetermined problem, the main task here is to find n-dimensional submanifolds Γ of plane sections (admissible submanifolds) such that the function f can be reconstructed from the restriction of the function $\widehat{f}$ to Γ.

The reconstruction of f from $\widehat{f}|_\Gamma$ is carried out by the general method of the operator κ, which was first applied in [**7**]. This method was explained in Chapter 2 using, as an example, the John transform in $\mathbb{C}^3$. In the case considered in this chapter, for any point $x \in X$, we assign to a function $\widehat{f}$ a closed differential form $\kappa_x \widehat{f}$ of degree n on the manifold of plane sections. The integrals of this form over various cycles are equal to the value of the original function f at the point x. This is a rare case where the method of the operator κ is developed for a nonlocal problem of real integral geometry.

Many formulas of this chapter can be written for arbitrary quadrics; however, the choice of appropriate cycles giving the inversion formulas is difficult and often impossible. For this reason we restrict ourselves to the case of a hyperboloid of two sheets and a sphere (which are projectively equivalent). The case of a hyperboloid of two sheets gives another proof of the inversion formulas in problems of integral geometry related to the hyperbolic space (see § 3). The case of a sphere is equivalent to the problem of integral geometry related to spheres in $\mathbb{R}^n$. The operator κ for quadrics was constructed by Gindikin [**15, 16**] and Goncharov [**24**]. The problem of sections of the quadric can be interpreted as a conformally invariant extension of the Radon transform. In this interpretation the Radon transform (Chapter 1) and the horospherical hyperbolic transform are conformally equivalent.

1. Integral transform related to the hyperplane sections of a hyperboloid of two sheets in $\mathbb{R}^{n+1}$

1.1. Definition. Let $\mathcal{L}^n$ be the upper sheet of the hyperboloid of two sheets in $\mathbb{R}^{n+1}$ given by

$$[x,x] = x_0^2 - x_1^2 - \cdots - x_n^2 = 1, \quad x_0 > 0. \tag{1.1}$$

Consider the sections of $\mathcal{L}^n$ by the hyperplanes $[\xi, x] = p$ in $\mathbb{R}^{n+1}$, where

$$[\xi, x] = \xi_0 x_0 - \xi_1 x_1 - \cdots - \xi_n x_n.$$

We recall that $\mathcal{L}^n$ is one of the models of the n-dimensional hyperbolic space, and corresponding to the hyperplane sections of $\mathcal{L}^n$ are various spheres in the hyperbolic space and their limit cases, namely, the geodesic hypersurfaces and the horospheres. Generic spheres occur as sections of $\mathcal{L}^n$ by the hyperplanes

$$[\xi, x] = p,$$

where $p \neq 0$ and $[\xi, \xi] \neq 0$. The sections related to the limit cases $p = 0$ and $[\xi, \xi] = 0$ are the geodesic hypersurfaces and the horospheres, respectively. Thus, the manifold of spheres is $(n+1)$-dimensional, and the manifolds of horospheres and of the geodesic hypersurfaces are n-dimensional.

Note that the horospheres and the geodesic hypersurfaces are the only n-dimensional families (in the manifold of all hyperplane sections) that are invariant with respect to the motion group $SO(1,n)$ of the hyperbolic space.

We introduce a differential n-form $\mu(x)$ on $\mathbb{R}^{n+1}$ satisfying the formula

$$d([x,x]) \wedge \mu(x) = dx, \quad \text{where} \quad dx = dx_0 \wedge \cdots \wedge dx_n$$

(i.e., $\mu(x)$ is an interior product of the differential forms dx and $d([x,x])$). Such a form on $\mathbb{R}^{n+1}$ is not unique; for instance, it can be given by the relation

$$\mu(x) = \frac{1}{2\,[\xi, \xi]} \sum_{i=0}^{n+1} (-1)^i \, \xi_i \bigwedge_{j \neq i} d\xi_j.$$

However, the restriction of this form to $\mathcal{L}^n$ is uniquely defined. In the coordinates $x_1, \ldots, x_n$ on $\mathcal{L}^n$, this restriction is

$$\mu(x) = \frac{dx_1 \wedge \cdots \wedge dx_n}{2x_0}.$$

We introduce the integral transform $\mathcal{R}$ taking a function f on $\mathcal{L}^n$ to the integrals of f over the sections of $\mathcal{L}^n$ by the hyperplanes $[\xi, x] = p$ in $\mathbb{R}^{n+1}$,

$$\mathcal{R}f(\xi, p) = \int_{\mathcal{L}^n} f(x)\, \delta([\xi, x] - p)\, \mu(x). \tag{1.2}$$

Assume that the functions f are compactly supported and infinitely differentiable on $\mathcal{L}^n$. Obviously, $\mathcal{R}f(\xi, p) = 0$ if the hyperplane $[\xi, x] = p$ is disjoint from $\mathcal{L}^n$.

REMARK. The integral transform $\mathcal{R}$ can be regarded as the Radon transform of the distribution $f(x)\, \delta([x,x] - 1)$ on $\mathbb{R}^{n+1}$.

It immediately follows from (1.2) that $\varphi = \mathcal{R}f$ is a function on $\mathbb{R}^{n+2} \setminus O$ satisfying the homogeneity condition

$$\varphi(\lambda\xi, \lambda p) = |\lambda|^{-1}\, \varphi(\xi, p), \qquad \lambda \in \mathbb{R} \setminus O, \tag{1.3}$$

and the differential equation

$$\frac{\partial^2 \varphi}{\partial \xi_0^2} - \frac{\partial^2 \varphi}{\partial \xi_1^2} - \cdots - \frac{\partial^2 \varphi}{\partial \xi_n^2} - \frac{\partial^2 \varphi}{\partial p^2} = 0. \tag{1.4}$$

Note that (1.4) is the wave equation.

1.2. Admissible submanifolds in the manifold of hyperplane sections of $\mathcal{L}^n$. An n-dimensional submanifold Γ of the manifold of hyperplane sections of $\mathcal{L}^n$ is said to be *admissible* if any function f on $\mathcal{L}^n$ can be reconstructed from the restriction of the function $\varphi = \mathcal{R}f$ to Γ. In Chapter 3 we obtained inversion formulas (for $n = 2$ and $n = 3$) reconstructing a function f on the hyperbolic space $\mathcal{L}^n$ from the integrals of this function over the horospheres or over the geodesics. In particular, we have established that the manifold of horospheres and the manifold of geodesics are admissible submanifolds of the manifold of all hyperplane sections.

We now present, for an arbitrary n, three examples of admissible submanifolds in the space of hyperplane sections of $\mathcal{L}^n$. In each of these examples, the proof of the inversion formula is reduced to the inversion formula of Chapter 1 for the Radon transform in the space $\mathbb{R}^n$.

EXAMPLE 1. Γ is the manifold of sections of $\mathcal{L}^n$ by the hyperplanes

$$[\xi, x] = p, \quad \text{where} \quad \xi_0 = 0.$$

Denote by $\psi(\xi_1, \ldots, \xi_n, p)$ the restriction of the function $\varphi = \mathcal{R}f$ to Γ, i.e.,

$$\psi(\xi_1, \ldots, \xi_n, p) = \int_{\mathcal{L}^n} \frac{f(x)}{2x_0}\, \delta(\xi_1 x_1 + \cdots + \xi_n x_n + p)\, dx_1 \wedge \cdots \wedge dx_n.$$

We see that ψ is the Radon transform of the function $\frac{f(x)}{2x_0}$ viewed as a function of $x_1, \ldots, x_n$ on $\mathbb{R}^n$. Hence, by using the inversion formula for the Radon transform (see 2.3 of Chapter 1), we obtain the desired inversion formula:

$$(1.5)\quad 2\,x_0 \int_\gamma \int_{-\infty}^{\infty} \psi(\xi_1, \ldots, \xi_n, p)\, (\xi_1 x_1 + \cdots + \xi_n x_n + p - i0)^{-n}\, dp \wedge \sigma(\xi) = c_n\, f(x),$$

where $c_n = \frac{(2\pi)^n}{i^n (n-1)!}$, $\sigma(\xi) = \sum_{i=1}^n (-1)^{i-1} \xi_i \bigwedge_{j \neq i} d\xi_j$, and $\gamma \subset \mathbb{R}^n$ is an arbitrary surface intersecting once every ray issuing from the origin O.

In particular, we have proved that the manifold Γ is admissible.

EXAMPLE 2. Γ is the manifold of sections of $\mathcal{L}^n$ by the hyperplanes

$$[\xi, x] = p, \quad \text{where} \quad \xi_0 = \xi_1.$$

Denote by $\psi(\xi_1, \ldots, \xi_n, p)$ the restriction of the function $\varphi = \mathcal{R}f$ to Γ, i.e.,

$$\begin{aligned}\psi(\xi_1, \ldots, \xi_n, p) &= \varphi(\xi_1, \xi_1, \ldots, \xi_n, p) \\ &= \int_{\mathcal{L}^n} f(x)\, \delta(\xi_1(x_1 - x_0) + \xi_2 x_2 + \cdots + \xi_n x_n + p)\, \mu(x).\end{aligned}$$

We find the inversion formula reconstructing f from the function ψ. To this end, we introduce a new system of coordinates on $\mathcal{L}^n$,

$$y_1 = x_1 - x_0,\ y_2 = x_2, \ldots, y_n = x_n,$$

and denote by $f_1(y)$ the original function f in these coordinates. It follows from the assumptions imposed on f that the function f_1 is defined on the domain $y_1 < 0$ and is compactly supported in this domain. Assume that f_1 is extended to the entire space $\mathbb{R}^n$ by setting $f_1(y) = 0$ for $y_1 \geq 0$. In the coordinates y, the differential form μ becomes

$$\mu(y) = -\frac{dy_1 \wedge \cdots \wedge dy_n}{2y_1},$$

and hence

$$\psi(\xi_1, \dots, \xi_n, p) = -\int_{\mathbb{R}^n} \frac{f_1(y)}{2y_1}\, \delta(\xi_1 y_1 + \dots + \xi_n x_n + p)\, dy_1 \wedge \dots \wedge dy_n. \tag{1.6}$$

Thus, ψ is the Radon transform of the function $F(y) = -\frac{f_1(y)}{2y_1}$. Therefore, it follows from the inversion formula for the Radon transform that

$$-2y_1 \int_\gamma \int_{-\infty}^{\infty} \psi(\xi_1, \dots, \xi_n, p)\, (\xi_i y_1 + \dots + \xi_n y_n + p - i0)^{-n}\, dp \wedge \omega(\xi) = c_n\, f(x),$$

where c_n, $\omega(\xi)$, and γ are defined in Example 1.

Returning from y to the original variables x, we obtain the desired inversion formula

$$\begin{aligned} 2\,(x_0 - x_1) \int_\gamma \int_{-\infty}^{\infty} & \psi(\xi_1, \dots, \xi_n, p) \\ & \times\ (\xi_i(x_1 - x_0) + \xi_2 x_2 + \dots + \xi_n x_n + p - i0)^{-n}\, dp \wedge \omega(\xi) = c_n\, f(x). \end{aligned} \tag{1.7}$$

REMARK. The manifolds Γ in Examples 1 and 2 have a simple geometric description; namely, they consist of the sections of $\mathcal{L}^n$ by the hyperplanes passing through the points at infinity with the homogeneous coordinates $(1, 0, \dots, 0)$ and $(1, 1, 0, \dots, 0)$, respectively. The first point does not belong to the projectivization of the hyperboloid, whereas the other point belongs to the projectivization.

EXAMPLE 3. Γ is the manifold of geodesic hypersurfaces in $\mathcal{L}^n$, i.e., the submanifold of sections by the hyperplanes passing through the origin, $[\xi, x] = 0$.

Denote by $\psi(\xi_0, \dots, \xi_n)$ the restriction of the function $\varphi = \mathcal{R}f$ to Γ, i.e.,

$$\psi(\xi_0, \dots, \xi_n) = \frac{1}{2} \int_{\mathcal{L}^n} f(x)\, \delta([\xi, x])\, \frac{dx_1 \wedge \dots \wedge dx_n}{x_0}.$$

One can view $f(x)$ as the restriction to $\mathcal{L}^n$ of a homogeneous function on $\mathbb{R}^{n+1} \setminus O$ defined by the relations $f(\lambda x) = |\lambda|^{-n} f(x)$ for any $\lambda \neq 0$ and $x \in \mathcal{L}^n$ and $f(x) = 0$ for $[x, x] \leq 0$. We also note that the differential form $(x_0)^{-1} dx_1 \wedge \dots \wedge dx_n$ in the integrand is the restriction to $\mathcal{L}^n$ of the differential form $\omega_{n+1}(x) = \sum_{i=0}^{n} (-1)^i x_i \bigwedge_{j \neq i} dx_j$ on $\mathbb{R}^{n+1}$. Hence (see 8.8 in Chapter 1), the function $2\,\psi(\xi_0, \dots, \xi_n)$ is the projective Radon transform of the homogeneous function $f(x)$. Thus, in view of the inversion formula for the projective Radon transform (see 8.8 in Chapter 1), we obtain

$$2 \int_\Gamma \psi(\xi_0, \dots, \xi_n)\, ([\xi, x] - i0)^{-n}\, \omega_{n+1}(\xi) = c_n\, f(x), \tag{1.8}$$

where $c_n = \frac{(2\pi)^n}{i^n (n-1)!}$.

1.3. Operator κ_x. Let us proceed with constructing a universal inversion formula for the integral transform $\varphi = \mathcal{R}f$. For any point $x \in \mathcal{L}^n$, we construct an operator κ_x from a function space on the manifold of hyperplane sections to the space of differential forms of degree n on this manifold. If $\varphi = \mathcal{R}f$, then the differential form $\kappa_x \varphi$ is closed. We also claim the following inversion formula for various cycles Γ : if $\varphi = \mathcal{R}f$, then

$$\int_\Gamma \kappa_x \varphi = c\, f(x).$$

We first introduce the following notation which is useful in what follows. We denote by $[a_0, \ldots, a_{n+1}]$ the determinant of order $n+2$ with columns a_i some of which consist of 1-forms. We expand the determinant from left to right and assume that multiplication of 1-forms is the exterior multiplication. Such a determinant is symmetric with respect to the columns consisting of 1-forms and may contain identical columns of 1-forms without being equal to zero.

For instance, under this notation, the differential form

$$\omega(\xi) = \sum_{i=0}^{n+1} (-1)^i \xi_i \bigwedge_{j \neq i} d\xi_j$$

becomes

$$\omega(\xi) = \frac{1}{(n+1)!} [\xi, d\xi, \ldots, d\xi],$$

or, briefly,

$$\omega(\xi) = \frac{1}{(n+1)!} [\xi, d\xi^{(n+1)}],$$

where $d\xi^{(n+1)}$ stands for the sequence of $(n+1)$ equal columns $d\xi$.

For convenience, in what follows we write ξ_{n+1} instead of p and $\varphi(\xi)$ instead of $\varphi(\xi, p)$, where $\xi = (\xi_0, \ldots, \xi_{n+1}) \in \mathbb{R}^{n+2} \setminus O$.

Let $\varphi(\xi)$ be an arbitrary infinitely differentiable function on $\mathbb{R}^{n+2} \setminus O$ satisfying the homogeneity condition $\varphi(\lambda\xi) = |\lambda|^{-1} \varphi(\xi)$ for $\lambda \neq 0$. To the function φ and to an arbitrary point $x \in \mathcal{L}^n$ we assign the following distributions on $\mathbb{R}^{n+2}$:

$$\Phi_i(\xi, x) = \frac{x_i \varphi(\xi)}{([\xi, x] - \xi_{n+1} - i0)^n} + \frac{1}{n-1} \frac{\epsilon_i \varphi'_{\xi_i}(\xi)}{([\xi, x] - \xi_{n+1} - i0)^{n-1}}, \quad i = 0, 1, \ldots, n+1, \tag{1.9}$$

where $\epsilon_0 = 1$, $\epsilon_i = -1$ for $i > 0$, and $x_{n+1} = 1$. Recall that the distributions $([\xi, x] - \xi_{n+1} - i0)^{-n}$ (see [**10**]) were introduced in Chapter 1 and used there in the inversion formulas for the Radon transform in the affine and projective n-dimensional spaces; see 2.3 and 8.8 in Chapter 1.

Note that $\Phi_i(\xi, x)$ can be given as

$$\Phi_i(\xi, x) = \epsilon_i \frac{1}{n-1} \Big\{ ([\xi, x] - \xi_{n+1} - i0)^{-n+1}, \varphi(\xi) \Big\}_i, \tag{1.10}$$

where the braces stand for the Poisson brackets

$$\{\varphi_1, \varphi_2\}_i = \varphi_1 \frac{\partial \varphi_2}{\partial \xi_i} - \varphi_2 \frac{\partial \varphi_1}{\partial \xi_i}.$$

To every point $x \in \mathcal{L}^n$ we assign the following operator κ_x from the space of functions $\varphi(\xi)$ to the space of differential n-forms whose coefficients are distributions:

$$\kappa_x \varphi = \frac{1}{n!} [\Phi(\xi, x), \xi, d\xi^{(n)}], \tag{1.11}$$

where $\Phi(\xi, x)$ is the column with the elements $\Phi_i(\xi, x)$ defined by (1.9).

Expanding the determinant (1.11) in the elements of the first column, we obtain the representation of $\kappa_x \varphi$ in the form

$$\kappa_x \varphi = \sum_{i=0}^{n+1} \Phi_i(\xi, x)\, \omega_i(\xi), \tag{1.12}$$

where

$$\omega_0(\xi) = \sum_{i=1}^{n+1} (-1)^{i-1} \xi_i \bigwedge_{j\neq 0,i} d\xi_j,$$

and the subsequent forms ω_i are obtained from ω_0 by a cyclic permutation of the coordinates $\xi_0, \dots, \xi_{n+1}$, and the sign depends on the parity of the corresponding permutation.

It follows from the definition that the differential form $\kappa_x\varphi$ is homogeneous of degree zero; moreover, this form is orthogonal to the fibers of the bundle $\mathbb{R}^{n+2}\backslash O \to S^{n+1}$, where S^{n+1} is the $(n+1)$-dimensional sphere. Hence, this form is defined on the manifold S^{n+1} of rays in $\mathbb{R}^{n+2}$ issuing from the origin.

THEOREM 1.1. *The differential form $\kappa_x\varphi$ is closed on $\mathbb{R}^{n+2}$ for any point $x \in \mathcal{L}^n$ and any function $\varphi(\xi)$ on $\mathbb{R}^{n+2}$ satisfying the homogeneity condition $\varphi(\lambda\xi) = |\lambda|^{-1}\varphi(\xi)$ (for $\lambda \neq 0$) and the differential equation*

$$\Box\varphi = \frac{\partial^2\varphi}{\partial\xi_0^2} - \frac{\partial^2\varphi}{\partial\xi_1^2} - \dots - \frac{\partial^2\varphi}{\partial\xi_{n+1}^2} = 0.$$

The proof of the theorem results from the following general lemma on determinants.

LEMMA. *If $a(\xi)$ is a column of length $n+2$ whose elements $a_0(\xi), \dots, a_{n+1}(\xi)$ are homogeneous functions of degree $-n-1$, then*

$$d\,[a(\xi), \xi, d\xi^{(n+1)}] = -\frac{1}{n+1}\Big(\sum_{j=0}^{n+1} \frac{\partial a_j(\xi)}{\partial \xi_J}\Big)\,[\xi, d\xi^{(n+1)}].$$

This lemma can be applied to the determinant (1.11) because the elements of the column $\Phi(\xi, x)$ are homogeneous of degree $-n-1$. Hence, by the lemma we have

$$d\kappa_x\varphi = -\frac{1}{n+1}\Big(\sum_{j=0}^{n+1} \frac{\partial\Phi_j(\xi,x)}{\partial\xi_j}\Big)\,[\xi, d\xi^{(n+1)}].$$

On the other hand, it follows from the definition of the functions $\Phi_j(\xi, x)$ in the form of Poisson brackets that

$$\sum_{j=0}^{n+1} \frac{\partial\Phi_j(\xi,x)}{\partial\xi_j}$$
$$= ([\xi,x] - \xi_{n+1} - i0)^{-n+1}\,\Box\varphi(\xi) - \varphi(\xi)\,\Box([\xi,x] - \xi_{n+1} - i0)^{-n+1} = 0.$$

Hence, $d(\kappa_x\varphi) = 0$.

COROLLARY. *If a function φ belongs to the image of the integral transform $\mathcal{R}$, then the differential form $\kappa_x\varphi$ is closed.*

In conclusion we present the proof of the lemma. It suffices to prove it for the case in which only one element of the column $a(\xi)$ is nonzero, say, the element $a_0(\xi)$. Then

$$u \equiv [a(\xi), \xi, d\xi^{(n)}] = n!\,a_0(\xi)\,\omega_0(\xi),$$

where the form $\omega_0(\xi)$ was defined above. Hence,

$$\frac{1}{n!}\,du = \sum_{i=0}^{n+1} \frac{\partial a_0(\xi)}{\partial \xi_i}\, d\xi_i \wedge \omega_0(\xi) + a_0(\xi) \wedge d\omega_0(\xi).$$

By the Euler theorem on homogeneous functions,

$$a_0(\xi) = -\frac{1}{n+1} \sum_{i=0}^{n+1} \frac{\partial a_0(\xi)}{\partial \xi_i}\, \xi_i.$$

Thus, since $d\omega_0(\xi) = (n+1)\, d\xi_1 \wedge \cdots \wedge d\xi_{n+1}$, it follows that

$$a_0(\xi) \wedge d\omega_0(\xi) = -\sum_{i=0}^{n+1} \frac{\partial a_0(\xi)}{\partial \xi_i}\, \xi_i\, d\xi_1 \wedge \cdots \wedge d\xi_{n+1}.$$

Substituting this expression into the formula for du, we obtain

$$\begin{aligned} \frac{1}{n!}\,du = {} & \frac{\partial a_0(\xi)}{\partial \xi_0}(d\xi_0 \wedge \omega_0(\xi) - \xi_0\, d\xi_1 \wedge \cdots \wedge d\xi_{n+1}) \\ & + \sum_{i=1}^{n+1} \frac{\partial a_0(\xi)}{\partial \xi_i}(d\xi_i \wedge \omega_0(\xi) - \xi_i\, d\xi_1 \wedge \cdots \wedge d\xi_{n+1}). \end{aligned}$$

It remains to note that the coefficient of $\frac{\partial a_0(\xi)}{\partial \xi_0}$ is $-\frac{1}{(n+1)!}[\xi, d\xi^{(n+1)}]$ and the coefficients of the other derivatives vanish.

REMARK. One can extend both the definition of the operator κ_x and the proof that the differential form $\kappa_x \varphi$ is closed to any nondegenerate quadric in $\mathbb{R}^{n+2}$.

1.4. Local and nonlocal operators κ. The distribution $(t-i0)^{-n}$ occurring in the definition of the operator κ_x can be represented as a linear combination of the distributions t^{-n} and $\delta^{(n-1)}(t)$,

$$(t - i0)^{-n} = t^{-n} + \frac{i\pi(-1)^{n-1}}{(n-1)!}\, \delta^{(n-1)}(t) \tag{1.13}$$

(see, for instance, [**10**]). Moreover, depending on the parity of n, one of the terms is an even distribution and the other is odd.

Accordingly, the operator κ_x can also be represented as the sum of local and nonlocal parts. Namely, define the operators κ'_x and κ''_x by

$$\kappa'_x \varphi = \sum_{i=0}^{n+1} \Phi'_i(\xi, x)\, \omega_i(\xi), \qquad \kappa''_x \varphi = \sum_{i=0}^{n+1} \Phi''_i(\xi, x)\, \omega_i(\xi), \tag{1.14}$$

where

$$\begin{aligned} \Phi'_i(\xi, x) = {} & \frac{i\pi(-1)^{n-1}}{(n-1)!}\Big(x_i\, \varphi(\xi)\, \delta^{(n-1)}([\xi, x] - \xi_{n+1}) \\ & \qquad - \epsilon_i \varphi'_{\xi_i}(\xi)\, \delta^{(n-2)}([\xi, x] - \xi_{n+1})\Big); \\ \Phi''_i(\xi, x) = {} & x_i\, \varphi(\xi)\, ([\xi, x] - \xi_{n+1})^{-n} + \epsilon_i \varphi'_{\xi_i}(\xi)\, ([\xi, x] - \xi_{n+1})^{-n+1}. \end{aligned} \tag{1.15}$$

Then it follows from (1.13) that

$$\kappa_x \varphi = \kappa'_x \varphi + \kappa''_x \varphi. \tag{1.16}$$

If the function φ satisfies the equation $\Box \varphi = 0$, then both differential forms $\kappa'_x \varphi$ and $\kappa''_x \varphi$ are closed.

When integrating the differential form $\kappa_x\varphi$ over various cycles, we shall restrict ourselves to the cycles $\Gamma \subset S^{n+1}$ symmetric with respect to the transforms $\xi \to -\xi$. For these cycles, the integral of the odd part of the form $\kappa_x\varphi$ vanishes.

Note that the form $\kappa'_x\varphi$ is even and the form $\kappa''_x\varphi$ is odd if n is odd. The picture is opposite if n is even. Thus,

$$\int_\Gamma \kappa_x\varphi = \int_\Gamma \kappa'_x\varphi \quad \text{for odd } n; \qquad \int_\Gamma \kappa_x\varphi = \int_\Gamma \kappa''_x\varphi \quad \text{for even } n.$$

1.5. Inversion formula. Recall that the differential form $\kappa_x\varphi$ is defined on $\mathbb{R}^{n+2}\setminus O$, and it can be pushed down (under the natural mapping $\mathbb{R}^{n+2}\setminus O \to S^{n+1}$) from $\mathbb{R}^{n+2}\setminus O$ to the manifold S^{n+1} of rays in $\mathbb{R}^{n+2}$ issuing from the origin.

THEOREM 1.2. *If $\varphi = \mathcal{R}f$, then the following inversion formula holds for each n-dimensional cycle $\Gamma \subset S^{n+1}$:*

$$\int_\Gamma \kappa_x\varphi = c(\Gamma)\, f(x), \tag{1.17}$$

where $c(\Gamma)$ depends only on the homology class of the cycle Γ. The integral over Γ must be understood as the integral over an arbitrary section of the bundle $\widetilde{\Gamma} \to \Gamma$, where $\widetilde{\Gamma}$ is the preimage of Γ in $\mathbb{R}^{n+2}\setminus O$.

All cycles Γ treated below belong to the same homology class, and we present the proof of the theorem for one of these cycles, namely, for the cycle Γ_0 that we are going to introduce. Since the form $\kappa_x\varphi$ is closed, the theorem will thus also be proved for each cycle $\Gamma \subset S^{n+1}$ which can be obtained from Γ_0 by a continuous deformation.

Let $\widetilde{\Gamma} = \{\xi \in \mathbb{R}^{n+2}\setminus O \mid \xi_0 = \xi_1\}$ and let $\Gamma = \Gamma_0$ be the image of $\widetilde{\Gamma}$ in S^{n+1}.

Denote by $\psi(\xi_1, \dots, \xi_{n+1})$ the restriction of $\varphi = \mathcal{R}f$ to $\widetilde{\Gamma}$; in other words, $\psi(\xi_1, \dots, \xi_{n+1}) = \varphi(\xi_1, \xi_1, \dots, \xi_{n+1})$. The inversion formula expressing f in terms of ψ was already obtained in 1.2 (see (1.7); recall that this formula for f immediately follows from the inversion formula for the Radon transform in $\mathbb{R}^n$). Thus, to prove the inversion formula (1.17) for $\Gamma = \Gamma_0$, it suffices to show that the integral $\int_\Gamma \kappa_x\varphi$ coincides with the left-hand side of (1.7).

Consider the expression for $\kappa_x\varphi|_\Gamma$ in (1.11). Since $\xi_0 = \xi_1$ and $d\xi_0 = d\xi_1$ on Γ, we see, after subtracting the second row of the determinant (1.11) from the first, that

$$\kappa_x\varphi|_\Gamma = \Big(\frac{(x_0 - x_1)\,\varphi(\xi)}{([\xi, x] - \xi_{n+1} - i0)^n} + \frac{1}{n-1}\,\frac{\partial\varphi(\xi)/\partial\xi_0 + \partial\varphi(\xi)/\partial\xi_1}{([\xi, x] - \xi_{n+1} - i0)^{n-1}}\Big)\,\omega_0(\xi)|_\Gamma.$$

Since $(\frac{\partial\varphi}{\partial\xi_0} + \frac{\partial\varphi}{\partial\xi_1})|_\Gamma = \frac{\partial\psi}{\partial\xi_1}$, we have

$$\kappa_x\varphi|_\Gamma = \Big(\frac{(x_0 - x_1)\,\psi(\xi)}{([\xi, x] - \xi_{n+1} - i0)^n} + \frac{1}{n-1}\,\frac{\partial\psi(\xi)/\partial\xi_1}{([\xi, x] - \xi_{n+1} - i0)^{n-1}}\Big)\,\omega_0(\xi)|_\Gamma. \tag{1.18}$$

For a section of the bundle $\widetilde{\Gamma} \to \Gamma$ we take a cylinder of the form $\{\xi \in \mathbb{R}^{n+1} \mid (\xi_1, \dots, \xi_n) \in \gamma\}$, where γ is an arbitrary surface in $\mathbb{R}^n$ intersecting once every ray issuing from the origin. The restriction of $\omega_0(\xi)$ to this cylinder is $\omega(\xi) \wedge d\xi_{n+1}$, where $\omega(\xi) = \sum_{i=1}^n (-1)^{i-1}\xi_i \bigwedge_{j\neq i} d\xi_j$.

Integrating (1.18) over this section and applying the integration by parts in the second term, we obtain

$$\int_\Gamma \kappa_x\varphi = 2\,(x_0 - x_1) \int_\gamma \int_{-\infty}^{\infty} \psi(\xi_1, \dots, \xi_{n+1})$$
$$\times\ (\xi_i(x_1 - x_0) + \xi_2 x_2 + \cdots + \xi_n x_n + \xi_{n+1} - i0)^{-n}\, d\xi_{n+1} \wedge \omega(\xi).$$

Thus, it follows from the inversion formula (1.7) that $\int_\Gamma \kappa_x\varphi = c_n\, f(x)$, where $c_n = \frac{2(2\pi)^n}{i^n(n-1)!}$.

The inversion formulas obtained by using the general formula (1.17) for various cycles Γ take into account not only the value on Γ of the function $\varphi = \mathcal{R}f$ itself but, as a rule, also the values of some derivatives of φ which cannot be expressed in terms of the restriction $\varphi|_\Gamma$. The terms with derivatives can be expressed in terms of $\varphi|_\Gamma$ for some special cycles Γ only. Just these manifolds Γ are admissible. Several examples of admissible manifolds can be found in the next subsection.

REMARK. The above definitions and formulas for the functions on $\mathcal{L}^n$ remain valid for functions f defined on the entire hyperboloid of two sheets H^n.

1.6. Examples. Consider the submanifold $\Gamma = \Gamma_a$ of sections by the hyperplanes passing through a point a with homogeneous coordinates $(a_0, a_1, \dots, a_{n+1})$; if $a_{n+1} = 0$, then a is a point at infinity. We first obtain the inversion formula for an arbitrary point a and then specify the formula for various points a that belong or do not belong to the hyperboloid.

Let $\widetilde{\Gamma} = \widetilde{\Gamma}_a$ be the preimage of Γ_a in $\mathbb{R}^{n+2} \setminus O$, i.e.,

$$\widetilde{\Gamma} = \widetilde{\Gamma}_a = \Big\{\xi \in \mathbb{R}^{n+2} \setminus O \;\Big|\; \sum_{i=0}^{n+1} a_i\xi_i = 0\Big\}.$$

The manifolds $\widetilde{\Gamma}_a$ corresponding to distinct points a can be obtained from one another by a continuous deformation; in particular, each of them can be continuously deformed to the submanifold considered in 1.5. Indeed, let g_t, $0 \le t \le 1$, be a continuous curve in $GL_+(n+2, \mathbb{R})$ such that $g_0 = e$ (the identity element), and let g_1 send a to b. Then $\widetilde{\Gamma}_a$ is transformed to $\widetilde{\Gamma}_b$ by the continuous deformation $\widetilde{\Gamma}_a \to g_t^*\,\widetilde{\Gamma}_a$, $0 \le t \le 1$, where g_t^* stands for the transform conjugate to g_t. Hence, the inversion formula (1.17) holds for each of these manifolds.

We find the explicit form of the inversion formula. If a is not a point at infinity, i.e., if $a_{n+1} \neq 0$, then one can set $a_{n+1} = 1$. We modify the expression for $\kappa_x\varphi$ in (1.11) by adding the ith row (of the determinant) multiplied by a_i to the last row for all i, $i \neq n+1$. We obtain

$$\kappa_x\varphi|_{\widetilde{\Gamma}} = \Bigg(\frac{(1 + \sum_{i=0}^n a_i x_i)\,\varphi(\xi)}{([\xi, x] - \xi_{n+1} - i0)^n} + \frac{1}{n-1}\,\frac{a_0\,\varphi'_{\xi_0}(\xi) - \sum_{i=1}^n a_i\varphi'_{\xi_i}(\xi) - \varphi'_{\xi_{n+1}}(\xi)}{([\xi, x] - \xi_{n+1} - i0)^{n-1}}\Bigg)\,\omega_{n+1}(\xi)|_{\widetilde{\Gamma}},$$

where

$$\omega_{n+1}(\xi) = (-1)^{n+1} \sum_{i=0}^n \xi_i \bigwedge_{j \neq i, n+1} d\xi_j.$$

Thus, the inversion formula for Γ_a becomes

$$\int_{\Gamma_a}\Big(\frac{(1+\sum_{i=0}^n a_i x_i)\,\varphi(\xi)}{([\xi,x]-\xi_{n+1}-i0)^n} + \frac{1}{n-1}\,\frac{a_0\,\varphi'_{\xi_0}(\xi)-\sum_{i=1}^n a_i\varphi'_{\xi_i}(\xi)-\varphi'_{\xi_{n+1}}(\xi)}{([\xi,x]-\xi_{n+1}-i0)^{n-1}}\Big)\,\omega_{n+1}(\xi)=c_n\,f(x). \tag{1.19}$$

If a is a point at infinity, i.e., $a_{n+1}=0$, and if $a_0\neq 0$, then one can set $a_0=1$. Adding the ith row (of the determinant (1.11)) multiplied by a_i to the first row for all i, $i\neq 0$, we obtain

$$\kappa_x\varphi|_{\tilde\Gamma}=\Big(\frac{(\sum_{i=0}^n a_i x_i)\,\varphi(\xi)}{([\xi,x]-\xi_{n+1}-i0)^n}+\frac{1}{n-1}\,\frac{\varphi'_{\xi_0}(\xi)-\sum_{i=1}^n a_i\varphi'_{\xi_i}(\xi)}{([\xi,x]-\xi_{n+1}-i0)^{n-1}}\Big)\,\omega_0(\xi)|_{\tilde\Gamma}.$$

Thus, the inversion formula for Γ_a becomes

$$\int_{\Gamma_a}\Big(\frac{(\sum_{i=0}^n a_i x_i)\,\varphi(\xi)}{([\xi,x]-\xi_{n+1}-i0)^n}+\frac{1}{n-1}\,\frac{\varphi'_{\xi_0}(\xi)-\sum_{i=1}^n a_i\varphi'_{\xi_i}(\xi)}{([\xi,x]-\xi_{n+1}-i0)^{n-1}}\Big)\,\omega_0(\xi)=c_n\,f(x). \tag{1.20}$$

Consider some examples.

EXAMPLE 1. Let $a=(0,\dots,0,1)$, i.e., Γ is the manifold of sections by the hyperplanes passing through the zero point, $[\xi,x]=0$. Recall that Γ is the manifold of geodesic hypersurfaces on $\mathcal{L}^n$. In this case, the inversion formula (1.19) becomes

$$\int_\Gamma\Big(\frac{\varphi(\xi)}{([\xi,x]-i0)^n}-\frac{1}{n-1}\,\frac{\partial\varphi(\xi)/\partial\xi_{n+1}}{([\xi,x]-i0)^{n-1}}\Big)\,\omega_{n+1}(\xi)=c_n\,f(x), \tag{1.21}$$

where $\omega_{n+1}(\xi)=(-1)^{n+1}\sum_{i=0}^n(-1)^i\,\xi_i\bigwedge_{j\neq i}d\xi_j$.

Along with the function $\varphi|_\Gamma$, the formula contains the derivative $\frac{\partial\varphi}{\partial\xi_{n+1}}|_\Gamma$ transversal to Γ.

To obtain an inversion formula depending on $\varphi|_\Gamma$ only, we replace the integral transform of functions on $\mathcal{L}^n$ by an integral transform on the full hyperboloid H^n; this transform takes a function on H^n to its integrals over the hyperplane sections of H^n. Both the definition of the differential form κ_x and the inversion formula remain unchanged in this case.

Let f be a function on $\mathcal{L}^n$, let f_1 be the even extension of f to H^n, and let $\varphi_1=\mathcal{R}f_1$. Since f_1 is even, it follows that the function φ_1 is an even function with respect to ξ_{n+1}, and hence $\frac{\partial\varphi_1}{\partial\xi_{n+1}}|_\Gamma=0$. Therefore, the second term in the inversion formula (1.21) for the integral transform $\varphi_1=\mathcal{R}f_1$ vanishes.

Moreover, note that $\varphi_1|_\Gamma=2\,\varphi|_\Gamma$. Therefore, the inversion formula for the transform $\varphi_1=\mathcal{R}f_1$ on H^n implies the following inversion formula for the original integral transform $\varphi=\mathcal{R}f$ on $\mathcal{L}^n$:

$$2\int_\Gamma\varphi(\xi)\,([\xi,x]-i0)^{-n}\,\omega_{n+1}(\xi)=c_n\,f(x).$$

Note that this inversion formula coincides with (1.8).

EXAMPLE 2. Let $a=(1,0,\dots,0,1)$; thus, Γ is the manifold of sections by the hyperplanes passing through the vertex of the upper sheet of the hyperboloid. The inversion formula is

$$\int_\Gamma\Big(\frac{(x_0+1)\varphi(\xi)}{([\xi,x]-i0)^n}+\frac{1}{n-1}\,\frac{\partial\varphi(\xi)/\partial\xi_0-\partial\varphi(\xi)/\partial\xi_{n+1}}{([\xi,x]-i0)^{n-1}}\Big)\,\omega_{n+1}(\xi)=c_n\,f(x). \tag{1.22}$$

Hence, the function f can be reconstructed from the restriction ψ of the function φ to Γ only,

$$\psi(\xi_0,\dots,\xi_n)=\varphi(\xi_0,\dots,\xi_n,-\xi_0).$$

Indeed, since $(\varphi'_{\xi_0}-\varphi'_{\xi_{n+1}})|_\Gamma=\psi'_{\xi_0}$, it follows that the inversion formula (1.22) can be written as

$$\int_\Gamma\Big(\frac{(x_0+1)\psi(\xi)}{([\xi,x]-i0)^n}+\frac{1}{n-1}\,\frac{\partial\psi(\xi)/\partial\xi_0}{([\xi,x]-i0)^{n-1}}\Big)\,\omega_{n+1}(\xi)=c_n\,f(x).$$

Using the integration by parts in the second term, we obtain

$$2\,(x_0+1)\int_\Gamma\varphi(\xi)\,([\xi,x]-i0)^{-n}\,\omega_{n+1}(\xi)=c_n\,f(x).$$

EXAMPLE 3. Suppose $a=(1,0,\dots,0)$, i.e., Γ is the manifold of sections by the hyperplanes passing through a point at infinity that does not belong to the projectivization of the hyperboloid. The inversion formula becomes

$$\int_\Gamma\Big(\frac{x_0\,\varphi(\xi)}{([\xi,x]-\xi_{n+1}-i0)^n}+\frac{1}{n-1}\,\frac{\partial\varphi(\xi)/\partial\xi_0}{([\xi,x]-\xi_{n+1}-i0)^{n-1}}\Big)\,\omega_0(\xi)=c_n\,f(x).\tag{1.23}$$

Along with the function $\varphi|_\Gamma$, the formula contains the derivative of φ transversal to Γ.

To obtain an inversion formula expressing f in terms of $\varphi|_\Gamma$ only, we pass to the integral transform of functions on the full hyperboloid H^n, as in Example 1.

Extend the function f on $\mathcal{L}^n$ to a function which is even with respect to x_0 on the full hyperboloid H^n, and let $\varphi_1=\mathcal{R}f$ be the integral transform on H^n of the function f thus extended. Since f is even with respect to x_0, it follows that φ_1 is an even function with respect to ξ_0, and hence $\frac{\partial\varphi_1}{\partial\xi_0}|_\Gamma=0$. Therefore, the term with the derivative $\frac{\partial\varphi_1}{\partial\xi_0}$ vanishes in the inversion formula for the integral transform $\varphi_1=\mathcal{R}f$. Since $\frac{\partial\varphi_1}{\partial\xi_0}|_\Gamma=2\,\frac{\partial\varphi}{\partial\xi_0}|_\Gamma$, it follows from the inversion formula for the integral transform $\varphi_1=\mathcal{R}f$ on H^n that the following inversion formula holds for the original integral transform $\varphi=\mathcal{R}f$ on $\mathcal{L}^n$:

$$2\,x_0\int_\Gamma\varphi(\xi)\,([\xi,x]-\xi_{n+1}-i0)^{-n}\,\omega_0(\xi)=c_n\,f(x).$$

Note that this formula was already obtained earlier; see (1.5).

EXAMPLE 4. Suppose $a=(1,1,0,\dots,0)$, and Γ is the manifold of sections by the hyperplanes passing through a point at infinity on the projectivization of the hyperboloid. The inversion formula becomes

$$\int_\Gamma\Big(\frac{(x_0+x_1)\varphi(\xi)}{([\xi,x]-\xi_{n+1}-i0)^n}+\frac{1}{n-1}\,\frac{\partial\varphi(\xi)/\partial\xi_0-\partial\varphi(\xi)/\partial\xi_1}{([\xi,x]-\xi_{n+1}-i0)^{n-1}}\Big)\omega_0(\xi)=c_nf(x).\tag{1.24}$$

Similarly to Example 2, this inversion formula can be written as

$$2\,(x_0+x_1)\int_\Gamma\varphi(\xi)\,([\xi,x]-\xi_{n+1}-i0)^{-n}\,\omega_0(\xi)=c_n\,f(x).$$

EXAMPLE 5. Let Γ be the manifold of horospheres, i.e., let the submanifold $\widetilde{\Gamma}\subset\mathbb{R}^{n+2}\setminus O$ be given by the equation $[\xi,\xi]=0$.

Note that the manifold Γ can be deformed to the manifold in Example 1 by the continuous mapping $\xi_0\to(1-t)\,\xi_0$, $0\le t\le 1$. Hence, the inversion formula (1.17) holds for the manifold Γ. Let us find an explicit form of this formula.

To do this, we modify the determinant expressions for $\kappa_x\varphi\,|_\Gamma$. Subtract the rows with indices $i = 1, \ldots, n$ multiplied by ξ_i/ξ_0, respectively, from the first row of the determinant. Since $[\xi, \xi] = 0$ and $[\xi, d\xi] = 0$ on Γ, we obtain

$$\kappa_x \varphi\Big|_\Gamma = \Big(\xi_0^{-1}[\xi, x]\,\varphi(\xi)\,([\xi, x] - \xi_{n+1} - i0)^{-n} + \frac{1}{n-1}\,\xi_0^{-1}\Big(\sum_{i=0}^{n} \xi_i \frac{\partial\varphi}{\partial\xi_i}\Big)\,([\xi, x] - \xi_{n+1} - i0)^{-n+1}\Big)\Big|_\Gamma\,\omega_0(\xi).$$

Note that the homogeneity condition yields

$$\sum_{i=0}^{n} \xi_i \frac{\partial\varphi}{\partial\xi_i} = -\xi_{n+1}\frac{\partial\varphi}{\partial\xi_{n+1}} - \varphi = -\frac{\partial(\xi_{n+1}\varphi)}{\partial\xi_{n+1}}.$$

Passing to the integral, we can integrate by parts with respect to ξ_{n+1} in the second term. This gives

$$\int_\Gamma \kappa_x \varphi = \int_\Gamma \xi_0^{-1}([\xi, x] - \xi_{n+1})\,\varphi(\xi)\,([\xi, x] - \xi_{n+1} - i0)^{-n}\,\omega_0(\xi).$$

Thus, the inversion formula related to the manifold of horospheres is

$$\int_\Gamma \xi_0^{-1}\varphi(\xi)\,([\xi, x] - \xi_{n+1} - i0)^{-n+1}\,\omega_0(\xi) = c_n\,f(x).$$

In particular, we have proved that the manifold of horospheres is admissible.

The reader can verify that this formula is equivalent to the inversion formula in Chapter 3.

2. Integral transform related to spheres in Euclidean space E^n

We consider here another overdetermined integral transform $\mathcal{R}$ that takes a function on n-dimensional Euclidean space E^n to the integrals of this function over all possible $(n-1)$-dimensional spheres in E^n. This transform can be regarded as an integral transform taking a function on the n-dimensional sphere $S^n \subset E^{n+1}$ to the integrals of this function over the hyperplane sections of the sphere S^n. Namely, let x and x' be two antipodal points of the sphere S^n. Then, when projecting the sphere S^n from the point x to the hyperplane $E^n \subset E^{n+1}$ tangent to S^n at the point x', the hyperplane sections of the sphere go to spheres on E^n.

There is a simple relation between this integral transform and the transform treated in § 1. Namely, under the projective transformation

$$(x_0, x_1, \ldots, x_n) \mapsto (1/x_0, x_1/x_0, \ldots, x_n/x_0)$$

the hyperboloid of two sheets $x_0^2 - x_1^2 - \cdots - x_n^2 = 1$ becomes the sphere $x_0^2 + x_1^2 + \cdots + x_n^2 = 1$ and the hyperplane sections of the hyperboloid become the hyperplane sections of the sphere. Due to this connection, one can obtain the main formulas for the integral transform on E^n related to spheres in E^n, by using the results of § 1. Here we give an independent proof of these formulas.

2.1. Definition. Let H be the manifold of $(n-1)$-dimensional spheres in Euclidean space E^n. It is assumed that H contains the hyperplanes, which must be viewed as spheres passing through a point at infinity.

We present the spheres by equations of the form

$$u(\xi,x) \equiv \xi_0 \sum_{i=1}^{n} {x_i}^2 + \sum_{i=1}^{n} \xi_i x_i + \xi_{n+1} = 0, \qquad \xi \neq 0. \tag{2.1}$$

Thus, the spheres are identified with points of the projective space P^{n+1} that are given by the homogeneous coordinates $(\xi_0, \xi_1, \dots, \xi_{n+1})$. Note that equation (2.1) defines an element of the manifold H if and only if

$$\sum_{i=1}^{n} {\xi_i}^2 \geq 4\,\xi_0\,\xi_{n+1}.$$

Every homogeneous equation with respect to ξ_i defines an n-dimensional submanifold of the manifold of spheres. The simplest examples are

(1) $\xi_0 = 0$, the submanifold of all hyperplanes;
(2) $\xi_{n+1} = 0$, the submanifold of spheres passing through the point 0;
(3) $\xi_i = 0$ $(i = 1, \dots, n)$, the submanifold of spheres centered at the coordinate hyperplane $x_i = 0$;
(4) $\sum_{i=1}^{n-1} \xi_i^2 = 4\xi_0\xi_{n+1}$, the submanifold of spheres tangent to the hyperplane $x_n = 0$;
(5) $r^2 \sum_{i=1}^{n} \xi_i^2 = (r^2\,\xi_0 + \xi_{n+1})^2$, the submanifold of spheres tangent to the spheres $x_1^2 + \cdots + x_n^2 = r^2$;
(6) $\sum_{i=1}^{n} \xi_i^2 = 4\xi_0\xi_{n+1} + r^2 a_0^2$, the submanifold of spheres of a chosen radius r.

Introduce the integral transform $\mathcal{R}$ taking any compactly supported C^∞ function f on E^n to its integrals over the spheres in E^n by the formula

$$\mathcal{R}\, f(\xi) = \int_{E^n} f(x)\,\delta(u(\xi,x))\,dx_1 \wedge \cdots \wedge dx_n. \tag{2.2}$$

For $\xi_0 \neq 0$, this integral coincides with the integral with respect to Euclidean measure on the sphere up to the factor $(2\,|\xi_0|\,r)^{-1}$, where r is the radius of the sphere. The function $\varphi = \mathcal{R} f$ thus defined is nonzero only in the domain $\sum_{i=1}^{n} {\xi_i}^2 \geq 4\,\xi_0\,\xi_{n+1}$.

It immediately follows from the definition that the function $\varphi = \mathcal{R} f$ satisfies the homogeneity condition

$$\varphi(\lambda\,\xi) = |\lambda|^{-1}\varphi(\xi), \qquad \lambda \neq 0, \tag{2.3}$$

and the differential equation

$$\Box\varphi \equiv \frac{\partial^2\varphi}{\partial\xi_0\,\partial\xi_{n+1}} - \sum_{i=1}^{n} \frac{\partial^2\varphi}{\partial\xi_i^2} = 0. \tag{2.4}$$

2.2. Operator κ_x. We construct (repeating the arguments of § 1) the universal inversion formula that recovers a function f from the values of the function $\varphi = \mathcal{R} f$ on various n-dimensional submanifolds of H.

Let $\varphi(\xi)$ be an arbitrary infinitely differentiable function on $\mathbb{R}^{n+2} \setminus O$ satisfying the homogeneity condition $\varphi(\lambda\xi) = |\lambda|^{-1}\,\varphi(\xi)$ for $\lambda \neq 0$. To the function φ and to an arbitrary point $x \in E^n$ we assign the following distributions on $\mathbb{R}^{n+2}$:

$$\Phi_0(\xi, x) = \frac{1}{2} \frac{\varphi(\xi)}{(u(\xi, x) - i0)^n} + \frac{1}{2(n-1)} \frac{\partial\varphi(\xi)/\partial\xi_{n+1}}{(u(\xi, x) - i0)^{n-1}};$$

(2.5) $$\Phi_i(\xi, x) = -\frac{x_i\varphi(\xi)}{(u(\xi, x) - i0)^n} - \frac{1}{n-1} \frac{\partial\varphi(\xi)/\partial\xi_i}{(u(\xi, x) - i0)^{n-1}} \qquad (1 \le i \le n);$$

$$\Phi_{n+1}(\xi, x) = \frac{1}{2} \frac{\left(\sum_{i=1}^n x_i{}^2\right)\varphi(\xi)}{(u(\xi, x) - i0)^n} + \frac{1}{2(n-1)} \frac{\partial\varphi(\xi)/\partial\xi_0}{(u(\xi, x) - i0)^{n-1}}.$$

Note that one can represent $\Phi_i(\xi, x)$ in the form

$$\Phi_0(\xi, x) = \frac{1}{2(n-1)} \left\{(u(\xi, x) - i0)^{-n+1}, \varphi(\xi)\right\}_{n+1},$$

(2.6) $$\Phi_i(\xi, x) = -\frac{1}{n-1} \left\{(u(\xi, x) - i0)^{-n+1}, \varphi(\xi)\right\}_i \qquad (1 \le i \le n),$$

$$\Phi_{n+1}(\xi, x) = \frac{1}{2(n-1)} \left\{(u(\xi, x) - i0)^{-n+1}, \varphi(\xi)\right\}_0,$$

where the braces are the Poisson brackets, i.e.,

$$\{\varphi_1, \varphi_2\}_i = \varphi_1 \frac{\partial\varphi_2}{\partial\xi_i} - \varphi_2 \frac{\partial\varphi_1}{\partial\xi_i}.$$

To each point $x \in E^n$ we assign the following operator κ_x from the space of functions $\varphi(\xi)$ into the space of differential n-forms whose coefficients are distributions:

(2.7) $$\kappa_x\varphi = \frac{1}{n!} [\Phi(\xi, x), \xi, d\xi^{(n)}];$$

here $\Phi(\xi, x)$ is the column with the elements $\Phi_i(\xi, x)$ defined by (2.5).

Expanding this determinant in the first column, we obtain the following representation of $\kappa_x\varphi$ in the form of a sum:

(2.8) $$\kappa_x\varphi = \sum_{i=0}^{n+1} \Phi_i(\xi, x)\, \omega_i(\xi),$$

where $\omega_i(\xi)$ are differential n-forms; for their explicit expressions, see § 1.

As in § 1, the form $\kappa_x\varphi$ is defined on the manifold S^{n+1} of rays in $\mathbb{R}^{n+2}$ issuing from the point 0.

THEOREM 2.1. *The differential form $\kappa_x\varphi$ is closed on $\mathbb{R}^{n+2}$ for any point $x \in E^n$ and any function $\varphi(\xi)$ on $\mathbb{R}^{n+2}$ satisfying the homogeneity condition $\varphi(\lambda\xi) = |\lambda|^{-1} \varphi(\xi)$, $\lambda \neq 0$, and the differential equation* (2.4).

PROOF. Since $\Box\, u(\xi, x) = 0$, it follows from the definition of the functions Φ_i as Poisson brackets that

$$\sum_{i=0}^{n+1} \frac{\partial\Phi_i}{\partial\xi_i} = (u(\xi, x) - i0)^{-n+1} \Box\, \varphi - \varphi\, \Box\, (u(\xi, x) - i0)^{-n+1} = 0.$$

Therefore, the assertion of the theorem follows from the lemma in 1.3.

COROLLARY. *If a function φ belongs to the image of the integral transform $\mathcal{R}$, then the differential form $\kappa_x\varphi$ is closed.*

2.3. Inversion formula. By analogy with § 1, the following inversion formula holds:

THEOREM 2.2. *If* $\varphi = \mathcal{R}f$, *then*

$$\int_\Gamma \kappa_x \varphi = c(\Gamma)\, f(x) \tag{2.9}$$

for any point $x \in E^n$ *and any* n*-dimensional cycle* $\Gamma \subset H$*, where* $c(\Gamma)$ *depends only on the homology class of the cycle* Γ.

We prove (2.9) for the submanifold $\Gamma = \Gamma_0$ given by the equation $\xi_0 = 0$, i.e., for the manifold of hyperplanes in E^n. Since the form $\kappa_x\varphi$ is closed, this will imply that the inversion formula (2.9) holds for any n-dimensional submanifold $\Gamma \subset H$ obtained by a continuous deformation of Γ_0.

In particular, using a continuous deformation, one can transform Γ_0 to a manifold $\Gamma^{(s)}$ of spheres that is defined by an arbitrary linear relation $\sum_{i=0}^{n+1} a_i\xi_i = 0$. Hence, corresponding to each of these manifolds is an inversion formula (2.9), where $c(\Gamma) = c(\Gamma_0)$.

Let $\psi(\xi_1, \dots, \xi_{n+1})$ be the restriction of the function $\varphi = \mathcal{R}f$ to Γ,

$$\psi(\xi_1, \dots, \xi_{n+1}) = \int_{E^n} f(x)\, \delta(\xi_1 x_1 + \dots + \xi_n x_n + \xi_{n+1})\, dx_1 \wedge \dots \wedge dx_n.$$

Since ψ is the Radon transform of f, the following inversion formula holds:

$$\int_\gamma \int_{-\infty}^{\infty} \psi(\xi_1, \dots, \xi_{n+1})\, (\xi_1 x_1 + \dots + \xi_n x_n + \xi_{n+1} - i0)^{-n}\, d\xi_{n+1} \wedge \sigma(\xi) = c_n\, f(x), \tag{2.10}$$

where $c_n = \frac{(2\pi)^n}{i^n (n-1)!}$ and $\sigma(\xi) = \sum_{i=1}^n (-1)^{i-1} \xi_i \bigwedge_{j\neq i} d\xi_j$.

It remains to show that the integral $\int_\Gamma \kappa_x\varphi$ coincides with the left-hand side of (2.10). It follows from the expression for $\kappa_x\varphi$ in the determinant form that

$$\begin{aligned}\kappa_x\varphi|_\Gamma &= \Phi_0(\xi, x)\, \omega_0(\xi)|_\Gamma \\ &= \frac{1}{2}\Big(\psi(\xi)\, (\xi_1 x_1 + \dots + \xi_n x_n + \xi_{n+1} - i0)^{-n} \\ &\qquad + \frac{1}{n-1} \frac{\partial\psi(\xi)}{\partial\xi_{n+1}} (\xi_1 x_1 + \dots + \xi_n x_n + \xi_{n+1} - i0)^{-n+1}\Big)\, \omega_0(\xi)\,\Big|_\Gamma.\end{aligned}$$

Passing to the integral over Γ and using the integration by parts in the second integral, we see that

$$\begin{aligned}\int_\Gamma \kappa_x\varphi &= \int_\Gamma \psi(\xi)\, (\xi_1 x_1 + \dots + \xi_n x_n + \xi_{n+1} - i0)^{-n}\, \omega_0(\xi) \\ &= \int_\gamma \int_{-\infty}^{\infty} \psi(\xi_1, \dots, \xi_{n+1})\, (\xi_1 x_1 + \dots + \xi_n x_n + \xi_{n+1} - i0)^{-n}\, d\xi_{n+1} \wedge \sigma(\xi).\end{aligned} \tag{2.11}$$

Comparing (2.10) and (2.11), we obtain (2.9).

2.4. Examples. Let us find the explicit form of the inversion formulas for the simplest cycles $\Gamma \subset H$.

EXAMPLE 1. Let Γ be the submanifold of spheres passing through the point 0. We write the equation of Γ in the form $\xi_{n+1} = 0$.

It follows from the expression for $\kappa_x\varphi$ in the determinant form that

$$\begin{aligned}\kappa_x\varphi\,|_\Gamma &= \Phi_{n+1}(\xi,x)\,\omega_{n+1}(\xi)|_\Gamma \\ &= \frac{1}{2}\Big(\sum_{i=1}^n x_i^2\Big)\psi(\xi)\Big(u(\xi,x)-i0)^{-n} \\ &\qquad + \frac{1}{n-1}\,\frac{\partial\psi(\xi)}{\partial\xi_0}\,(u(\xi,x)-i0)^{-n+1}\Big)\,\omega_{n+1}(\xi)\,\Big|_\Gamma,\end{aligned}$$

where ψ is the restriction of the function $\varphi = \mathcal{R}f$ to Γ. Passing to the integral over Γ and using the integration by parts in the second term, we obtain

$$\int_\Gamma \kappa_x\varphi = \Big(\sum_{i=1}^n x_i^2\Big)\int_\Gamma \varphi(\xi)\,(u(\xi,x)-i0)^{-n}\,\omega_{n+1}(\xi).$$

Thus, the inversion formula related to Γ becomes

$$\Big(\sum_{i=1}^n x_i^2\Big)\int_\Gamma \varphi(\xi)\,(u(\xi,x)-i0)^{-n}\,\omega_{n+1}(\xi) = c_n\,f(x). \tag{2.12}$$

In particular, this shows that the submanifold Γ is admissible.

EXAMPLE 2. Let Γ be the submanifold of spheres centered at the hyperplane $x_1 = 0$. We write the equation of Γ in the form $\xi_1 = 0$.

It follows from the expression for $\kappa_x\varphi$ in the determinant form that

$$\begin{aligned}\kappa_x\varphi|_\Gamma &= \Phi_1(\xi,x)\,\omega_1(\xi)|_\Gamma \\ &= \Big(x_1\,\varphi(\xi)\,(u(\xi,x)-i0)^{-n} + \frac{1}{n-1}\,\frac{\partial\varphi(\xi)}{\partial\xi_0}\,(u(\xi,x)-i0)^{-n+1}\Big)\,\omega_1(\xi)\,\Big|_\Gamma.\end{aligned}$$

Thus, the inversion formula related to Γ becomes

$$\begin{aligned}&x_1\int_\Gamma \varphi(\xi)\,(u(\xi,x)-i0)^{-n}\,\omega_1(\xi) \\ &\qquad + \frac{1}{n-1}\int_\Gamma \frac{\partial\varphi(\xi)}{\partial\xi_1}\,(u(\xi,x)-i0)^{-n+1}\,\omega_1(\xi) = c_n\,f(x).\end{aligned} \tag{2.13}$$

According to this formula, to reconstruct a function f, one must know not only $\varphi|_\Gamma$ but also $\frac{\partial\varphi(\xi)}{\partial\xi_1}|_\Gamma$, i.e., the derivative in a direction transversal to Γ.

The inversion formula can be simplified for functions $f(x)$ satisfying some additional conditions. We consider two cases of this kind.

Case 1: $f(x_1,\dots,x_n)$ is an even function with respect to x_1. Then the function $\varphi = \mathcal{R}f$ is even with respect to ξ_1, and hence $\frac{\partial\varphi}{\partial\xi_1}|_\Gamma = 0$. Thus, in this case the inversion formula becomes

$$x_1\int_\Gamma \varphi(\xi)\,(u(\xi,x)-i0)^{-n}\,\omega_1(\xi) = c_n\,f(x).$$

Note that $\varphi = \mathcal{R}f$ is an odd function of ξ_1, and hence $\varphi|_\Gamma = 0$ if $f(x_1,\dots,x_n)$ is an odd function of x_1. Thus, in this case the inversion formula becomes

$$\int_\Gamma \frac{\partial\varphi(\xi)}{\partial\xi_1}\,(u(\xi,x)-i0)^{-n+1}\,\omega_1(\xi) = (n-1)\,c_n\,f(x).$$

Case 2: the function $f(x_1,\dots,x_n)$ vanishes in the domain $x_1 < 0$. Let $f_1(x)$ be the extension of the function f to the domain $x_1 < 0$, which is even with respect to x_1, and let $\varphi_1 = \mathcal{R}f_1$. Then $\varphi_1|_\Gamma = 2\,\varphi|_\Gamma$ and $\frac{\partial\varphi}{\partial\xi_1}|_\Gamma = 0$. Therefore, applying the inversion formula to the function φ_1, we obtain

$$2\,x_1 \int_\Gamma \varphi(\xi)\,(u(\xi,x)-i0)^{-n}\,\omega_1(\xi) = c_n\,f(x).$$

EXAMPLE 3. Let Γ be the submanifold of spheres tangent to the hyperplane $x_n = 0$. For the equation of Γ we take $\sum_{i=1}^{n-1}\xi_i^2 = 4\,\xi_0\,\xi_{n+1}$.

We represent Γ as the union of the oriented submanifolds of spheres, Γ_+ and Γ_-, formed by the spheres tangent to the hyperplane $x_n = 0$ and belonging to the subspaces $x_n > 0$ and $x_n < 0$, respectively. Let us show that each of the submanifolds Γ_+ and Γ_- admits a continuous deformation (taking account of the orientation) to the manifold Γ_0 of all hyperplanes in E^n. This will yield

$$\int_\Gamma \kappa_x\varphi = 2\int_{\Gamma_+}\kappa_x\varphi = 2\,c_n\,f(x), \tag{2.14}$$

where the coefficient c_n coincides with that in the inversion formula for Γ_0.

We prove this assertion, say, for Γ_+. Using the stereographic projection, we can interpret the spheres and hyperplanes in E^n as hyperplane sections of the n-dimensional sphere $S^n \subset E^{n+1}$. In this interpretation, corresponding to the hyperplane $x_n = 0$ is the $(n-1)$-dimensional sphere $\sigma \subset S^n$ passing through the pole p of the sphere S^n, and to the manifolds $\Gamma_\pm$, the manifolds of $(n-1)$-dimensional spheres in S^n tangent to the sphere σ and belonging to the hemispheres $S^n_\mp$, respectively, into which the sphere σ partitions the sphere S^n. We contract the sphere σ to the point p in such a way that the hemisphere S^n_- contracts to p and the hemisphere S^n_+ becomes the entire sphere S^n. Then the manifold Γ_+ is deformed to the manifold of $(n-1)$-dimensional spheres on S^n passing through the point p, i.e., in the original interpretation, to the manifold Γ_0 of hyperplanes on E^n.

We modify the expression for $\kappa_x\varphi|_\Gamma$ in the determinant form. Take the first row $(i = 0)$, subtract from it the next $(n-1)$ rows (with indices $i = 1,\dots,n-1$), each multiplied by the respective factor $\xi_i/2\xi_{n+1}$, and add the last row multiplied by ξ_0/ξ_{n+1}.

Since $\sum_{i=1}^{n-1}\xi_i^2 - 4\xi_0\xi_{n+1} = 0$ and $\sum_{i=1}^{n-1}\xi_i\,d\xi_i - 2\xi_0\,d\xi_{n+1} - 2\xi_{n+1}\,d\xi_0 = 0$ on Γ, we see that

$$\kappa_x\varphi|_\Gamma = \Big(\Phi_0(\xi,x) - \sum_{i=1}^{n-1}\frac{\xi_i}{2\xi_{n+1}}\,\Phi_i(\xi,x) + \frac{\xi_0}{\xi_{n+1}}\,\Phi_{n+1}(\xi,x)\Big)\,\omega_0(\xi)\,\Big|_\Gamma.$$

Substituting here the expression for the functions $\Phi_i(\xi,x)$ (see (2.5)), we obtain

$$\kappa_x\varphi|_\Gamma = \frac{1}{2\xi_{n+1}}\Big(\frac{(u(\xi,x)-\xi_n x_n)\varphi(\xi)}{(u(\xi,x)-i0)^n} + \frac{1}{n-1}\,\frac{\sum_{i\neq n}\xi_i\,\varphi'_{\xi_i}(\xi)}{(u(\xi,x)-i0)^{n-1}}\Big)\,\omega_0(\xi)\,\Big|_\Gamma.$$

It follows from the homogeneity condition for φ that

$$\sum_{i\neq n}\xi_i\,\frac{\partial\varphi}{\partial\xi_i} = -\varphi - \xi_n\,\frac{\partial\varphi}{\partial\xi_n} = -\frac{\partial(\xi_n\varphi)}{\partial\xi_n}.$$

Hence,

$$\kappa_x\varphi|_\Gamma = \frac{1}{2\xi_{n+1}}\Big(\frac{(u(\xi,x)-\xi_n x_n)\varphi(\xi)}{(u(\xi,x)-i0)^n} - \frac{1}{n-1}\,\frac{\partial(\xi_n\varphi)/\partial\xi_n}{(u(\xi,x)-i0)^{n-1}}\Big)\,\omega_0(\xi)\,\Big|_\Gamma.$$

When evaluating the integral with respect to this differential form over Γ, one can use the integration by parts in the second term. We finally obtain the following inversion formula:

$$\int_\Gamma \frac{\xi_{n+1}^{-1}\,(u(\xi,x)-2\xi_n x_n)\varphi(\xi)}{(u(\xi,x)-i0)^n}\,\omega_0(\xi)=4\,c_n\,f(x).$$

In particular, we thus proved that the manifold of spheres tangent to the hyperplane $x_n=0$ is admissible.

EXAMPLE 4. Let Γ be the submanifold of spheres tangent to the sphere $x_1^2+\dots+x_n^2=1$. For the equation of Γ we take $\sum_{i=1}^n \xi_i^2=(\xi_0+\xi_{n+1})^2$.

As in Example 3, one can readily see that the inversion formula (2.14) holds for the cycle Γ.

We modify the expression for $\kappa_x\varphi|_\Gamma$ in the determinant form. Take the first row of the determinant, subtract from it each of the next n rows multiplied by the corresponding factor of the form $(\xi_0+\xi_{n+1})^{-1}\,\xi_i$, $i=1,\dots,n$, and add the last row.

Since $(\xi_0+\xi_{n+1})^2-\sum_{i=1}^n \xi_i^2=0$ on Γ, we see that

$$\kappa_x\varphi|_\Gamma=\Big(\Phi_0(\xi,x)-(\xi_0+\xi_{n+1})^{-1}\sum_{i=1}^n \xi_i\,\Phi_i(\xi,x)+\Phi_{n+1}(\xi,x)\Big)\,\omega_0(\xi)\,\Big|_\Gamma\,.$$

Substituting here the expressions for the functions $\Phi_i(\xi,x)$ (see (2.5)), we obtain

$$\kappa_x\varphi|_\Gamma=\frac{1}{2\,(\xi_0+\xi_{n+1})}\Big(\frac{P(\xi,x)\,\varphi(\xi)}{(u(\xi,x)-i0)^n}+\frac{1}{n-1}\,\frac{Q(\xi)}{(u(\xi,x)-i0)^{n-1}}\Big)\,\omega_0(\xi)\,\Big|_\Gamma,$$

where

$$P(\xi,x)=(\xi_0+\xi_{n+1})\Big(\sum_{i=1}^n x_i^2+1\Big)+2\sum_{i=1}^n \xi_i x_i,$$
$$Q(\xi)=(\xi_0+\xi_{n+1})\,(\varphi'_{\xi_0}+\varphi'_{\xi_{n+1}})+2\sum_{i=1}^n \xi_i\,\varphi'_{\xi_i}.$$

We transform the expression for $Q(\xi)$. Passing from ξ_0 and ξ_{n+1} to the new variables $\eta_0=\xi_0+\xi_{n+1}$ and $\eta_1=\xi_0-\xi_{n+1}$ and using the homogeneity property of the function φ, we obtain

$$Q(\xi)=2\Big(\eta_0\varphi'_{\eta_0}+\sum_{i=1}^n \xi_i\,\varphi'_{\xi_i}\Big)=2\,(-\varphi-\eta_1\varphi'_{\eta_1})=-2\,\frac{\partial(\eta_1\varphi)}{\partial\eta_1}.$$

When evaluating the integral over Γ with respect to the differential form $\kappa_x\varphi$, we can use the integration by parts in the second term. We finally obtain the following inversion formula:

$$\int_\Gamma \frac{(\xi_0+\xi_{n+1})^{-1}\,(\xi_0+\sum_{i=1}^n \xi_i x_i+\xi_{n+1}\sum_{i=1}^n x_i^2)\,\varphi(\xi)}{(u(\xi,x)-i0)^n}\,\omega_0(\xi)=4\,c_n\,f(x).$$

In particular, we have proved that the manifold of spheres is also admissible.

REMARK. In Examples 2 and 4, Γ is not a cycle, but the inversion formula still holds in these cases. One way to see it is to pass to sections of the sphere S^n, in which case Γ becomes a cycle.

REMARK. Using motions of the space E^n, one can transform the inversion formulas in Examples 1–4 into inversion formulas for the manifolds of spheres passing through a given point, or centered at a given hyperplane, or tangent to a given hyperplane or sphere, respectively.

Bibliography

[1] J. Bernstein and S. Gindikin, *Integral geometry on manifolds of curves*, Lie groups and symmetric spaces, Preprint, 2003.

[2] I. M. Gelfand and S. G. Gindikin (eds.), *Mathematical problems of tomography*, Amer. Math. Soc., Providence, RI, 1996.

[3] I. M. Gelfand, S. G. Gindikin, and M. I. Graev, *Integral geometry in affine and projective spaces*, Current problems in mathematics, vol. 16, VINITI, Moscow, 1980, pp. 53–224; English transl. in J. Math. Sci. **18** (1982), no. 1.

[4] I. M. Gelfand, S. G. Gindikin, and Z. Ya. Shapiro, *A local problem of integral geometry in a space of curves*, Funktsional. Anal. i Prilozen. **13** (1979), no. 2, 11–31; English transl. in Functional Anal. Appl. **13** (1979).

[5] I. M. Gelfand and M. I. Graev, *Geometry of homogeneous spaces, representations of groups in homogeneous spaces and related questions of integral geometry.* I. Trudy Moskov. Mat. Obshch. **8** (1959), 321–390. (Russian)

[6] ———, *Integrals over hyperplanes of basic and generalized functions*, Dokl. Akad. Nauk SSSR **135** (1960), no. 6, 1307–1310; English transl., Soviet Math. Dokl. **1** (1960), 1369–1372.

[7] I. M. Gelfand, M. I. Graev, and Z. Ya. Shapiro, *Integral geometry on k-dimensional planes*, Funktsional Anal. i Prilozhen. **1** (1967), no. 1, pp. 15–31; English transl. in Functional Anal. Appl. **1** (1967).

[8] I. M. Gelfand, M. I. Graev, and N. Ya. Vilenkin, *Generalized functions, Vol. 5: Integral geometry and related problems in the theory of representations*, Fizmatgiz, Moscow, 1962; English transl., Academic Press, New York–London, 1966.

[9] I. M. Gelfand and M. A. Naimark, *Unitary representations of the Lorentz group*, Izvestiya Akad. Nauk SSSR Ser. Matem. **11** (1947), 411–504. (Russian)

[10] I. M. Gelfand and G. E. Shilov, *Generalized functions and operations on them: Generalized functions*, Part 1, Fizmatgiz, Moscow, 1958; English transl., Academic Press, New York–London, 1964.

[11] S. G. Gindikin, *A universal integral formula for multicircular regions in n-dimensional complex space*, Sibirsk. Mat. Zh. **7** (1966), no. 3, 708–712; English transl. in Siberian Math. J. **7** (1966).

[12] ———, *Some solutions of the self-dual Einstein equation*, Funktsional. Anal. i Prilozhen. **19** (1985), 58–60; English transl., Functional Anal. Appl. **19** (1985), no. 3, 210–213.

[13] ———, *Between integral geometry and twistors*, Twistors in Mathematics and Physics, T. N. Bailey and R. J. Baston (eds.), Cambridge Univ. Press, 1990, pp. 30–35.

[14] ———, *Integral geometry on symmetric manifolds*, Amer. Math. Soc. Transl. Ser. 2, vol. 148, Amer. Math. Soc., Providence, RI, 1995, pp. 29–37.

[15] ———, *Integral geometry on real quadrics*, Amer. Math. Soc. Transl. Ser. 2, vol. 169, Amer. Math. Soc., Providence, RI, 1995, pp. 23–31.

[16] ———, *Real integral geometry and complex analysis*, Integral Geometry, Radon Transforms and Complex Analysis, Lecture Notes Math., vol. 1684, Springer-Verlag, Heidelberg, pp. 70–98.

[17] ———, *Integral geometry on $SL(2;\mathbb{R})$*, Math. Res. Lett. **7** (2000), 417–432.

[18] S. Gindikin (ed.), *Applied problems of Radon transform*, Amer. Math. Soc. Transl. Series 2, vol. 162, Amer. Math. Soc., Providence, RI, 1994.

[19] S. G. Gindikin and G. M. Henkin, *The Penrose transform and complex integral geometry*, Current problems in mathematics, vol. 17, VINITI, Moscow, 1980, pp. 57–112; English transl. in J. Math. Sci. **21** (1983), no. 4.

[20] S. G. Gindikin and N. D. Vvedenskaya, *Discrete Radon transform and image reconstruction*, Mathematical Problems of Tomography, Transl. Math. Monographs, vol. 81, Amer. Math. Soc., Providence, RI, 1990, pp. 141–188.

[21] A. B. Goncharov, *Three-dimensional reconstruction of arbitrarily arranged identical particles given their projections*, Mathematical Problems of Tomography, Transl. Math. Monographs, vol. 81, Amer. Math. Soc., Providence, RI, 1990, pp. 67–95.

[22] ———, *Integral geometry and manifolds of minimal degree in* $\mathbb{C}P^n$, Funktsional. Anal. i Prilozhen. **24** (1990), no. 1, 5–20; English transl., Funct. Anal. Appl. **24** (1990), no. 1, 4–17.

[23] ———, *Admissible double bundles*, 75 Years of Radon Transform, S. Gindikin and P. Michor (eds.), International Press (Boston), 1994, pp. 129–152.

[24] ———, *Integral geometry and* $\mathcal{D}$*-modules*, Math. Res. Letters **2** (1995), 415–435.

[25] I. S. Gradshtein and I. M. Ryzhik, *Tables of integrals, sums, series, and products*, Fizmatgiz, Moscow, 1963; English transl., Academic Press, San Diego, CA, 2000.

[26] S. Helgason, *The Radon transform*, Birkhäuser, Boston, MA, 1980.

[27] ———, *Geometric analysis on symmetric spaces*, Amer. Math. Soc., Providence, RI, 1994.

[28] F. John, *The ultrahyperbolic differential equation with four independent variables*, Duke Math. J. **4** (1938), 300–322.

[29] A. A. Kirillov, *On a problem of I. M. Gelfand*, Dokl. Akad. Nauk SSSR **137** (1961), no. 2, 276–277; English transl., Soviet Math. Dokl. **2** (1961), 268–269.

[30] F. Natterer and F. Wuebbeling, *Mathematical methods in image reconstruction*, SIAM, Philadelphia, 2001.

[31] J. Radon, *Über die Bestimmung von Funktionen durch ihre Integralwerte längs gewisser Mannigfaltigkeiten*, Ber. Verh. Königl. Sächs. Gesellsh. Wissensch. Leipzig, Math. Naturwiss. Kl. **69** (1917), 262–277.

Index

Titles in This Series

220 **I. M. Gelfand, S. G. Gindikin, and M. I. Graev,** Selected topics in integral geometry, 2003
219 **S. V. Kerov,** Asymptotic representation theory of the symmetric group and its applications to analysis, 2003
218 **Kenji Ueno,** Algebraic geometry 3: Further study of schemes, 2003
217 **Masaki Kashiwara,** D-modules and microlocal calculus, 2003
216 **G. V. Badalyan,** Quasipower series and quasianalytic classes of functions, 2002
215 **Tatsuo Kimura,** Introduction to prehomogeneous vector spaces, 2003
214 **L. Š. Grinblat,** Algebras of sets and combinatorics, 2002
213 **V. N. Sachkov and V. E. Tarakanov,** Combinatorics of nonnegative matrices, 2002
212 **A. V. Mel′nikov, S. N. Volkov, and M. L. Nechaev,** Mathematics of financial obligations, 2002
211 **Takeo Ohsawa,** Analysis of several complex variables, 2002
210 **Toshitake Kohno,** Conformal field theory and topology, 2002
209 **Yasumasa Nishiura,** Far-from-equilibrium dynamics, 2002
208 **Yukio Matsumoto,** An introduction to Morse theory, 2002
207 **Ken'ichi Ohshika,** Discrete groups, 2002
206 **Yuji Shimizu and Kenji Ueno,** Advances in moduli theory, 2002
205 **Seiki Nishikawa,** Variational problems in geometry, 2001
204 **A. M. Vinogradov,** Cohomological analysis of partial differential equations and Secondary Calculus, 2001
203 **Te Sun Han and Kingo Kobayashi,** Mathematics of information and coding, 2002
202 **V. P. Maslov and G. A. Omel′yanov,** Geometric asymptotics for nonlinear PDE. I, 2001
201 **Shigeyuki Morita,** Geometry of differential forms, 2001
200 **V. V. Prasolov and V. M. Tikhomirov,** Geometry, 2001
199 **Shigeyuki Morita,** Geometry of characteristic classes, 2001
198 **V. A. Smirnov,** Simplicial and operad methods in algebraic topology, 2001
197 **Kenji Ueno,** Algebraic geometry 2: Sheaves and cohomology, 2001
196 **Yu. N. Lin′kov,** Asymptotic statistical methods for stochastic processes, 2001
195 **Minoru Wakimoto,** Infinite-dimensional Lie algebras, 2001
194 **Valery B. Nevzorov,** Records: Mathematical theory, 2001
193 **Toshio Nishino,** Function theory in several complex variables, 2001
192 **Yu. P. Solovyov and E. V. Troitsky,** C^*-algebras and elliptic operators in differential topology, 2001
191 **Shun-ichi Amari and Hiroshi Nagaoka,** Methods of information geometry, 2000
190 **Alexander N. Starkov,** Dynamical systems on homogeneous spaces, 2000
189 **Mitsuru Ikawa,** Hyperbolic partial differential equations and wave phenomena, 2000
188 **V. V. Buldygin and Yu. V. Kozachenko,** Metric characterization of random variables and random processes, 2000
187 **A. V. Fursikov,** Optimal control of distributed systems. Theory and applications, 2000
186 **Kazuya Kato, Nobushige Kurokawa, and Takeshi Saito,** Number theory 1: Fermat's dream, 2000
185 **Kenji Ueno,** Algebraic Geometry 1: From algebraic varieties to schemes, 1999
184 **A. V. Mel′nikov,** Financial markets, 1999
183 **Hajime Sato,** Algebraic topology: an intuitive approach, 1999

For a complete list of titles in this series, visit the
AMS Bookstore at **www.ams.org/bookstore/**.